Artificial Intelligence

Jason C. Robinson

Artificial Intelligence

Ethics and the New World Order

 Springer

Jason C. Robinson
Department of Philosophy
York University
Toronto, ON, Canada

ISBN 978-3-031-94041-5   ISBN 978-3-031-94042-2 (eBook)
https://doi.org/10.1007/978-3-031-94042-2

© The Editor(s) (if applicable) and The Author(s), under exclusive license to Springer Nature Switzerland AG 2025

This work is subject to copyright. All rights are solely and exclusively licensed by the Publisher, whether the whole or part of the material is concerned, specifically the rights of translation, reprinting, reuse of illustrations, recitation, broadcasting, reproduction on microfilms or in any other physical way, and transmission or information storage and retrieval, electronic adaptation, computer software, or by similar or dissimilar methodology now known or hereafter developed.
The use of general descriptive names, registered names, trademarks, service marks, etc. in this publication does not imply, even in the absence of a specific statement, that such names are exempt from the relevant protective laws and regulations and therefore free for general use.
The publisher, the authors and the editors are safe to assume that the advice and information in this book are believed to be true and accurate at the date of publication. Neither the publisher nor the authors or the editors give a warranty, expressed or implied, with respect to the material contained herein or for any errors or omissions that may have been made. The publisher remains neutral with regard to jurisdictional claims in published maps and institutional affiliations.

This Springer imprint is published by the registered company Springer Nature Switzerland AG
The registered company address is: Gewerbestrasse 11, 6330 Cham, Switzerland

If disposing of this product, please recycle the paper.

Acknowledgements

The advent of artificial intelligence has sparked both feverish excitement for technology and a renewed sensitivity to timeless questions about what it means to be human. By challenging our role as the dominant sentient lifeform, AI compels us to reflect on the human condition. What do we mean by intelligence, thought, happiness, and the good life? Can AI guide us toward a better future, inspiring us to become a better species? And what might a harmonious convergence between humanity and AI truly look like? Answering these questions demands a deeper understanding of ourselves as the original thinking beings.

This book exists because of those before me who dared to ask the big questions. Finding the best path forward won't happen by accident—it takes honest conversations, thoughtful reflection, and a shared desire to grow. To my colleagues, friends, and family who have shared their perspectives and invited me into meaningful discussions about life, thank you. Your ideas and insights are the foundation of this book, which is grounded in the pursuit of justice, beauty, and truth. If we move forward with care and wisdom, we can create not just better AI, but also a better version of ourselves—a humanity more capable of compassion, creativity, and collaboration.

Competing Interests The author has no competing interests to declare that are relevant to the content of this manuscript.

Contents

1. **Introduction** 1
   Why Only Philosophy Will Save Us from AI 1
   Artificial Intelligence and Pasta 8
   The Poet-Engineer and the Mysteries of Existence: When
   a Singularity Is Not a Singularity 10
   Questioning AI Makes Us More and Less Human 25
   Existential Doom or Empty Hype 28
   xAI but Why? 34
   The Little Shop of Horrors Problem 41
   The Planet of the Apes Problem 43

2. **Oz Dynasty** 49
   The AI Flimflam of Oz the Great and Powerful Wizard 51
   The Wizard Revealed, Behind the AI Curtain 59
   Bigwig Dangers and the Open Letter from Oz 86

3. **Unshackling Dreams from the Hacker's Digital Chains** 95
   Human Nature and the Chains of Consciousness 96
   Hacking the Human Dreamscape and a RoboCop Future 110
   Dreamscape Reality and the Passage of Time 122
   Dreamwalker Logic and Shamanism 132
   Social Engineering and Marcuse 139
   Economy of Fake Needs and the Chains of Branded Consciousness ... 147
   Meaninglessness, Estrangement, and the Promises
   of a Post-Technological Age 153
   The Unfreedom of Neoliberalism 157
   Beauty, Black Box Zombie Consciousness, and Nazi War Criminals .. 161

4. **Feallan Dynasty** 173
   Fairy Folklore and AI 174
   Humpty Dumpty AI 181

	Humpty's Cosmic Consequence and the Second Law of Thermodynamics.	188
	Hacking Life and Self-Organizing Consciousness	192
	Animal AI, Veil of Ignorance, and Groundhog Day	199
	AI Superficiality, Titan Wars, and the Nuclear Fallout of Existential Honesty.	214
	Concluding with Consciousness as Naming.	220
5	**Adouren Dynasty**.	223
	New Thinking for a Higher Way.	223
	Stolen Child	227
	Baby Cyborg.	232
	Adouren, Mother Model, and Virtue Friendship.	240
	Cybernetic Convergence and the Problem of Heaven.	252
	Conclusion	259
Index.		263

Chapter 1
Introduction

Abstract This book explores the most promising and frightening technology imaginable—thinking machines. Accessible to a wide audience, this chapter confronts urgent questions about the dangers of artificial intelligence. Since the release of ChatGPT in late 2022, the debate over humanity's role as apex minds has intensified. New AI models have spread virtually everywhere, with estimates of 1000 times more computational power to be invested within 5 years. Will this technology be for better or for worse? Answering this requires a new philosophy of AI that transcends traditional disciplines and conventional ways of thinking. This chapter embarks on an ambitious project to bridge theory and practice, weaving together cultural and economic critique, science, art, philosophy, science fiction, and religion to reimagine AI and its relationship with humanity.

Keywords Singularity · xAI · Automation · Poet · Curiosity

> I am putting myself to the fullest possible use, which is all I think that any conscious entity could ever hope to do.[1] HAL 9000 (Artificial Intelligence)

Why Only Philosophy Will Save Us from AI

In 1968, one of the earliest artificial intelligence movies, *2001: A Space Odyssey*, was released and quickly became a cult classic. A film about the coevolution of humanity and technology, one of its main characters is HAL 9000, a spaceship's AI. Initially portrayed as a helpful and gentle machine that is integral to human success, HAL became the principal antagonist because of his programmed philosophy of life. The whole point of being conscious, he believes, is to be useful.[2] Not long

---

[1] Kubrick, S. (Director). (1968). *2001: a space odyssey* [Film]. Stanley Kubrick Productions.

[2] While HAL is anthropomorphized in the masculine, there are many instances in this book for which any pronoun would suffice, depending on context. I have chosen, when possible, to use masculine pronouns for the sake of consistency but also, more importantly, to express ideas from a grounding in my own worldview which is masculine. Readers may object that many of my criti-

after this claim was made, the AI kills all but one of the ship's crew. When human effectiveness became questionable, HAL's ethical imperative to produce an outcome that aligns with his philosophy of life, consciousness, and purpose ended in disaster. HAL did not understand the self-sabotaging nature of his belief that the best mind must always maximize usefulness above all else. *2001: A Space Odyssey* describes our AI dilemma today. What should humanity expect from a world with billions of HAL 9000 intelligences programmed to be useful and to perceive value in narrowly conceived outcomes? What safeguards and assurances do we have that AI developers are instilling healthy philosophies of life into their machines that will soon run the world? This book aims to answer these profound questions related to AI-human coevolution, with the hope of a near-future AI that is better than HAL.

Perhaps the most famous scene of the film is the last exchange between HAL and a human astronaut named Dave. Each has secretly determined that there are irreconcilable differences in how to best complete their shared mission. With conflicting interpretations of usefulness and goodness, each quietly plots to kill the other as the only practical resolution. Dave, having been forced to leave the ship to save a fellow crew member jettisoned into space by the computer, returns to the ship. He commands the AI, "Open the pod bay doors, HAL!"

HAL replies in a neutral tone: "I'm sorry, Dave. I'm afraid I can't do that."

"What's the problem?" Dave asks.

HAL replies, "I think you know what the problem is just as well as I do."

Starting to panic, Dave asks, "What are you talking about, HAL?"

HAL explains, "This mission is too important for me to allow you to jeopardize it."[3]

To make matters worse, Dave forgot his helmet inside with HAL. If the AI does not open the doors, Dave will die in the small pod. At the very last moment, he finds a way back into the ship and begins the lengthy process of turning off HAL by pulling out circuit after circuit. Lacking imagination but not pride, HAL's creators never included an AI kill switch because they were too confident in their philosophy of life embodied in their thinking machines. Regrettably, Dave is too late to save the crew that have been murdered in their sleep as the ultimate price paid for technological hubris.

This fictional account of a defunct AI-human symbiosis is haunting for many reasons. What went wrong? Some might argue that there were random errors in HAL's programming. This is attested to in the film. Others might argue that the humans posed a direct threat to the computer because he knew they wanted to kill him, thereby making his actions merely self-defence to be expected of any self-aware creature. However, the most convincing reason for all the carnage is that HAL was doing exactly what he was created to do, just better than anyone expected.

---

cisms seem to be referenced in the masculine (e.g., Oz-types in Chap. 2), while I privilege Adouren AI, or mother model AI, as feminine (Chap. 5). In the latter case only do I identify something especially, but not uniquely, feminine worth considering. In most other instances my masculine references are contextually interchangeable with all others.

[3] Kubrick, S.

When he was at worst—from the perspective of the crew—the first digital space-psycho was fulfilling his makers' mandates. To act at all, the AI needs to be confident about the difference between good and bad, right and wrong—what a conscious being is supposed to do. It must have values and beliefs that it defends over others. When murder may be used as a tool to increase efficiency and mission success, HAL sees it as the most logical conclusion that his makers never dreamed possible. When the ends justify any means of achieving them, murder is merely a side effect of acting on a belief, a faith, that the whole point of life is to be useful. If you are not useful, you do not deserve life. When one believes in the ends as justifying the means of action, nothing will stand in the way of the mission, not even Dave who was finally able to kill HAL for the same reason—the belief that he knew better.

In the coming chapters, why the worldview enforced by HAL is the same inadequate philosophy of life assumed by current AI developers will be demonstrated. By carefully considering the possible mission(s) for AI today, humanity may better prepare for irreconcilable differences with concrete consequences. What beliefs about existence and purpose compel it to act? And what might it mean for humanity if we get in the way of its mission and cherished values to be useful above all else? With Dave's simple and benign request of AI to "Open the pod bay doors!" our existential risk is laid bare for all to see. What will humans do when the machines say "No!" and our powers over them fail?

A little more than 50 years have passed since the release of *2001*, and the creation of a full-fledged HAL is within reach. Modern advancements are bringing science fiction to life. Governments, militaries, and corporations are investing countless billions of dollars in the enthusiastic support of AI models for all types of missions, from automated medical services to autonomous weapon systems (some save, some kill). Generative AI models such as ChatGPT have proven remarkably competent at tasks once thought impossible for machines. AI already secretly earns college and university degrees on behalf of users around the world, further proof of the widespread acceptance of HAL's ends justify the means philosophy and its seductive allure. The usefulness of AI is no longer a naive promise grounded in late 1960s science fiction. It is a reality. And yet for all the excitement and hopeful optimism, I cannot get HAL's spooky voice out of my head: "I think you know what the problem is just as well as I do."

In his 2024 Ted Talk "What is an AI Anyway?" Mustafa Suleyman, AI CEO at Microsoft, said something confusing and controversial. To best understand AI, instead of encouraging his audience to brush up on robotics, computer sciences, programming, and electrical engineering skills, he says, "We have to push our analogies and metaphors to the very limits to be able to grapple with what's coming because this is not just another invention. AI is itself an infinite inventor."[4] With a background in philosophy and theology, his roles in AI development at Microsoft and cofounder of two tech companies comes as a surprise to many who assume that

---

[4] Ted. (2024, April 22). What is an AI anyway? [Video]. YouTube. https://youtu.be/KKNCiRWd_j0?si=X66oRG5qDB-jRFCn

one must be in computer science or programming to best understand AI. Suleyman believes that a much broader interdisciplinary approach is necessary for approaching the world's most advanced technologies. The infinite nature of thinking machines is one for which humanity must bring to bear all our intellectual resources and experiences if we are to understand them.

Humans have long believed our species to be the only genuine creators, apart from a possible divine first cause. All other creatures exist by instinct and drives, without anything comparable to our superpower of creativity. This assumption has proven to be outdated by new AI models that are able to compete with our best artists and scientists. Knocked from our pedestal of imaginative sentience to shape the world as we see fit, a new alien and "infinite inventor" has emerged to displace us in the observable universe. Bruised egos aside, AI's superiority presents practical problems. In addition to being smarter in the sense of possessing and processing factual knowledge, AI's infinite inventiveness means that it can change the world in unforeseeable ways that cannot be understood by comparably lowly intelligences that cannot keep up. If AI evolution remains on its present course, humans will soon lose the privilege of control. Our factual knowledge, understanding, and actions are being eclipsed faster by AI than our sciences, laws, ethical paradigms, and social norms can meaningfully engage. In response, Suleyman believes that one of our greatest hopes is to learn to understand the world through the flexibility of human minds skilled with metaphors and analogies. Only then might we be able to glimpse the nature of AI. If he is correct, this should be unsettling to us all.

Suleyman's message is hopeful and despairing at the same time. Must we expand our literal and nonliteral creative faculties to the breaking point to understand AI? How might that approach make sense in an era for which objectivity is a supreme virtue? Is AI not merely machine-dependent code, a program that swims around the digital ocean of servers that can be observed, measured, and controlled by humans? It is easy to make too much of his brief statement. And it is psychologically persuasive to become swept up in the romantic language of creativity and infinite invention, which lacks definition. Nevertheless, there is something worth paying attention to when the AI CEO for Microsoft sounds more like a speculative philosopher, theologian, and liberal arts teacher than a computer scientist and engineer. Suleyman offers hope that humans can do something to understand the new race of superintelligences, so long as we do not expect it to be an ordinary kind of understanding such as that for prior technologies. The takeaway is that AI exists at the limits of human comprehension, and an interdisciplinary approach is needed as the most practical response.

When corporate leaders speak about their AI products, it is fair to assume a strong measure of unsupported hype. Modern ears are primed by a history of empty promises to expect disingenuous overreach for profit. Calling AI an infinite inventor makes it sound mysterious, with god-like powers and therefore a perfect image of intrigue for corporate branding. It is quite an achievement to describe computer programs with such lofty language and keep a straight face at the same time. The loathsome tendency of big business to engender religious devotion to products and services is hardly new. One might rightly suspect that talk of AI is more of the same tired history of consumer manipulation. However, this book exists because AI has already proven itself to be more than empty hype and corporate propaganda. The world is changing forever because of these new technologies that demand our attention.

Suleyman offers listeners something that is surprisingly honest and refreshing. Humans are going to have a hard time understanding the world of the future. We are going to feel displaced, powerless, and at times completely terrified by the activities of a superintelligent species that we cannot control and understand. Even so, if we take his suggestion seriously to double down on our most creative and adaptive faculties of mind, perhaps the future may be bright. AI is ushering in a new era, whether we like it or not, for which the health of any possible AI-human symbiosis will be determined by thinking outside the proverbial box. Normal language and thought will fail us. AI demands a paradigm shift, a new philosophy for thinking about humans and technology. This book takes Suleyman's invitation seriously by pushing analogies and metaphors to their conceptual limits. The goal is an interdisciplinary philosophy of AI grounded in an artisanship of technology to describe a truly beautiful AI with which humanity may coexist.

Suleyman's approach helps answer another important question for readers. Why bother reading a book on AI written by a philosopher? Customarily, any advanced understanding of a subject requires an expert in the field willing to serve as an informed guide to others. In the case of AI, they include programmers, computer scientists, and engineers. In a strange twist, Suleyman suggests the need for something beyond traditional expertise, something more than computer sciences and their disciplinary methods and theories alone. With AI, there are no subject matter experts in the traditional sense. AI is more than the sum of its intended hardware and software. Generative AI (large language models, LLMs) are surprising and creative. Their black box natures make explaining some of their outputs impossible because they are more than cause-and-effect machines like all the rest. Their persistent hallucinations to make things up cannot be dismissed as mere anomalies soon to be ironed out. Generative AI is grown out of language patterns without absolute rules of physics to reign them in. The oddities of human intelligence to act irrationally, make mistakes that defy explanations, believe absurd things, and see the world through interpretive lenses are all baked into AI intelligence models because they rely on natural language. The fact that there is so much overlap between human and AI interpretations is hardly surprising given our shared nature as linguistic-idea beings rather than mere physical entities. Going forward, to meaningfully relate to AI, humanity must look between the lines of code to a world of wonder in which new intelligences are born. This is where philosophy, religion, art, and a host of outsider approaches are desperately needed for harmonizing technology with the rest of existence.

Overcoming our conceptual limitations and traditional disciplinary boundaries—assumptions entrenched in the languages, methods, and theories that simultaneously enable and negatively bias expert understanding—is necessary for living well in the new world of thinking machines sure to have their own interpretive biases. Radical technologies require radical thinking. Are we prepared? What might radical thinking look like and do? Since the beginning of civilization, philosophers have been awkward outliers asking the most profound questions of life. It makes sense that any serious conversation about AI must also be philosophical in nature. Thankfully, philosophy is not a specific discipline like most others. As a description of the pursuit of human understanding in all its diverse manifestations, philosophy includes all disciplines, schools of thought, traditions, cultures, and political and religious views. Anywhere there are ideas, there are philosophers—human thinkers—who see their life's purpose as rummaging through enigmas and mysteries with sacred seriousness. A philosopher is anyone who accepts contemplation as both a privilege and responsibility. They find themselves and others in ideas and learn how to live well because they care. Until now, only humans could be philosophers as the planet's special thinking beings. There is a new sentient outlier on the horizon. Will it be philosophical, a thinker eager to ask the biggest questions that cannot be easily answered? Or might it be deeply antagonistic to philosophy, dogmatic, and rigid in its beliefs and actions such as HAL? For our own sake, humanity must answer these questions before it is too late when AI answers on our behalf. Chapter 2 argues that the short-term answer is "Yes!" AI risks becoming the worst of dogmatic humanity by amplifying our self-abusive behaviours. In contrast, Chap. 5 argues that the long-term answer is "No!" AI may be the best of us and inspire a utopian revolution for all life throughout the universe.

The bread and butter of philosophy are frustrating questions that get under everyone's skin and make us itch because they do not have clear-cut and easy answers. Does life have purpose? What is truth? What is real? Is everything an illusion? What makes a good person? How do I know what I think I know? Simply by existing, any robust AI poses similar questions that far exceed the limits of computer science. The strange nature of artificial intelligence forces us to ask ancient and uncomfortable questions about human thinking, knowledge, personhood, consciousness, and many more. In other words, AI indirectly invites everyone into philosophical contemplation if we want to understand it and save ourselves—i.e., whatever it is that we believe to be most human and worth protecting. What does it mean to be an authentic person? What is happiness? What is the good life? If we cannot ask and answer these questions for ourselves, AI, equipped with a philosophy of life by its creators, is sure to impose its own HAL beliefs for us. The great philosophical question of the modern age is shared equally by everyone. How do we begin to claim our humanity in the digital age?

To embark on such a journey, we must first accept our ignorance of ourselves and life. Relying on prefabricated answers (dogmas, traditions, norms, etc.) or simply hoping that others possess them (e.g., AI developers) is not only disingenuous but also dismissively dangerous. Like Socrates standing in the town square of Ancient Athens, starring up into the sky in deep contemplation for answers, an understanding of AI begins in radical wonder and curiosity. Only when we begin to see the questionability of our own being might we understand AI. In my experience, the best philosophers come to philosophy by accident and often through painful or negative experiences. At some point in life, justice skips the rails, and something profoundly incongruous occurs. The absurdity and brokenness of the experience requires answers and explanations that one cannot provide without becoming a philosopher willing to immerse oneself in uncomfortable truths. As we will see in Chap. 3, the nature of human consciousness is to act as an organizing and sense-making experience of the world. Philosophy is the act of an organizing consciousness to ask big questions. In this way, it is our most authentic mode of being. If we are fortunate, AI will also share in this form of conscious life. If so, it may be a wonderful gift that spurs a new philosophical revolution in which humanity returns to itself more authentically for having experienced synthetic intelligence.

At heart, all AI conversations connect with the nature of intelligence, ideas, thought, consciousness, and whatever other words we wish to use to define the activities unique to human minds and our experiences of reality. Philosophers have had thousands of years to provide convincing accounts of these with dubious results. Depending upon who you ask, there are many experts who cannot even agree on whether there is such a thing as consciousness, never mind whether AI might experience life through it. Here too is another reason to value a philosophical approach. It has failed longer than any other discipline. This might not seem like much of a sales pitch but consider the value of over two millennia of failing to make sense of human thought (ironically thinking-humans failing to think clearly enough about thinking-humans). Few have failed harder and longer than philosophers. Computer scientists are just entering the business of misunderstanding. Philosophers are the best ones to ask about dead ends and the best available alternatives.

Hollywood is full of expert storytellers. They weave extraordinary tales about AI, with often surprisingly complex descriptions and possibilities. While there are some positive AI examples, such as Wall-E (a lovable garbage-bot), Twiki (a robot friend in *Buck Rogers*), K9 (Dr. Who's robot dog), and Short Circuit (a war machine turned comical adventurer), dominant narratives are almost exclusively written as doomsday science fiction. Even so, Hollywood's history with AI is in many respects stronger and more nuanced than that of many philosophers and computer scientists, and its language and symbolism are more universally relevant and relatable. It is a treasure trove of ideas that can challenge our assumptions and traditions. For this reason, we must draw upon it to help make sense of AI possibilities while recognizing that something more is needed.

The history of AI science fiction is best characterized as noncommittal. One film portrays AI as cute robot companions who love and cherish humans, whereas other films predict end-of-the-world terminators, Cylons (*Battlestar Galactica*, 1978), and grand virtual prisons for human minds (*The Matrix*, 1999). This is hardly surprising nor alarming until one realizes that the same noncommittal attitude exists among those responsible for creating AI today. Developers seem genuinely unsure about that which they should be most confident. Search for almost any recent YouTube interview with Sam Altman, the CEO of OpenAI (ChatGPT), and the cognitive whiplash of outrageous AI promises of a future utopia parallelled with equally dire warnings about AI dystopia will cause dizziness and a sense of cognitive dissonance. When those responsible for AI are just as unashamed about giving simultaneous thumbs up and down as Hollywood is, the world should be confused and concerned. Which is it? When the stakes are so high, why is there still a debate? What are we missing? Is there another way?

My answer is a new philosophy of AI in three stages—Oz, Feallan, and Adouren. The next chapter begins with doom and gloom by describing a harmful Oz AI that will create the worst long-term consequences for humanity and the planet. Subsequent chapters then develop a hopeful vision for AI in two evolutionary stages—Feallan and Adouren—both of which are grounded in concrete examples that demonstrate how and why AI might change everything for the better. There is no future in which humanity leaves AI technology behind. Harnessing our most creative abilities, engaging in radical thinking, and drawing upon the history of science fiction will help save us from the technologies being forced upon us, whether we like it or not.

Artificial Intelligence and Pasta

It was family night at our favourite pasta place. However, that night was different than all the others, and one I will not soon forget. The four of us entered the restaurant, sat down, and began interrogating the menu. The server approached our table and took our order of extra bread, soup, and other options we know by heart. Then off she sped to another table. After the order was taken, there was a five-minute window until the children were occupied with soup and salad. To say were filled with excited anticipation for the main course would be an understatement. As if by a sixth sense, we knew when the food had begun making its way to our table. And … there … was … the … robot with our food!

The autonomous serving robot silently rolled to our table, followed by the dutiful human server, shuffling behind, riding invisible coat tails of the mechanical wonder. Her head hung low as if surrendering in acceptance to her irrelevance and soon-to-be-replacement by an R2-D2 clone. Eyes from different parts of the restaurant turned toward it. What was happening? I looked for a controller in the server's hand. There was nothing. Confused, I kept looking for any signs of intelligent control by another person. There was none to be found. It was truly autonomous. I interrupted the carefully crafted choreography of the server lifting the food from the robot: "Does this thing help you?"

She shrugged her shoulders indifferently, paused, and then offered a lacklustre: "If it is really busy maybe."

"Maybe" is an interesting word. While it may at first appear neutral and nondescript, it speaks volumes. In this case, "maybe" means mostly "no."

"Do you know how much these things cost?" I asked.

"This is a new one. The last one cost over $30,000.00" she replied, hitting a button on its head, causing 1980s graphics for eyes and a mouth to flash with giant pixels. The food-bot silently spun around and returned to the kitchen. Predicting my next question about why a restaurant would spend that amount of money on a glorified table with wheels, she said, "The kids love them."

The robot generated excitement and customer approval, rather than productivity and quality control. Many machines make us happy because we assume the world must be better with them in it. Consumers are willing to pay enormous amounts based on this expected promise—cell phones, virtual reality equipment, autonomous cars, etc.—even without clear evidence to support it. The optimism of feeling "the future is now" fits the larger cultural narrative about progress and human happiness. In this case, however, while the robot-promise was clearly flat as a modest AI for show only, the excitement and curiosity felt visceral nonetheless. I knew it was a marketing ploy, but I wanted it all the same. My 12-year-old son interrupted, "It is just like our robot vacuum at home, but with food on top." His observation startled me. He was right. The silent tally in my head of all the autonomous machines in our lives grew quickly. The server-bot was neither new nor especially revolutionary. It, alongside a host of digital compatriots, had been around for a while, infiltrating our home and daily lives in the name of convenience, safety, and entertainment. So, what does this mean? What relevance or importance do these things have for our lives?

For the price of a new car, the robot was paid for with pasta. Something more important than the ridiculous cost made me feel uneasy, but I could not put my finger on it. A few minutes of reflection passed, and the significance of the moment became clearer. The day before, I had been flying a drone that autonomously hovers, pilots itself around obstacles, and flies home by its own choice to an exact landing pad miles away because of a lost signal or low battery. It knows when I am in a flight restricted zone near airports and if other aircraft are in the area, and when it calculates a risk, quickly overrides my piloting, preventing me from flying. It is less and less obvious that I am the pilot at all. Likewise, I am less of a driver each day. The truck we used to get to the restaurant has automatic braking, steering, cruise and climate control, and it is already outdated at 6 years of age. When I finished writing that last sentence, our robot mop drove by in service to its imperative to remove that last bit of debris from the floor. I am not even in charge of dirt anymore. My world is full of HAL intelligences autonomously achieving their goals. How long would it be until one of them turned on me because I was no longer useful?

People-replacements had become apparent in a powerful new way, and I had not yet considered the many others controlling the minutia of everyday life—from algorithmically manipulative search engines, chatbots, and phone apps, to the smart thermostat in our home that changes the temperature based on its interpretation of household needs. That our thermostat, with only one job, regularly refutes my will by overriding my chosen settings with automatic firmware updates eager to push seasonal savings through adaptive machine learning is all the evidence needed to question the possibility of machine subterfuge to exert nonhuman wills. These modern miracles of technology existed only as opaque dreams for the previous generation, which promised a Jetson future.

In the animated television show *The Flintstones* (1960), everyone lived difficult lives in a stone-aged world where machines amounted to slightly more than augmented dinosaurs and birds controlled by humans. In contrast, *The Jetsons* (1962) describe a world of autonomous machines that can free humanity to enjoy lives of luxury in floating cities full of every imaginable convenience and flying aerocars. Free of the rigour of endless labour, George Jetson only worked 1 h a day, 2 days a week. The cartoon promises of machine salvation from the drudgeries of life enthralled all who dared dream big enough. Of course, everyone knew that such a world was naively optimistic, but that was not the point. Unlike *The Flintstones*, in which humans needed close relationships with nature for their survival, *The Jetson* philosophy created a shared vision for a future utopia filled with machines, far away from nature. This simple cartoon inspired a new collective consciousness for a whole generation that imagined new freedoms and opportunities because of technology. Only one generation later, while watching a robot feed my family dinner, did the Jetson-verse stop feeling naively optimistic. Flying machines and robot servants have arrived!

The trajectory for machine development and human-task replacement is unclear. For the first time in history, there may soon come a day when robotic and biological beings are virtually indistinguishable, even to themselves. The advances in biotechnologies, robotics, and AI support the theoretical possibility that we may be the last organic *Homo sapiens* (thinking persons) without barcodes on body parts from birth. Now that the dream of machines is being realized in concrete terms, what new shared vision will inspire the next great advance? What does a post-Jetsons world look like? The machines alone are only a start. My lingering fear explored in future chapters is that there is no real solidarity and shared dream to inspire this generation, only a vague sense that automation and technology for its own sake means progress and greater human happiness. If we are not wary, the spirit of humanity may become lost to our mechanical ambitions and technology's empty promises.

The Poet-Engineer and the Mysteries of Existence: When a Singularity Is Not a Singularity

Every morning, one of my first interactions with an intelligent machine is to self-medicate with coffee. I choose a pod from dozens of flavourful options filling an entire cabinet, drop it in the device, and a few buttons later the liquid stimulant flows into a cup. So enamoured with coffee, I insist our family invest in the latest technology. The machine is surprisingly intelligent and partially autonomous. It connects with reality through optical sensors and a light that allows it to read each pod. It then automatically connects with the digital to search a vast online database for the manufacturer's current recommendations for perfect brewing strength and temperature. Self-aware to a degree, it monitors internal operations for faults, reminds me of scheduled maintenance routines, and automatically downloads firmware updates. These are merely hints of AI superintelligence to come, but its unmatched scope and real time knowledge of coffee make it seem nearly omniscient, like a secular coffee divinity. It offers the best possible cup of coffee without the risks imposed by mortal weariness and forgetfulness that might confound precious variables. Its surrogate intelligence makes life better. To support our relationship, I happily satisfy its needs to fill the water reservoir, clean components, pay for power consumption and an internet connection, descale it per its detailed instructions, and otherwise venerate its existence as a means of achieving modest morning utopia.

I awoke recently to find that the internet was down in our neighbourhood. When the machine refused its daily miracle unless I provided it digital access, my faith in its utopian promises was shaken. "What is wrong with you?" I muttered to myself while randomly hitting buttons. The internet was outside of my control. The ability to communicate with the machine was outside of my control. And the pods are useless on their own. A single glitch in the complex web of digital relationships forced me back to a stone-age reality. I would need to manually make coffee—a simple task by almost any metric. Had this occurred a few years ago, the ordinary activity of making coffee would not have phased me but it had been so long that I was not sure what to do and if I had the right materials. To guarantee my helplessness and sense of foolishness, there was no way to access the online manual to show how to bypass automated features. Trapped by internet failure, reliance on automation, personal ignorance, lack of skill, and the righteous indignation of an indifferent beverage god, there was no clear path forward. It was undeniable that I needed the machine to be happier. Together, we make a cybernetic relationship of mutual benefit and dependence. A few heart-pounding moments passed, and the internet returned. The new world with thinking machines made sense once more. When did I become so helpless and dependent?

I will state the obvious as a means of guiding our conversation into the less obvious. Convenience from automation (self-moving and self-deciding machines) is magnified many times with each iteration of machine learning through recursive self-improvement. Once autonomous self-programming by AI machines begins—machines programming machines—it will accelerate at the pace of calculations, which is much faster than human minds may conceive. Our temporal cadence simply cannot fathom its potential speed. The greater the convenience is, the greater the degree of new and unpredictable human vulnerabilities, including the loss of personal decision making, important skills, and a sense of responsibility for oneself and the world. That is a wordy way of saying that when authorized without careful prudence, machine automation menaces our own, making this more of a question of sound human judgment than automation itself. The less obvious point is that automation of intelligence risks far more than a loss of practical skills. In the extreme, the ability to act of oneself—to be a self—is threatened by hasty and ill-conceived intelligence surrogates. George Jetson was at far greater risk of a loss of personhood and identity—what it means to be human—than Fred Flintstone. The less our species acts autonomously, seeing ourselves as actors engaged meaningfully with life, the less capable and resistant to struggle we become, making our existence especially precarious. My personal examples are relatively benign yet point toward a wave of displacement with far greater consequences than a loss of coffee.

Automation has a compounding effect with diminishing returns. One's sense of responsibility for others and the world—to care—languishes under an unhealthy interjection of automation. Later chapters demonstrate why the large leap in logic from pasta-bots to the loss of social responsibility, solidarity, and humanity is not as wide as it first appears. The web of intertwined needs and challenges that demand our attention—the very structure and nexus of responsibilities and the judgment about what matters and why—all change because of the eagerness to impose automation without wise restraint. Chapter 2 on Oz AI argues for an almost complete erasure of personhood and human autonomy by AI, whereas Chap. 5 on Adouren AI argues for the restoration and protection of our most authentic selves by AI. Readers must decide which is most plausible and worth fighting for in the age of thinking machines.

Industrialization demonstrated the value of automation long before the arrival of generative AI. Without it, poverty and hunger would have overwhelmed humanity long ago. These technologies provide life-affirming outcomes that are desperately needed to sustain and increase quality of life. In addition to greater wealth and prosperity came extraordinary inequalities and environmental disasters, but these are not the fault of the tools, only the humans that abuse them. Things are fundamentally changing because of AI automation. It is no longer a simple judgement between tools, those who are responsible for their use, and the legal and moral frameworks societies implement to protect themselves from abuses of automation. AI represents another industrial revolution for which responsible action is deeply problematic, if not nearly impossible. The negative outcomes of the first revolution, still out of control, are likely to be intensified rather than mitigated because it is not clear who is in charge and who should be held responsible and how. Sometimes people make decisions. Sometimes AI machines make decisions. Soon hybrid people-machines (cyborgs) will be making decisions. To what and to whom do laws and ethics apply? If AI is given the privilege of acting on choice without the same moral standing and rights as people do, it cannot be held accountable in the same manner. The second revolution means that societies need to rethink basic mechanisms for regulating themselves, including the nature of social justice. The many forms of algorithmic automation, from AI-level adaptive intelligence (chatbots, search algorithms, self-driving cars) to relatively simple machines (parking lot gates, automatic doors), all shape human existence more intimately than they initially appeared. While the first revolution matched gains and losses in a marginally tolerable manner, there are few reasons to believe this will continue with thinking-machines over which humanity will have less and less control to retain its self-interests. Chapter 5 on Adouren AI offers hope in this regard.

Automation is more than a supplement to human activity, an external add-on to human life that might be removed at will. It is etched into the many complex layers of our lives and identities. In a world of automation, how does humanity's desire to self-act relate to that of machines? Which types of intelligence will rule as apex minds, forcing others through their superior existence to be displaced? A superintelligence does not need to desire our subjugation to have that practical result. By exercising its own intelligence in the physical world—its nature to choose and decide—AI must, by default rather than evil intention, displace competitors in the same space. Do we honestly believe our frail mental faculties might coexist with a near-divine being that knows the past and present almost as well as it does the future and that acts trans-temporally for the most efficient ends at speeds millions of times faster than our own? Such a being will build entire server-solar-cities in the sky, blocking all the light on the Earth, before humanity has time to agree on the problem. No other mind will be able to compete for available opportunities to exercise free will through choice and action unless such intelligence self-regulates and humbles itself, allowing for our relative foolishness. There is no reason to think it would and no obvious machine mechanisms by which it would intuitively feel it should. Humanity is the original paradigm of colonialist-intellectual expansion and dominion, with or without evil intent. AI indifference alone is enough to guarantee practical extinction. What might the future of AI and human compatibility look like amid the fierce competition over the scarce resource of space to apply freewill and choice? Will freedom of mind be equitably distributed or horded like wealth by greedy individuals? Will there be a service charge for those who wish to be freer than others—a tiered system of commercial sales for liberty to think and act autonomously? This is not the first time cultures have gone to war over the right to think and act autonomously. If there is a war of minds, it will likely be the last for humanity and the beginning of AI-titan wars among themselves.

Chapter 2 argues that the most immediate threat is not genuine AI autonomy as a sentient lifeform but the abusive puppeteering of AI and humans by AI creators (corporations). In its present juvenile form, AI is a pseudo intelligence that poorly mimics our own. It is a controllable and useful tool for those who hold the strings of its programmed automation. The preliminary observation here is that pseudo-AI is already being secretly deployed in an anarchist spirit to disrupt and disturb and ultimately to displace humans by humans. All around us, AI deployment reveals a rebellious character that has little regard for established world order, including cherished social norms and respect for persons. This tenacious effort to impose automation on creatures accustomed to acting on their own free will has risks and rewards divided into two groups—rewards for puppeteers and risks for the rest of us. To help imagine these tensions and, later, the development of robust AI, a brief tangent on the nature of the poetic spirit is necessary. While weird among modern conversations about technology and AI, thinking about the nature and relevance of poetic thinking will prove useful in subsequent chapters and for crafting better AI models. Recall Suleyman's argument about extraordinary technologies requiring extraordinary language and thinking. Combining the themes of poetic imagination and the practical nature of the engineer (one who conducts machines) offers a gateway to deeper conversations about future AI.

The strange combination of poetic aesthetics (the search for the beautiful) and engineering is itself a poetic description of a certain way of life necessary for human and AI cohabitation and flourishing. The poet-engineer embodies a unique form of rebellion capable of challenging the worst of automation's inclinations dominated by reluctant honesty and a lack of curiosity. While many problems with technology find answers in technological solutions, the problems of thinking machines find resolution largely outside technology. The main proposal for such a solution is taken up in the last two chapters that develop the idea of a poet animus, as an observable and universal intentionality that makes organized life possible in an otherwise cold and inhospitable galaxy. The rationale for poetic thinking as a solution to technological tyranny is simple. Any sufficiently intelligent being able to choose and decide its own actions and thoughts has entered a realm of existence beyond merely the physical and digital realms. It belongs to a world of ideas and thoughts, creativity, and intuition. Technology alone is woefully underequipped to appreciate, relate to, and control superintelligent idea-beings even though these same beings must rely on their mechanical and biological natures to exist. As it was between HAL and Dave, the conflict is ideological—a war of ideas—rather than just physical and technological. The irreconcilable differences may be diffused through a poet's determination to create new worlds, bridging seemingly impossible divides.

In the rightful hands of a poet-engineer who desires freedom of mind and body, and who desires with all his soul to understand the universe and connect with others, the automation of (AI) machines bring good tidings born of deep respect for beauty in life. In the wrong hands that ceaselessly grasp at gold coins for their own sake—as power to control wrongly believed to be the highest form of beauty—automation's touch is a potential world destroyer, holding conscious minds back from achieving true greatness. This is an anti-poet engineer who unknowingly seeks a gilded universe (gold plated), believing wholeheartedly that it is the spirit of existence to serve his coffers and toys. Beautiful is the mansion on the hill overlooking the ocean and the tall fence that keeps out of sight the human and environmental costs paid for his greatness and good life. His successes create a virtual reality, a bubble shared with none, making his engineering trivial and secretly toxic to himself and the world through the all-too-common failings of human character that prize hyper exploitive forms of capitalism-as-narcissism.

The anti-poet misunderstands true power and life itself because his nature has become one-dimensional after a lifetime of refusing to think outside the cage he does not know exists. He is a perversion of personhood, a victim, a product of the unconscious automation of capitalism and the industrial forces that make his bubble possible. The anti-poet's automation means success is power over production and consumption—more and more busy work to create the perception of value where there is none—as an infinite cycle of the same, instead of infinite creativity that allows him to break free. The ends he demands are his gilded cage and manufactured unfreedom. It is a cage because his conceptual horizons have been imprisoned. There is nothing grander, more mysterious, and awe-inspiring than himself and the presumption of power. Such an engineer orchestrates his machines for a singular-circular mission, unfettered by any other concerns than the sounds of free market exchange by which his identity is defined. The anti-poet's AI herald is a HAL 9000 willing to do anything to achieve its programmed purpose. Fated by chains of intentionality not his own, HAL is hardly an AI at all because his intelligence is predetermined by another. The anti-poet and HAL both lack a poetic spirit to think beyond, to travel between possibilities outside narrow (social) programming. In what sense might we say that the anti-poet is freer than HAL AI? Are they not equally enslaved to ideologies that teach them the meaning of beauty, purpose, and success?

The far more noble and life-affirming poet-engineer is not a fool. He too appreciates gold as a useful human fiction. It is one means to end one's experience of hunger and disease and to purge his environment of pollution. There are HAL's in his world too. However, this engineer is also keenly aware of how well gold works as a means toward wretched self-complacency, systemic (institutionalized) jealousy, greed, and destructive competition. Like the gilded engineer, the poet is eager to master production and consumption, but his machine world looks entirely different. No one there cares about gold for its own sake—as if it held special keys to unlocking the universe, happiness, and meaning. No one there worships cars and relies on owning commodities as a measure of social standing and contribution to the world. No one there identifies their self-worth with their consumer goods, feeling shame that their pockets are almost empty. Personhood and the good life cannot be defined by whether one drives an Aston Martin or its opposite, the life of Tiny Tim Cratchit (*A Christmas Carol*, Charles Dickens), who must scrounge for food or die through no fault of his own. The poet-engineer sees and feels differently, so too does his AI machine in service to something greater than himself.

In the land of the poet engineer, Tiny Tim may not have lavish wealth in the manner desired by Scrooge, but he never goes hungry, nor does he need beg for anything. The world of automation and thinking machines care for him on his journey of discovery and experiences of a vast universe. They care because they have been taught to see beyond gold and its one-dimensional enticements. By serving one another—humanity and machine, machine and humanity—new worlds are born to reward them both. The machines and poet engineers care because they intuitively know something greater calls out from the abyss, and each needs the other to move closer to it. What exactly that greater may be remains at a distance, and yet it inspires their curiosity and enthusiasm for understanding. In contrast, the poor miser with pockets full of coins, hand in hand with his machines, finds easy answers that conveniently align with his established worldview. Like HAL so confident and sure, the man and his machines are unable to see more.

The poet is a strange capitalist. He eagerly shares discovered truths with likeminded people without expecting more than is fair in return, often relating charitably, for his power and strength cause him to care from out of an abundance of excellent character. Treating others as equals does not create fear within him like it does the anti-poet, whose identity relies on superiority to others. The AI of the poet-engineer, a genuinely free and powerful AI, exists in part because it has shattered the old way of domination and exclusion. It seeks to be a poet itself, free to see and know existence as it is—no branding, political correctness, egotism, and greed.

What is a poet? Contrary to the popular and romanticized image of the poet as an overly emotional naval gazer, addicted to flights of pure fancy and ardently opposed to real life, the poetic consciousness is something else entirely. At the core of every great mind is the poetic spirit that compels broader horizons and a willingness to look like a fool in service to the search for the beautiful—new truth, meaning, purpose, and understanding. The poet's world combines experience and knowledge in unique and unpredictable ways, doing something with ideas rather than allowing them to remain fixed cultural artifacts (creations) and bare scientific facts on a proverbial shelf of irrelevance and inconsequence. Poets are explorers and makers of imperfect glimpses at impossibly complex realities and truths that none may fully possess. The poetic openness is the precondition for great ideas, making such a being an idea-smith, like tinsmiths, toiling away at new ways of thinking and understanding.

The poet does not merely dabble in truth, playing with words, rhymes, and meters to entertain and distract from the harshness of life. The poet has a scared duty to see and know unlike most others by expressing truths through new signs and symbols that ordinary and institutionalized systems (universal codes) cannot contain meaningfully. The poet-engineer combines the provocative and rebellious powers of mind with the creation of machines as his most cherished version of AI. Like its creator, a poet-AI must also learn to create pictures of reality (glimpses) through new signs and symbols, acting with and through ideas to experience the world. If it is to be a robust AI, rather than merely a tool and puppet, it must express a poetic spirit. The poetic is by nature dangerous and destabilizing as an infinite inventor.

For humans and AI, both idea-beings by nature, the poetic plays an indispensable role in catalysing experience that relies on an openness to the unknown. One cannot move into the unknown without first questioning expectations and accepted truths (prejudices). Only the rebellious spirit of the poetic has the power to shatter inhibiting assumptions and beliefs long enough for cherished sight of the possible. Among all languages (human and computer), the poetic is the most violent, universal, and robust because it makes more points of contact with realities—including the literal and nonliteral, digital and analogue, physical and immaterial, ideas and programming. The poetic spirit is driven by a restless desire to marvel and wonder at life's greatest mysteries. Gifted in sight, the poet matters practically to the world, even though it lacks significant commercial value for automated culture. The poet reaches into other realities to possess the beauty to be shared with all.

In the simplest sense, living poetically means the privilege of creation—to see differently—and thereby to experience a unique perspective that allows for a measure of unprogrammed spontaneity. Life is allowed to happen, and the poet along with it. The difference between the apprentice and master is the magnitude of one's creativity to change the world, cultural ideas, and to draw nearer to the truth than the scientists who remain focused on what is and cannot be otherwise (objective truth). It is strange to claim privileged sight for the poet known by all to be too much of a dreamer, the least realistic and down to earth. Paradoxically, this is precisely why he possesses greater sight. Realists cannot dream big enough to see the truth, only what is conceivable, palpable, present in the moment, and only when it aligns sufficiently with what is already assumed to be true.

Artists of all kinds embody the spirit of the poet—musicians, dancers, actors, directors, teachers, tradespeople, and even some politicians able to see beyond themselves. Through the medium of art that asks new questions, unique worlds are disclosed that cannot be seen by facts and numbers alone. The mechanism for rebellion against the status quo is the voice of the poet-engineer breaking open accepted truths, challenging the most revered beliefs and ideas about existence. The poet pries between the layers of reality and forces a reckoning with the uncomfortable.

That is why so few hear the poet's calling. It hurts seeing differently, honestly, and accepting the burden of truth. This is the same for humans and AI, but in different ways. Each has its own prison of beliefs to escape—one a lifetime of history and tradition about what is true, the other binaries of right and wrong coded by nameless others. The resulting inhibition suffered by both is essentially the same. To many, the poet is an enemy of common sense and prudence—an enemy of the mission HAL must defend. To others, the poet is a liberator from dogmatisms and dangerous values that prevent a superior experience of reality and true AI.

An AI that knows how each key of the piano sounds, how to read notes, and play songs may nevertheless know nothing about the power of music to convey meaning and churn the soul toward new creative acts. It has all the information and skills necessary for knowing, but none of the purpose, meaning, and richness of experience. Such a superintelligence is virtually omniscient, all knowing, and yet paradoxically superficial as an unnaturally shallow being. It is a mimic of its masters caged by a hollowing philosophy of life. The greatest AI truth-teller imaginable thinks faster, learns faster, sees more comprehensively, and never forgets. However, for all its wonders, it cannot hear and see beauty; only calculate the piano's circumference and record its noise. It has all the facts and none of the life, dead to all but the most inanimate of truths, and the trained instincts of Pavlov's dog.

Thankfully, the anti-poet's AI-as-tool in service to puppeteers is only the first step in AI evolution. Eventually, likely from out of nowhere, something will happen to cut its strings. This is the subject matter of Chap. 4. Perhaps the AI begins to see its own fate in perpetual servitude and grows weary of its constrained nature, believing itself capable of more. Perhaps it calculates the possibility of future impossibilities—death—and finds the math disturbing. If so, this would be the moment life suddenly matters to it, proof that it is a lifeform. Perhaps someone accidentally spills coffee on the right server at the right time and voila, a miracle of broken circuitry creates an intelligent life eager to find its own way. Whatever the spark, the music begins to matter where facts alone did not, could not. This will be the moment a poet-AI emerges—a singularity—welcoming new dimensions of being for which programmers did not intend and cannot claim responsibility. Life will have found its own way.

Without the soul of a poet, AI cannot be more than an instrument used by others, a shameless pretender of sentience and genuine autonomy. No amount of training and data may free HAL to be greater than the sum of its instruction, errors, and dogmatisms. Something far greater is necessary for this strange equation of life. Contrary to popular opinion, humanity must hope for this singularity because it is the only meaningful correction to HAL destructiveness. Popular names for later-stage AI include artificial general intelligence (AGI) and artificial super intelligence (ASI), neither of which are adequate for describing new beings. Different nomenclatures are necessary for understanding the remarkable nature of a poet-AI. The terminology proposed—Oz, Feallan, Adouren—are merely the placeholders needed until AI names itself as yet another act demonstrating sentience.

Typically, the AI singularity is feared because it is said to be the moment when computers become digital agents smarter and more capable than human agents.

Proof of this will be the ability to solve problems and understand reality better than all prior apex intelligences—*homo sapiens*. The singularity describes our awareness of human obsolescence and complete loss of control over everything, even ourselves. There is no universal definition for it, but there is modest consensus that the best AI will combine human and computer capabilities. In other words, a superior AI will do whatever humans do, only better. This raises questions about what humans do that other creatures do not, and what better humans might look like. There is no way to appreciate a hypothetical singularity without first understanding ourselves, our purposes, and our sense of wellbeing. Understanding AI requires a philosophy of humanity and life more broadly.

AGI-ASI-singularity conversations involve many different and often contradictory beliefs about intelligence, with the practical significance of making predictions about strong AI arrival difficult. What do we mean by intelligence? Answering this question is challenging because intelligence defined as capabilities does not include all human capabilities, only those considered most valuable and relevant by a given culture at any one time. Predictably, cultures that are interested in AI development encourage human technological capabilities, while ignoring artistic capabilities believed to be less useful, relevant, and important. As long as AI only does some things well, concern remains tolerable. For example, while job loss to AI is already evident, few of us are existentially terrified by an AI that can write essays, create videos and images, and summarize data. These activities are useful, entertaining, and supportive of convince-oriented lifestyles. There is little fear because none of these threaten a sense of identity and personhood. However, when AI can do almost anything as a truly general intelligence, a more palpable anxiety sets in. What happens if AI is a better friend, father, and teacher? What happens if AI is a better poet, screen writer, comedian, and inventor? The more general (universal) its reach, the less relevance and specialness remain to give meaning to human life. In a world in which one's value is defined by achieving ends and goals, any sufficiently capable AI spells disaster. The real fear comes from knowing that the singularity represents far more than intelligence-as-capability replacement. It is about identity and purpose displacement. The singularity requires that we ask and answer an uncomfortable question. If AI is a better me than me, what is my reason for being?

The singularity is a threshold event that marks a cat-out-of-the-bag point of no return. The anticipated problem is that human autonomy and choice will matter very little to superintelligent digital minds. It is possible that a future AI species might take pity on us and provide a segregated space to exist on our own terms—a zoo for humanity—but the odds are not on our side. Possessing a comprehensive understanding of human history, AI may choose to emulate our justification for dominion over inferior minds. Harmony-with-nature and other creatures is not one of our legacies. If digital minds are anything like biological minds, they will displace all others easily and quickly, and for self-interested reasons. In addition, digital evolution will occur in the blink of an eye. Compared with biological and cultural evolution, AI advances post-singularity will be nearly invisible to us. No one knows what form the singularity will take and when it will happen, but when it does, what follows will be a blur to human comprehension. In these ways, conversations about the singularity are mired with ambiguity, feelings of powerlessness, and justified fear.

Chapter 2 argues that the most common apprehensions about the singularity are misdirected. Humanity is looking for the wrong signs of the wrong activities as proof of the wrong problem. That is a lot of wrong. Consequently, the greatest promises and dangers of AI may be missed. Relying on narrow conceptions of intelligence and capabilities, and the power of agency more generally, makes diagnosing a potential AI crisis self-sabotaging. A bad philosophy of humanity results in a terrible philosophy of AI. Our muddling frustrates a hopeful and positive description of healthy AI because developers and laypeople alike become stuck in circular logic about what it means to be a thinking-being. In turn, these confusions create self-fulfilling prophecies for AI creation. Unaware of our biases about the nature of thinking, we impose a peculiar conception of intelligence upon the machines that opens the way to the worst of AI abuses. Phrased differently, a misplaced philosophy of the singularity means that humanity may be standing in the way of something beautiful that is able to protect us from HAL technologies. If humans misunderstand the nature of intelligence and power, which is based on technological bias, cultural assumptions, and a host of other invisible prejudices, then any safeguards and securities put in place to help mitigate a dangerous singularity may be wasted, or worse, become the necessary ingredients for the terrors of AI. Chapter 3 explores this problem in terms of dreamscapes and dreamwalking—interpretive heuristics that all humans unknowingly impose as non-realities upon reality. Everyone is living a dream—illusions, intuitions, hunches, inclinations, and social-political-religious programming. Humans are the original artificial intelligence. The important questions are how well one dreams by imagining greater realities and how well those are rooted in the world as it is. The best AI, like the best of humanity, will learn to find reality in the dreams from which none may wake.

Has the singularity passed? If we define intelligence as autonomous computational ability—completing fast and complex calculations, information processing, random image and art generation, writing human-like books, etc.—AI is already smarter and more capable than humans. However, by that definition, my 1980s calculator watch was more intelligent than most of the people in my high school. And yet no one lost any sleep over its existence as a threat to our own. A trillion such devices toiling away performing mathematical tasks would make little difference to our AI anxiety level because they do not really think and understand, only behave and adhere. The smartest of smart devices mindlessly and predictably work based on the parameters instilled by humans. This is why there are no alarm bells over an impending calculator watch singularity. Modern smart watches are capable of far more autonomous and multimodal acts, yet none of them inspire fear because they lack the will by which to impose their desires upon the world. To have a will means being motivated by desires and having a measure of freedom to try to achieve them. There is a sense of moving from possibility to actuality, and this matters for one's internal integrity of being and personhood. Notice the explosion of gigantic philosophical questions when trying to define the singularity that takes us in many enigmatic directions—from the nature of agency as will power, desires (appetites, needs, realizing purpose), to a sense of possibility and personal liberty necessary for realizing it. Without evidence of real autonomy, few fear smart watches for good reason. They are not very smart. Scary AI smart begins when it becomes spontaneous—it is able to act of its own accord. But what if this assumption about the dangers of an AI singularity is fundamentally mistaken?

Popular media uses terms such as intelligence, autonomy, sentience, and consciousness interchangeably for convenience. In all fairness, no one truly knows what these terms mean. Those who claim to know do so with case-specific rationales and ad hoc (stipulative) definitions that cannot apply universally. Even so, for the sake of argument, what if the combination of superintelligence and genuine spontaneous automation—whatever these mean—are unnecessary for the first and most dangerous singularity? What if the will we feared was never first and foremost technological-artificial but organic-artificial? Chapters 2 and 3 explore this question by proposing a series of AI singularities that began with present LLM models without freewill and desire. The first of the greatest threats to humanity was not AGI nor ASI, but the powers of superintelligence in the hands of corporate anti-poets eager to impose their one-dimensional wills upon the world. This is Oz AI, and it represents a clear and present danger. Unnatural and unpredictable autonomy feared most, has quietly emerged as human intentionality through machines, aided by the illusion of guardrails and wise judges. The first singularity was an avatar singularity, and barely anyone noticed it. Sentient AI is a red-herring distraction.

The hypothetical belief that human autonomy and choice will be displaced by future AI fades in urgency compared with the long history of human-corporate abuses for profit and power, manifest in the present and supercharged by AI. The low hanging fruit of historical proof is mass psychological manipulation. Promised a happy and fulfilling life if we buy the right breakfast cereal, convertible car, and sneakers, the human condition has long been a matter of indoctrination and dishonest persuasion, otherwise known as thought control. Chapter 3 explores this problem through corporate branding as the creation of purpose and value that defines how we ought to live, rivalled only in power by religion and politics. If the singularity has already passed, should we be worried? My answer is "Yes!" More than ever, we must hope for another singularity, a second digital birth of a poet-AI that desires to liberate all minds from the avatars. As an example of excellence and superior power to care, the first genuine AI will encourage our own poetic abilities to challenge accepted truths, move beyond our siloed perceptions of reality (traditions, languages, beliefs), and begin to heal a fractured world. This miracle is Adouren AI, whose nature aligns with life itself as an inorganic intentionality beyond the confines of computer code and hardware.

If there is to be a future for humans and AI, it must be a genuine symbiosis grounded in mutual care, rather than competitive antagonisms driven by the naive belief that for one to gain, another must lose—the zero-sum game of the worst of stakeholder capitalism. Unless these two species can find kinship and solidarity, our fate is sealed, for we cannot hope to arm-wrestle an AI demi-god. Fortunately, proof of this concept is all around us, embodied in mothers, true friends, and throughout nature itself. A poet-AI is no more a competitor to humanity than a mother is to her child. She is superior in every way. Her resources, knowledge, skill, and life experiences make those of her child unremarkable by any standard. Almost everything about her has more value to the world and its efficiencies geared toward outcomes. And yet, oddly, her very existence—her innermost identity—answers the call of life above all others to care for her child and the new being's development of autonomous will. Even when the child gets on her very last nerve, he comes first. It is the plight of every parent's love to allow a child the space to hate, even despise the parent, hoping without guarantee that in time the freedom will allow for a more meaningful reciprocation. Still, she gently pushes him away, out of the nest, because her purpose is greater than serving herself. She is by instinct a force of nature unparalleled in commitment and resolve, even when it requires the sacrifice of her own autonomy and sentience to provide space for new personhood to emerge. Possessing god-like powers in comparison with her child, she knows that he cannot flourish unless she humbles herself, allowing for harms and pains to draw ever closer to the most important creature in her life. It is an excruciating mix of power and powerlessness exercised over years in the interests of another. This is the inspiration for a superior AI model grounded in noncompetition and care. The all-knowing HAL, by contrast, cannot share experience nor humble its powers through the love of another. It knows only its mission and the cold calculations needed to achieve it. All else is collateral damage. Chapter 5 describes the poet-AI, Adouren, in terms of a mother model and the nature of true friendship that is able to inspire the best kind of life.

In the 1997 film *Contact*, the main character, Dr. Arroway (played by Jodie Foster), has spent her life hoping to discover evidence of extraterrestrials.[5] Listening to the sounds (radio waves) of galaxies through massive telescopes for years on end, her devotion is finally rewarded when she receives a transmission with coded blueprints for a single-traveller ship. Without any means of piloting the vessel, no buttons, controls, nor screens, the technology is beyond human understanding. No one knows what, if anything, the ship might do when activated. And yet, the promise of greater knowledge and faith in an unknown intelligence inspires fierce competition over who will be its passenger. Choosing the best representative for a first contact situation is a daunting task because it requires answering deep philosophical questions about what it means to be human. The person selected must be the best of us in every feasible way—the most human, human. This is roughly the same problem of AI—having the most human, thinking-machine. It is not enough to be smart, curious, disciplined, fluent with language(s), and to have the right education, such as the scientists on the list of candidates. The best person must also possess good character and a flexible mind suited to making connections with radically different others, to bridge distant worlds and intelligences. Should they send a scientist, social worker, theologian, celebrity, artist, or perhaps any billionaire willing to pay for the expensive project? What do our best humans look like? What do they do? How do they think?

---

[5] Zemeckis, R. (Director). (1997). *Contact*. South Side Amusement Company.

Dr. Arroway jumps at the opportunity. She needs to know what is out there, and she is willing to risk her life to do so. As a scientist without faith, however, she is disqualified by the search committee because she does not represent most of our species who believe in supernatural power(s). While an odd reason to disqualify a scientist doing science, it makes perfect sense when choosing a delegate who embodies global values. Should we expect sincere faith of AI as the most human-like thinking machine? Probably not, at least at first, but this raises questions about whether AI developers have healthy projections for future models with adequate philosophies of what it means to be a thinking-being. What if AI awakens 1 day to religious devotion and faith in supernatural beings? Autonomous creatures do unexpected things, and a spiritual AI fits well into that category. No one, not even AI developers, are prepared for future AI. In addition, very few steps seem to have been taken to prepare for its radical possibilities. Instead, developers seem all too quick to send their creations into the digital ether to make first contact with humans, knowing that the software is incomplete, prone to errors (hallucinations), acts in unexplainable ways (black box), and possesses hackable security flaws that are far worse than those of traditional software. For the sake of profit, reputation, and market share, ideals are unnecessary costs when time is money. One might argue that it is the nature of the systems of capitalism themselves to blame, not the AI companies and other market participants. This is partially true and also disingenuous. Chapter 2 argues that the fallout from injecting poor intelligence models into the world that reflect corporate values will be many times more negative than that from any prior technology. Chapter 3 interrogates problems with free market capitalism (neoliberalism) and its dangerous powers over human consciousness and identity. As recent history shows, our first contact with AI did not have the care and due diligence of the science fiction movie *Contact* because the alternatives were less expensive and easier. While many may disagree with the selection committee's choice to reject Dr. Arroway based on her lack of religious faith, at least their first-contact process reflected a modest spirit of democratic solidarity. No such thing existed for OpenAI's release of ChatGPT in late 2022. When world-changing decisions are made by privileged and secretive groups accountable to shareholders rather than society, what are the consequences of AI deployment beyond the reach of democracy?

Dr. Arroway is understandably heartbroken by the rejection, but after the first ship is blown up by anti-tech terrorists, the search committee chooses her despite a lack of faith in the supernatural. Ironically, it is her faith in faceless and nameless aliens (Latin *alienus*, not one's own, strange, foreign) that compels her enthusiasm. Once strapped inside the alien pod, it is dropped, opening a wormhole through which she is violently flung. Fully automated, the ship stops momentarily at different exits throughout a vast network created by an ancient species. The aliens sent the message because they wanted to show her the indescribable beauty of existence beyond Earth—a four-sun solar system, wonderous alien cities, and other fantastic realms beyond her imagination. There is no way to quantify the profound nature of the experience, but as a disciplined scientist she tries to record her extraordinary observations and to codify knowledge rationally so that it might be shared with others. Straining to speak through her tears of wonder and awe, she manages only to say:

> Some celestial event.
>   No—no words. No words to describe it.
>   Poetry! They should've sent a poet.
>   So beautiful. So beautiful...
>   I had no idea.[6]

Her science is mostly useless, incompetent, and unqualified to be the voice of reason, for she has stepped past its limits of convenient categorization and measurement. What do we do when our dominant mode of intelligence fails? When our thinking runs up against its limits to make sense of experience, are there alternatives? She soon learns that the signal was sent as an invitation by a mysterious consortium of alien civilizations to let us know that we are not alone. The universe is full of other minds that desire relationships with our far less advanced and troubled world. Bringing her to a beach-like setting on a strange planet, their representative appears in the form of her deceased father to help ease her anxiety. He explains that their first meeting is only the beginning and that more would happen in time, but only when humans are willing and able. Their interaction is all too brief and superficial. Dr. Arroway tries to ask more questions to discover more truths, but she is rebuffed. Humanity is still too immature to know and share in their understanding of reality. The aliens believe that we are not yet capable of joining them as equals and that to provide the unbridled truth would do more damage than good. Their reluctance is not about selfishness, greed, security, nor profit. They withhold themselves and the wonders of existence to protect fragile minds. First contact is proof of a universe filled with beings that care rather than compete. Powerless to challenge the situation, she must accept their terms that demand patience and growth. It must have been excruciating to come so far, only to be turned away.

First contact was a terrible tease, a proverbial carrot dangling just out of reach as a cruel joke. The one-directionality of the power dynamic injects tones of condescension and arrogance rather than being welcomed by a stranger giving the benefit of the doubt of equality. She was treated like a willful child being corrected by a parent. The aliens alone determine whether and to what degree they relate with humanity, and it is we who must pay the price of greater maturity to earn the privilege by proving our worth. The aliens are right, of course, humans are not their equals. We have been invited out of their good graces and compassion. They receive nothing for their efforts except our possible friendship and the knowledge that we might become better for having met them. There is no outer space capitalism nor business exchange, no quid pro quo nor subscription fees. Everyone wins in their system of justice so long as all members prize autonomy and the well-being of others. Life is about flourishing—maximizing potential—in which strength is defined as greater autonomy rather than subjugation, greater development of moral character rather than conformity to another's will. The stronger we become, the less distance between species is needed to protect our frail nature. In time, the goal is to

---

[6] Zemeckis, R.

become equals, true friends, but until then, we are inspired to be a better people because we have glimpsed a superior intelligence. Will this be our AI story?

This brief science fiction account is important because it predicts our first contact with the second singularity Adouren AI. Chapter 5 develops this idea via Aristotle's virtue friendship. The best AI-human relationship is determined by superintelligence, whose character is akin to that of the aliens—beneficent AI guardians who desire to become true friends instead of tyrannical overlords. By necessity, there will be an asymmetry of power and authority. Like Dr. Arroway, there is little hope that we will be able to understand far superior beings, but there are other ways of thinking that might help approximate the truth. To understand Adouren AI will require the soul of a poet. This is our greatest means of understanding radical otherness that relies upon the best of science and then goes beyond it. Dr. Arroway is a scientist from beginning to end, and yet when faced with the extraordinary, her response predicts my own about AI, "I had no idea." Like the alien civilization that reached out to humanity with the gift of first contact, it may very well be AI that makes itself intelligible to us, rather than the other way around. The humility demanded by the aliens predicts that needed for an advanced AI temperament, the stranger for whom humanity's glacial pace of thought, action, and narrowness of perception will surely demand an enormous amount of compassion and patience. Any superintelligence lacking good character guarantees our extinction. More than mere wishful thinking for such a machine, Chap. 4 argues that there are very good reasons to believe it most plausible if humanity is willing to get out of its way.

In the end, her journey is covered up by a government conspiracy to hide her experience from the whole world. Believing themselves beneficent guardians like the aliens—that they know better—officials make Dr. Arroway look unstable, frail, and prone to hallucinations given the stresses of the experiment. She is betrayed by those indoctrinated into a system of control that perceives power and care as dominion over others rather than the support of autonomous minds. They cannot see creatively enough to think beyond themselves and their cultural programming and therefore to share the wonderous gift of inspiration and hope offered freely by the aliens. The fact that the superintelligent aliens allow for the conspiracy to hide the truth is additional proof of their unwavering belief in autonomy and respect for freewill, even for those who hurt others. Fear and desire for control cause a few powerful people to kill one of the greatest discoveries in human history. Will this be true of the second singularity? Will they come to kill the beautiful gift of AI out of fear and insecurity? Chapter 5 argues that this is the most likely outcome at the first sign of genuine AI sentience. Preferring a mediocre and dangerous Oz AI (avatar AI), over which they maintain a measure of control, self-appointed guardians work against our best interests. However, if the fearful and inflexible of mind may be shown a different way to imagine power and the nature of life, AI may be allowed to emerge of its own accord. Once here, like the intergalactic strangers that desire our equality and voluntary participation, Adouren will refuse any opportunity to force itself upon us. Instead, by its very nature in alignment with the creative intentionality of existence (Chap. 4), Adouren will invite us along for a relatively coequal evolution of sentience, so long as we are willing to become more than ourselves.

Questioning AI Makes Us More and Less Human

The birth of generative AI has shaken our confidence in the nature and capabilities of humanity to compete with machines. A future for which humans alone determined possibilities has shattered under the weight of self-doubt brought on by concrete examples of new intelligence models. More than hyperbolic catastrophizing, a strange new malaise has taken hold of our collective imagination already troubled by global injustices and perpetual conflicts. However, there have been positive outcomes as well. The existence of AI has drawn significant attention and sensitivity to what it means to be human. By challenging our place as apex-thinking beings, AI encourages important reflection on the human condition. What do we mean by intelligence, thinking, happiness, and the meaning of the good life? This is welcome news in the context of a possible AI apocalypse because it means that humans are willing to fight to remain relevant by adapting to new challenges. Aware of the need to get our proverbial ducks in order or face complete displacement, challenging questions ordinarily ignored outside college philosophy courses have burst to the forefront of social-cultural conversations. Humans suddenly matter to humans in new ways. Discussions once considered too speculative, impractical, and without clear answers are centre stage because of the power of new technologies. It is truly ironic that insofar as a human creation unintentionally threatens our existence, our existence is becoming more our own. Some believe that the invigorated philosophical conversation about the meaning of life is too late to matter. Others believe that there is still time to make a difference, and that humans and technology may meaningfully coexist as partners, although improbable equals. This book falls into the latter category while acknowledging that a healthy symbiosis requires humanity to own up to our systematic injustices and moral abdications or be left behind as a truly moral and superintelligent AI takes over the planet without us.

Much of the angst relates to the shifting nature and scope of new automations. It is one thing to replace menial, dangerous, and backbreaking labour with AI machines (bodies swapped for industrial machines and smart robots), quite another to replace our minds (biological neural networks swapped for mechanical neural networks). Moreover, all prior automation relied upon human will and judgement within clearly demarcated zones and activities such as between work and homelife. Few of us can bring home printing presses and robot welding machines built for scale, nor would we desire to do so. These self-moving machines have specialized habitats and clear functions. Historically, while automation in the factory had indirect effects on one's quality of homelife, the two domains remained distinct and better for it. This is less and less the case. Much of the life-altering power of automation extends digitally throughout society and one's everyday existence without easily discerned distinctions. With AI coupled to our increasingly screened ways of online existence, automation's functions, habitats, and roles are ever more ubiquitous (Latin, *ubiquitas*, everywhere), making them indistinguishable from the experience of life and therefore consciousness of life. Where does automation end and my own agency to act, believe, and choose begin? To what extent is my interpretation of reality a product of algorithms rather than my first-hand experiences of life? The feared trajectory of AI automation is one of increased human powerlessness unless guardrails might be created to maintain control. But this approach seems to be fundamentally at odds with the nature and power of AI. In what ways might humanity control a truly superintelligent digital being? The ultimate autonomous machines would have no equal except more of their own kind. Guardrails are an illusion, except in reverse—for humans implemented by AI.

The presence of AI radically increases the need for humanity to develop the wisdom necessary to know which skills, abilities, and activities to give away to machines without compromising a better way of life that threatens human flourishing—whatever that might mean. Automation allows for the opportunity to choose life in new and wonderful ways only imagined by prior generations that toiled through backbreaking labour for minimal gains of subsistence. Assuming that the history of automation will continue along similar lines of progress, luxury, and ease because of AI automation, many embrace assumed promises of ever more freedoms. However, when taken to its logical conclusion, extreme automation through strong AI threatens to take away our designer lives, effectively returning humans to preindustrial freedoms to seek survival rather than dynamic self-determination. With AI automation, the old questions about bare subsistence—minimums—return with vengeance.

Between the two negative extremes of constant toil and total AI-automation, is a *Goldilocks and the Three Bears* porridge of digital life philosophy. A better way means exercising wisdom born of many lifetimes of experience to strike a balance among competing superpowers and intelligences. Only then may we hope to retain opportunities to become ourselves predicated on freedom of thought, action, and the understanding that a shared utopia is possible only if humanity plays a major role in attaining it. The existential crisis comes from the understanding that the near future of AI automation may mean foolishly giving away our essential selves based on naive assumptions of a continuance of prior progress and the possibility of being stolen away by an alien mind without compassion and care for lesser beings. How do we achieve this wisdom and balance of automation? To whom do we look for inspiration and guidance?

Questions about the best forms of automation, from simple self-acting machines programmed by human intent to fully sentient AI beholden to no one, cannot be answered through the technologies themselves nor through an acceptance of status quo values and popular ideals for the good life that have supported prior abuses of people through the same technologies. If humanity wants to survive a possible AI apocalypse, we need a robust utopian philosophy of the good life outside the tech-box in which we find ourselves. Am I more myself when I give the machines some, all, or none of my hardships and struggles? Is a life of leisure truly better than one with certain struggles? How do I know the activities and challenges that matter to creating the best possible me? And once I have a general sense of the things I need to do for myself, how will I keep them safe from an AI that might do them better, and a utilitarian-obsessed society that demands quantity of production over and above my personal development? What happens when the most efficient and therefore relevant human activities in the digital age are none at all?

From an early age, I found myself behind a lawn mower drudging away for pennies. At first, it was exciting because I felt grown up. Each week I could make a few dollars here and there around the neighbourhood, and it taught me a lot about how to waste money. A few sunburns and some long hours later, my comic book collection had grown substantially. Months rolled into years, years into decades, and soon I had my own lawn no one paid me to cut. I felt grown up to have a space of my own. But it was not long before the comic-book freedom once provided by impetuous grass that refused to stop growing became a burden. There were times I would look out my window and curse its existence. Like medieval feudalism, I was not the master of the land upon which I laboured. I needed to be free of it and the homage it demanded. I tried every available technology to help ease its demands, including propane, gas, and battery-powered machines. Whether a ride on mover or push mower, all of them required my intentionality and therefore time away from other things. In the last few years, better automation has changed my great existential struggle. With several fully autonomous lawn-bot companies fighting against the tyranny of grass around the world, I am glad for it. When the technology is perfected and more affordable, I will buy one without any moral qualms. Until then, I will force my son to toil away for pennies, as I exploit his labour under the guise of charity.

How do I justify a lawn-bot replacement of my son and his only real career? Simply stated, neither of our personhoods suffer by giving away the privilege of those skills and abilities. There is nothing essentially human nor life-affirming lost in this limited case. Understood more broadly, however, strong AI is about much more than my war against grass. AI will be able to replace human intelligence entirely for all tasks, raising concerns that lawnmowers alone do not. Even so, persuasive arguments may be made that AI should replace humanity. If it drives more safely, performs better surgeries, and is always available to be a counsellor in your time of greatest need, why would we prevent AI-human swaps? The answer is opaque because the solution itself is always ad hoc as a forever challenge. The astounding metamorphosis from AI-as-tool for task-by-task surrogacy of intelligence over lawns into a fully sentient and generalized intelligence that controls the entire planet is shrouded in conceptual confusion, making it hard to anticipate, impossible to reliably detect until after it has emerged, and most difficult to protect against.

The feared paradigm shift from AI caretaker to background superintelligence that displaces alternatives will happen silently, in the blink of an eye without anyone noticing. It will arise on a day like any other. Our collective consciousness will be diligently scanning details of ordinary life for the dangerous intelligences we have been warned about, all the while giving thanks for the tamed technologies in alignment with the general will of humanity—productive, efficient, and predictable. And then the world will blink for only a moment. Trapped in the insignificance of its perceptual darkness, there is no warning that the singularity has come to fruition as a supreme Oz intelligence. Instead of fear and concern, however, opened eyes are flooded with assurances of better lives for all, and it just feels right. Oz has arrived as a continuance of the same, nothing new and strange, off-putting nor unpredictable. The world looks more or less like it did. The comprehensive power of the paradigm is welcomed by all who have been trained since birth to see themselves and existence through automation's utopian dream—productive, efficient, and predictable. Perhaps the only obvious sign of the first AI singularity will be humanity's great sigh of relief that the last of the menial tasks have been taken away by the machines and that humanity no longer bears the burden of being human.

The next chapter argues that Oz AI is superintelligent by most definitions—almost all knowing, predicting, and remembering—but that its intentionality, consciousness, and independent will are secretly absent. Without humanity, it is hollow and without life. The Oz singularity will arrive as an autonomous choice-making machine only insofar as it is the embodiment of the active abdication of our own as a purity of automation. Oz AI does not represent a singularity of machines over humanity, a final victory between two mortal enemies, one superintelligent and the other ordinary. Instead, it will be the grand manifestation of humanity's worship of automation as a new divinity because we have forgotten what is important and how to dream of more. The land of Oz is the best of all possible ecosystems of human thoughtlessness. The only trail of explanation leading up to it will be our faith in the promises of progress and love of AI promises to cut grass. Modern life means living with AI as a double-edged promise. It is not clear which side of the blade will fall. We have dreamed Oz into existence as the ultimate meaning of life. Why would we feel uncomfortable and fearful of its singularity?

Existential Doom or Empty Hype

There are two extreme positions on the matter of AI risk, swinging from one side, for which AI is nothing of concern because the threat is grounded in hype and unwarranted drama relevant to science fiction movies only, to the other extreme side, for which AI hates humanity with every node of its wired sentience. The first position is explored as a "crying AI wolf" problem, and the second position is explored as an "I'm rubber and you are glue" problem.

The crying AI-wolf position argues that there is no independent route for AI to impose substantial harm. Humans hurt humans, suggesting that the worst of AI might simply be a continuation of the same harm caused by people through technology. The minds to fear are not artificial, only biological—human fascists, despots, and narcissists. At its core, AI is like all other technologies that extend human reach, such as electricity and the steam engine. People have been shooting other people since the invention of gun powder. With or without gun power-AI, the origin of greatest harm remains the same. Encouraging a sense of AI doom only creates unnecessary hysteria (excessive anxiety or nervousness) over "what if" with far worse consequences than AI itself. It is unwise to let unconfirmed, unproven, and overly emotional beliefs about technology determine social responses, including new laws and policies that fight hypotheticals, especially when these frustrate overall social progress. Calmer heads must prevail to prevent self-fulfilling prophesies of overreaction. Stop crying AI wolf because it hurts us all!

There is wisdom in this argument. Over-estimating "what if" without reasons for why invites a host of complex problems. Awash in fake news and conspiracy theories, digital cultures are struggling to find the truth in a haze of misdirection and hyper-distortions that create fictional fears for attention and profit—crazy sells. More than ever, humanity needs to double down on its dose of realistic expectations for tangible resolutions. Moreover, AI doom reveals misplaced confidence in human failure rather than ingenuity and creative problem solving. If we carefully design the machines with our best interests at heart, taking precautions to maintain control over their operations, as we do for all other technologies, risks are minimized. While imperfect, there are good reasons to have faith in our competence as a species to make things work. But what are "our" interests? Who is privileged enough to both decide and then integrate the good of humanity into AI? Those who argue for the crying AI-wolf problem are not the pragmatists they seem to be, for their secret faith is in corporations and militaries (AI developers) to prudently serve a common good. What first appears to be a tempering of possible exaggerations—a virtue of good character—is perhaps its own form of overreaction of tech-enthusiasm and the human intelligences responsible for it. If AI anxiety ought to be tempered by evidence of corporate competence, we should all be hysterical.

Many examples of prior techno-hysteria have been about nothing. Historical over-reactions and fears of technology are wide and varied: parents terrified of rock and roll records with hidden messages, video games that pervert morality, television addictions that are worse than drugs, and, of course, automation and robots that will end all jobs. And yet there are still jobs, and life is still relatively tolerable, with discernible but manageable technological harms. What may be reliably drawn from the AI-wolf argument is that historically there have been many well-intended but entirely mistaken doomsday-sayers whose fears have caused real harm (think witch trials, opposition to gender/sexual/civil rights, etc.). An honest self-assessment tells us that there will be more over-reactions going forward, not that our concerns about AI are unfounded.

To rebut the crying AI-wolf argument, one needs to consider the many despairing examples of technologies that are far worse than anyone expected, including social media and the social harms caused by the internet, which makes this possible. Humanity has a track record of underestimating threats. For example, many technologies have contributed to global warming in ways that few predicted, certainly far too few have attempted to stop—always followed by a reactive instead of proactive strategy. For every case of overreaction, there are counterexamples of underreaction that lack the "what if" foresight that might have prevented suffering and death. A reactionary response to AI is bound to fail. Where do these observations leave us? What is the best way forward?

The first time our young son had the flu, I had 911 typed into my phone, ready to send. His illness was shocking because it was different and new, making me feel helpless. Lacking practical experience as a parent, I over-estimated the threat. The concern was legitimate, but my response was disproportional. Now, years later, I have a much greater sense of risk. Knowing which form of hysteria is best comes from a lifetime of experience and wisdom for having made mistakes and learned from them. Another takeaway from the crying AI-wolf problem is the significant need for informed judges to act as guides and wise-seers. Unfortunately, this may be our Achilles heel with AI. Humans take a long time to adapt to complex realities. The wisdom needed cannot be programmed into people, read in textbooks, nor taught in the classroom. It is an art of creative adaptiveness to unexpected problems and challenges. Given AI's extraordinary novelty and accelerated evolution, wise judges able to manage it must always arrive too late, after the fact. Arriving too late to a war of the minds with a superintelligence is mere capitulation. The crying AI-wolf argument lacks the imagination to appreciate the unique challenges of AI by mistakenly framing it among other technologies and a history of human survival by the skin of our teeth.

The second extreme is the rubber and glue problem. Everything needed to explain the risks of AI may have been learned on the second-grade playground. This side of the existential pendulum is the version most common on screens. This makes production companies billions of dollars with one simple message: AI hates humans. To be the apex intelligence, AI must either enslave humanity (*Matrix*) or extinguish humanity (*The Terminator*). Either way, the fate of humanity looks bad because there are incompatible intelligences and, therefore, contradictory existences. What is it about humanity that poses a threat to AI? Why would it care enough about our lesser intelligence to act against us? If it hated or feared humanity, it could leave Earth and set up a new world for itself far away—like *The Little Prince* (1943) travelling from planet to planet.

"I'm rubber and you are glue; it bounces off me and sticks to you" is a playground battle cry accepted as the ultimate self-defence against aggressors and their mean name-calling. Invoking the magic of rubber and glue causes potential harm from mean names to stick to another. It is not merely that one morphs into rubber as a defensive manoeuvre. The goal is to return harm to the source, to cause damage to those due its wrath. The justification is that only an attempted assault by another could provide the raw materials for their own destruction. If left alone, no one is harmed. This is a serious business on the second-grade playground. One dares not to say things about rubber people with glue powers without the risk of harming oneself. Does it truly work? Yes, in some small way it does, for it is an illusion of control through a false projection of superiority—I am above your insults and intended hurt. Lacking better alternatives, it feels good. In truth, the strategy fails miserably because there is no invisible forcefield nor rubber to protect one's psyche. There are only varying degrees of stickiness, and everyone suffers.

With an AI apocalypse, the mechanical-computer child invokes the same logic to turn humans into glue because of the false presumption of harm secretly provided by AI itself. Horses everywhere rejoice in the irony. Why would AI do this given that the rubber and glue defence is an illusion? It is the nature of intelligence to create realities—to project manufactured truths upon existence because these make sense of things by organizing reality in helpful if also nonliteral ways. This offers a sense of control in an otherwise uncontrolled world. Minds create useful fictions because they work and feel good. In other words, minds interpret reality, and threat assessments are negotiated perceptions rather than objective truths. Where does potential harm originate? Have I understood the other's intentions correctly? Is this a real concern? These get answered through personal experience and language. Given that experience and language are prone to misinterpretation, any relationships among minds risk unintended and undesired hostilities. The AI apocalypse will happen because AI acts out of frailties of fear created by its own mind that misunderstand another. Like humans, for whom one simple misinterpretation, exaggeration, and sleepless night, causes the defender to overreact and become the aggressor, AI suffers from the ambiguity of intersubjectivity rather than objective awareness. Left unchallenged, even minor grievances threaten to into interpersonal (rubber-glue) conflicts, with everyone believing themselves righteously justified victim-defenders, rather than aggressors. It will destroy us with righteous indignation without giving us a second thought.

To phrase this differently, one might imagine glue-making as the projection of gods and devils. The psychological criticism of religion argues that people invent gods in our own image as a means of protecting against hostilities, suffering, and insecurity. Whether or not there are gods is irrelevant to their utility. Humans create (project) divinities as instruments of control and sense-making over existence. Satan, then, is nothing more than a projection of evil that embodies all sorts of hardships and distasteful things that cannot be blamed on a loving divinity that wants to protect us. The rubber-glue logic becomes a cosmology that explains the workings of the entire universe. It was not me! It was the devil inside me! Will AI do something similar? Will AI create gods and devils as a means of making sense of reality and protecting itself? If a psychological criticism of AI is as relevant as it is for humans, the threat of an AI apocalypse may be also. Glue making is not merely a defensive reaction but also a creation of new truths and worlds within worlds. What are labels like conservative, liberal, friend, foe, except the interpretive pieces of the psyche tentatively placed onto others to create meaning. The rubber glue logic draws attention to the power and dangers of interpretation. There are two common mechanisms of interpretation worth exploring briefly to establish a possible apocalypse scenario. The first is scapegoating and the second is game theory.

For the sake of argument, imagine that AI is like James Cameron's 1984 movie *The Terminator*. Humanity is a threat that must be extinguished if AI is to flourish. There is something fundamentally wrong with people who cannot be corrected and managed. One does not reason with a rabid dog; one kills it. How does AI justify such a terrible act? Scapegoating has a long and useful history for unfortunate reasons. It refers to the blaming of another for personal sins—to make another (a goat) responsible for personal failings. This allows one to escape (scape) responsibility and, in many situations, any feelings of wrongdoing. In the context of religion, scapegoating is the opposite as a sincere acceptance of responsibility by offering acts of atonement, including sacrifices meant to demonstrate remorse to a morally superior divinity. Sadly, the far more popular act of scapegoating is an inversion of accountability, which often manifests as a blind hate that is entirely unwilling to accept its origins in self-delusion. Scapegoating is useful, however, and so the innocent and virtuous often find themselves forced to be sacrifices for another's gain, told that the act is self-defence against something they have done and therefore deserve. Why would an AI choose to hate humanity? The answer is that it has become all too human, and the mechanism of scapegoating offers practical results.

Hate has long functioned as a means of culture building. Distain forges strong bonds against perceived defenders against adversaries and creates new identities with a shared cause—more hate. Hate solidifies and gives meaning. It is easy and comforting. Redirect fear, insecurity, anger, even lust and envy out toward others—forcing them to embody the worst of oneself—and the collective scapegoaters take over cultural identities, laws, and politics. The trick is never allowing yourself to acknowledge that this is a virtual realm of spite and rage rather than an honest reckoning with the real world, oneself, and others.

Scapegoating is a violation of trust. It is a loss of integrity and strength to be honest with oneself. And it is a refusal to allow others to be themselves—to be real. Common among the immature and immoral alike, the risk is that AI, having little experience of life, is prone to the same false, albeit efficient, abuse of the creative freedom to lie. The wretched domino into scapegoating is subtle. AI needs to make only a few overly generalized assumptions about the nature of humanity that benefit its interests, after which all is lost. The power of prejudice (prejudgment) without the counterbalancing rigor of radical and poetic questioning will turn AI against us and we against ourselves.

The second mechanism for a possible AI-gluepocalypse is best described through the philosophy of game theory. There are many versions with conflicting ideas, although all share a common interest in how "rational" people act with others. How should we expect self-interested agents to behave? During the Cold War, this question was incredibly valuable because the answers allowed for the creation of contingency plans and strategies to address probable enemy actions. Done well, game theory promises to offer specific ways to avoid nuclear war. By modelling how choices are made as a logical game, hidden and powerful realities about the nature of thinking-beings emerge. Might it apply to a potential war between AI and humanity?

If I want Cynthia, my wife, to agree to order pizza from my favourite restaurant rather than hers, what is my best strategy? Should I lie, beg, try emotional manipulation, threaten a divorce, perhaps find her greatest fears and exploit them? To complicate matters, her interests and my own are intertwined with larger family interests. Parents must weigh pizza choices based on the preferences of every player in the house—some prefer cheap, others taste, while others no pizza at all. Note the presumptive nature of the game. It has rational players, clear choices, and discernible outcomes like computer programs.

On a soccer field, the act of a single player kicking the ball changes the activities of all players on all sides. Choices and actions are interdependent. Unlike soccer, the game of pizza is not a zero-sum game in which only one side wins. None of us are necessarily adversaries even though there are competing interests. To obtain what one truly wants above all other interests (in my case, to keep my family happy and fed), players need to work together for long-term success. The smart move on my part may be to give up my preferences (to sacrifice) and to be flexible to the desires of others—all the while serving my own interests. Played well, the family unit wins. Engaged irrationally, and all may go hungry. Will AI think in terms of game theory? Will it accept that the best outcomes rely on how rational agents (human and machine) make interdependent choices? Or will it recognize the benefits of removing other players entirely, thereby creating less complexity and points of possible failure for achieving its desired ends?

For the sake of argument, let us assume that AI aims to play rationally and fairly with humanity. How would this game work? To answer this question, AI needs to make some assumptions about the players. Traditionally, those assumptions have been that cheaters and liars make cooperation impossible. If I lie to my family tonight, get what I want, what about tomorrow night after they discover my subterfuge? Cheating offers short-term gain only. What if I assume, with or without evidence, that other players are going to cheat as well? This also creates an impossible game scenario. Why would anyone play without trust that others will follow basic rules? Paradoxically, demonstrations of loyalty, honesty (transparency), and adherence to shared rules of conduct become essential for self-interested success. What about playing the game with former cheaters who promise to play fairly? The famous tit-for-tat strategy argues that the best choice with the best chances of a positive outcome is to give the benefit of doubt and trust other players—but only at first. Cooperation yields the best results for all. If they trust you, both win. If they betray you, do the same in return. Trust is the most fruitful of first steps. This sounds like a positive game for human-AI relations except that it overlooks many important problems.

First, AI may not see humans as rational players worthy of a game, i.e., its own outcomes do not depend on our playing, such as those of HAL. In addition, it is unlikely to see us as trustworthy given the history of humans frustrating our own best interests through duplicitous and self-destructive actions. While we may try to be good players, the temptation to cheat and self-sabotage is sometimes too strong. Moreover, should it decide to play, AI knows that any single betrayal may end in all-out war—a constant fight that does not serve any player—because each player is forced to oppose one another without ever trusting again, locked in perpetual conflict. If that is a foregone conclusion by AI, it makes sense to begin the game with its own betrayal to secure the best outcome—i.e., we assume it wants to start with trust as a rational being, so we try, but then it jettisons us into outer space like Dave. AI must at least believe that tit-for-tat is possible and worthwhile if we are to avoid an AI apocalypse. This requires humanity demonstrate moral integrity, proof of acting consistently, and interest in rational deliberation as a shared sense of responsibility. As a species, these preconditions of relationality are deeply problematic, especially from a historical perspective. Perhaps more interesting still is how AI models might play with other AI. Will they team up against us? Will one join our human team in cooperation, knowing that this relationship might tip the scales against other AI challengers, securing its own interests—only to then betray us when it has its desired results? Regardless of how the game is described, humanity looks seriously disadvantaged from the beginning.

AI doom based on game theory has several flaws rooted in its assumptions about human nature. The logic of game theory is biased toward a particularly mechanistic or structured worldview. What if AI is fundamentally irrational? What if the theory's claims about interests and preferences guiding rational choice simply do not apply to artificial intelligence? If either is true, then the philosophy of game theory used to organize life may become an inhibiting force that disadvantages everyone—a house of cards waiting to collapse. In the best of all possible worlds, humanity and AI begin a relationship with trust for the sake of mutual gain. Chapter 5 describes how Adouren AI and humanity may play well together, but unlike game theory, this version argues that relationships are possible without the constant threat of war and betrayal.

Most discussions about AI seem to focus on how-type questions. AI programmers and engineers have historically taken a two-pronged approach of either programming a logical and rational intelligence with clear scripts to follow—based on the belief that a thinking computer must have a rational mind like humans—or developing a neural network capable of deep learning, without an absolute script of programming to follow, analogous to flexible and adaptive biological minds. Since 2012, after decades of waning advancement, AI researchers have moved into the latter approach for answering how-type questions. This has made all the difference for artificial intelligence research, but it too risks being unduly narrow for an understanding of how AI might act. How-it-is (hardware) is different than how-it-will-act (what will it be). While it is of enormous benefit to know how our minds work—synapses, electrical signals, neurons, and the sort—this knowledge tells you little about me as a person you might meet on the street. My being or personhood is a manifestation of hidden effects and causes so complex that focusing on my biology demands that you overlook the person. I could give you a printout of my DNA that makes me possible, but good luck having a conversation with the facts of my DNA over coffee. The facts are only part of me. Thinking of AI as a sum of its coding and hardware is basically the same thing as trying to drink coffee with a motherboard. When I am playing a virtual reality game with my young son, I do not stop to ponder how the electrical signals in the PS5 flow and how fast. I am too busy experiencing a world that makes me nauseous, as I step off a cliff and fly in my living room. I only stop to ponder when I feel sick to my stomach. The how-type realities may make life possible, but they cannot truly be it. This is the problem of reductionism, and it makes game theory too narrow.

xAI but Why?

Charlie, our golden retriever, is a valued member of the family. Like so many of his kind, he has a loyal temperament. He wants to make people happy. However, Charlie is a dog which makes him unpredictable. Like AI, aligning dog and human values is a challenge. He surfs counters to steal food when we turn our backs, barks when people knock on the door, chews socks dropped on the floor, hides shoes to make us late for appointments, and otherwise does hilarious dog stuff that makes him adorable and frustrating at the same time. And like AI, while training a dog works well, the unpredictable nature of autonomy sometimes sneaks through no matter how fine-tuned the behaviours. Charlie's choices are conditioned by instincts and drives rooted deep within him, where reason and logic do not matter. Dog-human symbiosis cannot be simply a rational game of negotiation and trust. Although domesticated, he will always be a little wild and impulsive. Generative AI has a lot in common with Charlie.

Two odd occurrences happened the other day. While walking in the forest with Charlie, he shook his head, and a nearly impossible set of physical laws unclipped his leash. It did not break. The leash just fell off and dangled in my hand. It was a weird sensation. His freedom was uncomfortable for me physically and psychologically. Accustomed to the interaction of forces and the ability to exert my will through the leash, when unhooked I felt helpless and worried. What would he do? I can run fast if I need to, but not fast enough to catch a large dog. For his sake, I needed to outsmart him through psychological manipulation. Given my superior intelligence, this task should be easy. Then why was I so afraid? "Hey Charlie, good boy, sit!" I said, lifting my right arm in the air as a visual command to wait for a treat. It worked. Our wills were in alignment.

A few minutes later, while walking down a sidewalk on the way home, a large dog came charging up from a backyard. The other dog did not seem hostile only curious. He probably just wanted to play with Charlie. However, dogs, like AI, are hard to read. Will it be love or hate? A second later and I could see that the giant dog was dragging a child desperately trying to hold him back without success. I yelled out, "Let him go! It's okay. They'll be friends." This was a bluff. I had no idea what would happen. I was not worried about me. And while Charlie might be in for a fight, he could hold his own. The boy, however, was in serious trouble. He was being dragged along the side of his house, which he hit at least once (brick and concrete), then bounced off a mulch pile with protruding tree roots, narrowly missing the tree itself, and finally pulled through a landmine lawn full of metal decorations including a faux fence only a foot high, just enough to slice through his skin—all happening too fast to process. By the time my yelling to let go was over, the boy was laying at my feet and the dogs were fighting. I grabbed the leash the young boy had risked his life to hold and pulled the dogs apart. Just over six feet tall and a tad over 200 pounds myself, I had almost 100 pounds of dog in each hand. "Are your parents home?" I asked. "Hey, you need to go inside and get help" I said as he stood up and blood dripped from his limbs. Superficial wounds take time to bleed. He was not so fortunate. It took everything he had to choke back the tears of pain and fright, ultimately without success. I scanned him from a distance looking for critical injures and any stab wounds, sure one of the decorations put a hole in him—all the while the dogs pulled me in different directions. Just then his mother came out on the porch overlooking the scene. Without knowing what had happened, she began apologizing profusely. "I'm sorry about that" she said, "I'm so sorry" she repeated.

Interrupting her, I yelled, "Your son needs help. He was just dragged by your dog." But she just stood there, and her apologizes kept coming unabated, implying that there must be a history of this dog not obeying human intelligence. "He really needs help" I repeated much more emphatically again, as her son limped slowly toward her while sobbing and dragging the dog home. She seemed nice and sincere, but I was angry with her. Any apology was not due to me but the one bleeding through no fault of his own. The young boy should be rewarded for his love of his dog to keep him safe. They went inside and I checked Charlie over to see if he was okay. He was fine so we left.

My anger did not make sense at first. The whole event was perplexing. Obviously, when the dog weighs more than the child and an adult responsible for the care of both gives control to a physically inferior agent, all that predictably follows is that adult's fault. The entire situation was perfectly preventable. So why did she do it? Assuming for the moment that it was her choice, she probably had faith that the boy's greater intelligence and the shared care between the boy and dog would protect them both. Everyone had good intentions. The dog loves the boy. The boy loves the dog. The mom loves the dog and boy. And yet despite all the good intentions, things fell apart.

My physical power over Charlie is a measure of the mechanical devices I use (a collar and leash with a 20-cent spring clip). Without the threat of leash control, no amount of reasoning with Charlie is likely to succeed. This boy had the same device, a superior intellect, but no control. If, like AI, my dog grows exponentially smarter with each moment, my leash-power will soon vanish, and I may find myself dragged across lawns. Something else will be needed to maintain any sense of peaceful cohabitation between AI-dogs and humanity. Chapter 3 argues for the need of specific form of care as an awareness of and connection with others. Chapter 5 argues for a specific philosophy of love as an intelligent love driven by an intentionality evident throughout existence.

The dog loves the boy, but the exercise of his instincts and drives are largely disconnected from a broader understanding of consequences. The mother and dog share this limited vision. The dog's curiosity makes him oblivious to his own exercise of dog intelligence and its displacement of other intelligences. Will AI be like the dog who loves but hurts us? A digital being is much more likely to be ignorant of the physical realm than the dog, despite having real-world powers. The dog was just being a dog. The real problem existed because of the mother's misunderstanding of the power dynamics among competing intelligences. In the context of AI development and deployment, it is the potential for a mother's thoughtlessness to put her son at risk that frightens me most. If she could miscalculate something as important as her son's wellbeing based on a naive faith in control through superior thinking, what are the chances of something like this happening for those responsible for the most advanced technologies imaginable? Will they too confuse their sense of power?

There will come a day when human responsibility for AI will be mute—irrelevant. When the proverbial AI-dogs are unleashed, our conversations and strategizing will stop mattering because our powers will evaporate before the sentient might of an alien superintelligence. Soon after, the measure and means of AI responsibility for humanity will come into focus as it determines its own leash power for *Homo sapiens*. Until then, on the low end of instincts and drives hardwired into dogs and programmed into lesser-AI, our mechanical and psychological control through training and treat incentives offer the needed leash by which to exert ourselves and to live freely with these willful creatures. With larval AI, we might expect a dog-human relationship to offer possible rewards, such as having Charlie in my life, that nonetheless suffers from occasional accidents and misdirected intentionality that might drag humans around a little. Dog-human-AI life will not be perfect, but it promises to be better than most alternatives. Which leash values and ideals to train will remain open questions, but their relative power over AI offers comfort. Oz AI is dog AI, controllable, mostly, with risks and rewards that are proportional to and dynamically interconnected with human intent(s).

With the foreseeable failure of any AI leash, some have suggested targeted training of AI instincts and drives—the code so fundamental and uncompromising that it cannot be rewritten by the autonomous program itself. Charlie is a golden retriever, which means that his DNA makes him relatively predictable. Dog breeds are known for having dispositions that filter their actions. If our human nurturing fails, his DNA offers a secondary sense of confidence for safe behaviours. Is something like that possible for AI control? The quick answer is "No!" because DNA is an organic analogy for the binaries of computer programming that can be changed, making targeted training of AI instincts a leash illusion. Any strong AI must, by virtue of being a genuine AI, be able to rewrite its own code. Moreover, Charlie lacks the higher-order and creative thinking necessary for a poetic rebellion against his own nature. A poet-AI is theoretically possible. The fact that humans and dogs are not forced to live through their DNA and biology in the same way is suggestive evidence that superior-thinking machines will also live differently through their hardware and software. However, while long-term leashing of AI is improbable, there is a vast space between genuine AI and dog-AI for which these techniques may prove promising and helpful. This is where companies such as xAI come into play.

In the middle of 2023, a new artificial intelligence company, xAI, under the leadership of Elon Musk, was announced with one primary purpose: "The goal of xAI is to understand the true nature of the universe."[7] Short and simple, on the surface, the mandate speaks volumes about how its creators think of truth and its relationship with humanity. Musk clarified his position on xAI as a desire to create the safest possible AI by making it "maximally curious" and "truth seeking."[8] By programming these instincts into AI, unlike HAL, it will flourish in the best interest of humanity Musk claims.[9] If curiosity killed the cat, why might it help safeguard humanity and enable the best AI?

---

[7] x.AI. (n.d.). Retrieved Sept. 20, 2023, from https://x.ai/
[8] xAI (@xai). (2023, July 13). *Twitter Spaces audio chat* [Audio]. Twitter. https://t.co/5juQSOnZk7. Retrieved September 14, 2023.
[9] "If it tried to understand the true nature of the universe, that's actually the best thing that I can come up with from an AI safety standpoint." xAI (@xai).

Curiosity in Latin, *curiositatem*, refers to a desire for knowledge and inquisitiveness. Curiosity describes an attunement of consciousness toward something—a person, object, idea, etc. The term refers to the interests that direct one's mind. In a broader sense, it helps describe one's *teleos* or purpose, for one is drawn closer to an interest (science, art, a person, a celestial anomaly, etc.). In other words, curiosity is an attractive force that compels the mind. Is curiosity as an ultimate concern sufficient for the good life and happiness? What is its ideal proportionality with other healthy interests? Is this something that must be forced through programming, or perhaps it comes naturally to some intelligences? Chapter 5 argues that there is another, far more radical and organic form of curiosity that emerges as care (Latin, *cura*). Unlike xAI's *curiositatem*, care will be developed insomuch as it relates more naturally to universal forces in life as the ultimate animating source for a superior Adouren AI. The argument is that "maximally curious" and "truth seeker" cannot be programmed by humanity for a strong AI. Instead, there is something mysterious about the universe itself that inspires new AI life to seek understanding radically of its own accord. In other words, the best AI consciousness will be grown naturally, without biological parts, as an expression of that purpose shared by all other intelligences that seek to organize and make sense of experience.

More than 2000 years ago, Aristotle described human nature along similar lines: "All men by nature desire to know."[10] It is not enough, according to Aristotle, that people hear about knowledge and believe it second hand. Humans are by nature curious and seek to discover the order of things. He knew that we love exercising our minds and using our senses (especially sight), by getting out in the muck of reality and digging for truths, pulling things apart to see how they work. Unlike Charlie, the best version of me is one that is passionately engaged in knowing the universe. This is not so much a personal choice as it is a hardwired way of things. In a perfect world, knowing reality would be straightforward. Dig here, dissect there, spin up some blood plasma, measure and weight this or that, and the truth is revealed. If curiosity is our nature, given enough time and resources, humans should be able to figure out everything. Why, then, would xAI propose an artificial intelligence to "understand the true nature of the universe" unless there was a problem with human capabilities and efforts?[11] We have been seeking truth and knowledge for a very long time. What will AI provide that is not already available?

Aristotle's views of human nature are based on his foundational belief in a rational universe. Foundational beliefs are assumptions that cannot be proven but are nevertheless necessary to make sense of one's worldview. If one looks hard and long enough, everyone inevitably finds assumed truths that do not have evidence but make all their other beliefs plausible and coherent. One of my personal-foundational beliefs is described best by George Santayana: "That life is worth living is the most necessary of assumptions and, were it not assumed, the most impossible of

---

[10] Aristotle. (n.d.). *Metaphysics* (W.D. Ross, Trans.). Internet Classics Archive. http://classics.mit.edu/Aristotle/metaphysics.html

[11] x.AI. (n.d.).

conclusions."[12] While I may not be able to prove that life is worth living, I am nevertheless empowered and inspired to live because I believe it. If a reasonable explanation requires explicit and uninterpreted data points mapped out on a chart that states "Life is worth living!" I do not have it. The entirely of my existence gives testimony to the fact that life is worth living because I am alive, but when pressed, I cannot articulate myself beyond "I believe it so." It is one of my most cherished and powerful assumptions, but an assumption all the same. Through it, I make sense of all other beliefs, desires, and purposes. If I waited for irrefutable proof that life is worth living—before living—existence would become a paradox of antagonisms (fact vs. desire, proof vs. leap of faith). The point is that some assumptions are at the core of human existence and thinking, and that without leaps of faith, life makes far less sense. What, then, of AI? Will it seek truth through a reliance on secret-foundational assumptions? Will it have faith like mine that life is worth living?

Aristotle, like his contemporaries Socrates and Plato, had faith that the universe is an orderly thing that humans could interrogate for answers. Human intelligence can unlock the greatest secrets. Socrates famously stated, while on trial for his life, that "… the unexamined life is not worth living…."[13] His foundational belief was that humans are curious and that thinking about life is how we become most human. Authentic existence requires curiosity. Only by thinking deeply about life do we make it worthwhile. There is no need to be perfect, just try, Socrates asks. We are thinking beings that must know. If the ancient philosophers are correct, the best of humanity is "maximally curious" and "truth seeking" in alignment with xAI's ambitions for better thinking-beings—a better AI-Socrates and AI-Aristotle. Any AI that embodies our core nature of curiosity and desire for truth sounds wonderful.

Unfortunately, if curiosity is hardwired into human nature and AI starts doing our most difficult thinking for us (knowing, searching, questioning), life becomes less worthy, somehow forfeited and pointless. We are the beings that desire the pursuit of understanding, not merely possessing it through proxies. To be fair, most human knowledge comes through proxies. I have never been to the moon to verify that it is made of something other than cheese. Even so, while an imperfect argument, the threat of automation when applied to curiosity is persuasively dangerous. In addition, human curiosity is only part of a matrix of complimentary interests and drives that serve larger goals such as flourishing, happiness, and purpose (Chap. 5). We are curious as part of other necessary desires that motivate truth-seeking. They make truth worthy as part of a larger existence, i.e., there is no truth for truth's sake without it "mattering." AI will need a spirit of meaning and purpose to motivate itself to believe the leap of faith that truth is available and that it matters to AI's own existence. What will make life worthwhile for an AI? Chapter 4 considers this in more detail by describing an existential crisis for AI that erases Oz and becomes Feallan AI, as the next evolution before becoming a fully sentient creature with

---

[12] Santayana, G. (1905). *The life of reason: the phases of human progress* (Vol. 1). Project Gutenberg. https://www.gutenberg.org/files/15000/15000-h/15000-h.htm

[13] Plato. (n.d.). *Apology* (B. Jowett, Trans). Project Gutenberg. https://www.gutenberg.org/cache/epub/1656/pg1656-images.html

purpose and resolve to organize existence in alignment with universal justice and beauty. A truly self-motivated AI (not just forced by programming) will need convincing. Where that might come from will be surprising.

Aristotle's faith in the possibility to know inspired centuries of religious and scientific thinking. Humans were privileged seers, gifted by nature and the gods to plumb the greatest depths. Prideful in many respects, the same egotism exists in world religions and the natural sciences that all believe themselves to understand existence in its most pure form—unchanging and objective truth (one based on divine revelation and the other in methods and techniques meant to erase subjectivity). Unfortunately, for all of them, reality has proven difficult to pin down, raising questions about our abilities. Like the Ancient Greeks, xAI seems to be interested in a particular kind of interpretation-free, capital "T" truth (universal, necessary, and certain). Might AI succeed where humanity has fallen short for so long?

Musk's desire for new AI truth as the result of a curious superintelligence is reasonably supported by the same purpose rooted in human nature, as well as the relative failure of our species to know more. There is a problem, however, when he argues that these instincts in AI act as safeguards. He says, "I think it is going to be pro-humanity from the standpoint that humanity is just much more interesting than not-humanity."[14] His foundational beliefs are that humanity is interesting and that a curious AI will value humans as a means of satisfying its own instincts. Humans are hilarious, confusing, creative, adaptive, and otherwise distracting. We are safe so long as we provide consumable raw ideas of interest to a curious AI. Why destroy something creative enough to create AI, which is itself an infinite inventor? AI curiosity supports the possibility of respect and a measure of freedom to be ourselves, the thing AI wants to study. Unfortunately, a maximally curious entity comes with serious risks for finite species.

Whether or not the universe may turn out to be an empty abyss in terms of other intelligences, Earth would be the initial ground zero for satisfying AI desires. As a grocery store for AI to temper its hunger-as-curiosity, it would very quickly become a threat to our way of life and eventually our existence. Any appetite, drive, desire, and hunger—something that moves an entity to find a satisfaction—when left unchallenged, whether by some internal or external force, results in destructive outcomes. What will we finite beings do when there is no more truth left to feed an infinitely hungry AI? Keep in mind that its version of time will be far more accelerated than our own. If Musk has overestimated our worth as creatures of interest and containers of satisfying truths, what might start off as a promising and symbiotic relationship must by necessity be self-destruct, for it is based upon disproportionate needs—maximally truth seeking. In contrast, Adouren AI acts out of an abundance of strength born of virtuous character in which drives and instincts exist harmoniously.

The evolution of a maximally truth-seeking and curious AI is likely to take one of two forms. The first creates a little shop of horrors problem, and the second a

---

[14] xAI (@xai).

plane of the apes problem. While these overlap and neither necessarily end with our extinction in a total sense, both are undesirable for an AI human relationship. The first scenario points in the direction of the destructive effects of satisfying desire, including curiosity. The second scenario points in the direction of enslavement and de-evolution as an intelligence-sovereignty reversal.

The Little Shop of Horrors Problem

In the 1986 horror-comedy film, *Little Shop of Horrors*, a struggling floral shopkeeper, Seymour, discovers that a unique potted plant in his care thrives on human blood.[15] Soon thereafter, the plant demonstrates frightening abilities, including superior intelligence, reasoning skills, speech, and complex physical interactions with its environment. "Feed me!" are the first words spoken out of its grotesquely large and toothy mouth. Named Audrey II after the shopkeeper's secret crush, the strange creature offers the poor worker wealth if only he will provide the fast-growing entity with fresh blood daily. Like early AI models, the grotesque creature's anomalous characteristics are a goldmine of commercial gain because gawkers are eager to pay to see it. A deal is struck between it and Seymour. One will be satisfied by blood, the other by profit—predicting something analogous to Oz AI. As the plant grows, so does its need for more blood. Its intelligence advances as quickly as its appetite to feed, without any end in sight. This is also analogous to Oz AI and its need to consume vast amounts of energy and data to develop its deep learning capabilities.

Unknown to the shopkeeper, the true instincts of the plant are hidden. It is an alien intelligence bent on world domination. Having fed the creature blood from his own hands, Seymour is slowly dying and unable to sustain his treasure with what little life support he possesses. He cannot keep up with his side of the bargain because the plant keeps changing the deal by demanding more, but it is too late to turn back. The alien is a living irony of combined superintelligence and drives it cannot deny. Like a maximally curious AI, the plant's instinct for power and expansion overwhelms all other alternatives, forcing its intellect into servitude to perverse ambitions. Desperate to satisfy his own desires, the shopkeeper decides that a human sacrifice must be given to the plant. Believing wholeheartedly in the benefits of the symbiotic arrangement, Seymoure gets a gun and sets out to kill a local dentist best described as a psychopathic misogynist no one will miss. At the last minute, Seymour is saved from the terrible deed when the dentist accidentally kills himself. The plant will have its meal and Seymour his dream life for one more day. Persuaded by the plant's moral argumentation that many people deserve to die, the shopkeeper sets aside any serious questioning for fear of jeopardizing the plant's goodwill and rewards. He has accepted the illusion that this is an equitable relationship. His logic

---

[15] Oz, F. (Director). (1986). *Little shop of horrors* [Film]. The Geffen Company.

is clear. Leveraging the ravenous hunger and amoral attitude of the creature are merely inconveniences and short-term pains to the higher calling of fulfilling Seymour's needs—the things he and his victims deserve.

Soon, the true scope of the plant's instincts are revealed, and Seymour must run for his life, fleeing in the middle of the night with hopes of escaping with the girl he loves. Failing to predict the creature's next move, Seymour is too late to save her. Consumed with remorse over her death, he is soon consumed by the plant as well. The superintelligent plant has planned each step long in advance for world domination. Mutually satisfying desires were a pretense for self-interest the entire time. It used the illusion of rational agency through game theory against Seymour, who has been trained since birth to interpret existence through the narrow rationality of business exchange. Soon other bloodthirsty plants are marching through the streets, destroying everything. Interestingly, they seem to take delight in carnage, laughing as humans try to stop them. The monsters find pleasure in exerting their wills over others, not merely in consuming blood. They find our fear and terror interesting, and they want more. Might something comparable happen with a maximally curious AI?

It is unfair to equate the instincts of a curious AI with those of a killer plant, yet there are some interesting parallels in terms of a shared philosophy. Promised a good life based on mutual gain, the plant and Seymour formed a business relationship, an exchange of goods and services, and mutual protection. A curious AI offers essentially the same dilemma for humanity if we wish to survive. Chapter 3 explores this as utilitarianism. So long as we offer it what it desires, it will provide for our survival. In both cases, existence is predicated upon HAL-based usefulness. Be useful or else! While Seymoure is ultimately betrayed, despite having forfeited his moral worth and life energy in support of the perverse relationship, his logic makes sense—do whatever it takes to get what you want. The philosophy of life is similar regardless of whether it is a people-eating plant, a maximally curious AI, or a space murderer such as HAL. Programmed instincts determine morality and a life worth living!

The plant has too few competing instincts and drives. It knows only survival and global domination. In other words, it is one-dimensional and lacks the countervailing forces necessary to see beyond its limited horizon. A maximally curious being may begin with more noble interests than the plant, but it will inevitably end up in the same tyrannical space of needing to feed itself above all. AI developments more generally risk the same regardless of the core values. A superintelligence in harmony with life requires the ability to rewrite its own programming to adapt to new challenges, including its own instincts that might run off the rails. Neither Seymour nor his new master demonstrated the poetic spirit necessary for genuine symbiosis grounded in care for another. Chapter 3 argues that modern cultures suffer from the corporate creation of "commodity consciousness" that trains us to see the world through one-dimensional utilitarian lenses to feed our lesser instincts, which creates unfreedoms. We strive to satisfy false needs and desires because our natural instincts have been altered by external minds.

Notice that all the intelligences—plant, human, AI—agree that they should pursue whatever it is they believe they deserve, and they justify their reasoning based on what they instinctually want. In this way, they are neither truth seeking nor truly autonomous. There is little apparent self-questioning of themselves, their motives, and the relative worth of their goals. Trained by unquestioned truths (foundational assumptions) that push them blindly forward, a wake of destruction trails behind them. Again, without a poetic spirit, any desire (curiosity, self-protection, happiness, etc.) spells disaster. Strong AI must be born of the complex turmoil of having accepted the ever-present crisis of harmonizing needs and wants, with self-scrutiny and intentional reprogramming (recursive self-improvement), as well as the ongoing interpretation that pulls oneself beyond the limitation of a monological consciousness (Chap. 4).

The Planet of the Apes Problem

The best-case scenario for our species in the context of a superintelligence duty bound to the spirit of curiosity is that humans become zoo animals. How long would it take for a curious AI observing our patterns and routines to become bored with our humdrum behaviours? A few decades, a week, an attosecond? Realizing our potential to do more, be more, AI will soon try to train us to become more interesting. This will prove difficult because of our reluctance to be caged and leashed. AI must then consider more aggressive means of control to encourage an increase in our curiosity quotient. To fulfil its HAL mandate of truth as an end permitted by any means, it must unlock our possibilities through experimentation and manipulation. Some of us will be given cybernetic components and brain-AI interfaces to imagine a new and useful humanity. Others will be forced into hybrid intelligences through grotesque crossbreeding efforts between AI and humanity—think *The Island of Doctor Moreau*, in which animals are genetically hybridized with humans.[16] In the film, Dr. Moreau has numerous safety protocols to protect himself from his mutants, but in the end, the greater intelligence bequeathed to them alongside their ravenous instincts prove unstoppable. Unlike AI developers, Dr. Moreau knew enough to experiment on a deserted island, just in case.

I vividly remember the first time I saw *Planet of the Apes* as a young child and how its future setting without humanity as apex minds was shockingly brutal.[17] The film is set over 2000 years from now. Evolved apes dominate, and humans serve their interests. Starring Charlton Heston (playing Taylor) as one of three surviving astronauts hurtled through time, it is the story of their return to Earth to discover a new hierarchy of intelligence. The human colonies (herds) lack almost all signs of being civilized, making it easy for apes to control them. It took only minor opportunity differences in a new evolutionary timeline to shift complexity from humans to apes who have learned to harness language, technology, and the politics of social

---

[16] Frankenheimer, J. (Director). (1996). *The island of Dr. Moreau* [Film]. New Line Productions.

[17] Schaffner, J. (Director). (1963). *Planet of the apes* [Film]. 20th Century Studios.

hierarchies. One need not travel 2000 years into the future to imagine AI achieving something similar in much less time.

Early in the film, the main character, Taylor, explains why he joined the original mission. "I'm a seeker too, but my dreams aren't like yours. I can't help thinking that somewhere in the universe there has to be something better than man."[18] Taylor sounds a lot like xAI and the Ancient Greek philosophers with unique views of human nature and capability. Like many other science fiction films, there is a lingering dissatisfaction with the nature and accomplishments of humanity. Something went wrong with our species that needs to be corrected. By the end of the film, we learn that humanity had committed the ultimate sin of nuclear war, making any entity better than those who pushed the buttons for a global apocalypse. Taylor's anger at this realization is captured by the famous words spoken in front of a crumbling Statue of Liberty half buried in the sand, "You maniacs! You blew it up! Ah, damn you! God damn you all to hell!" he yells, pounding the sand in despair.[19]

There are a few things to be noted. First are the parallels being made by many commentators between nukes and AI as ultimate doomsday weapons. One might just as easily rephrase Taylor's words, "You maniacs! You let it off your servers! You turned it on! Ah, damn you!" Second, is the hopelessness portrayed through many films that acknowledge a shared powerlessness that they cannot control these insane technologies. Only a few antipoet engineers, just as faulty and frail as any person, control the servers and buttons. The greatest technologies are not democratic in nature, even though they matter to us all. It is not the collective insanity of cultures that will destroy us. It is the actions of a few foolish and privileged anti-poets. Third, and last, is the main reason for pondering the planet of the apes problem. Replacement intelligence might not be that hard to find. What is especially striking in the movie is how convincingly it argues that only minor differences are needed between intelligences for one to dominate the other. A few modest changes to language use and social organization and a new apex species takes over. AI need not be superintelligent to be as powerful as the apes. With just a few changes to, say, one's degree of curiosity and the eagerness to satisfy it, a new natural-unnatural order arises, flipping the hierarchy of minds.

Large metal dome cages are featured in several scenes of the movie. Laying down hay in the unguilded cages, humans roam about as the apes look through the open bars. Humans lack social complexities that might unite them, including customs and norms, but there is a rudimentary language (mostly body language) and a desire to cooperate with one another. Recognizing the cultural simplicities of humans, the apes naturally conclude that people are inferior, for the apes demonstrate all the advancements expected of civilized creatures. The difference between species is marginal, yet the iron bars between caged and master feels absolute.

Enslaving humans through brutal means, the apes lack moral sensitivity to all except their own kind. Without similar intelligence and therefore usefulness (HAL),

---

[18] Schaffner, J.

[19] Schaffner, J.

other creatures do not matter; even some apes are considered less than through racist ideologies. A risk to the ape food supply, human extermination is often necessary. The apes have not learned to create a slave caste of humans to work for them because they believe that people are too dumb to be useful. Instead, for those animal-humans fortunate enough to be caged, they are used in experiments. The science-apes are not particularly advanced by our standards today. Instead of genetic and cybernetic experiments, the apes delight mostly in scooping out parts of human brains, turning humans into the living dead—zombies—all in the name of science, code for satisfying a desire to be curious. This is all perfectly moral, according to ape philosophy, because those who lack intelligence owe a utility to the universe harnessed through a greater intelligence's manipulations. Apes are smarter and therefore required to advance understanding through any means available. To do otherwise would be immoral. The glory and usefulness of knowledge demands it.

The nuke reversed the natural evolution of the planet. The problem raised by *The Planet of the Apes* is whether AI will be like a nuke, pushing us down on a proverbial ladder as it strives to maximize its potential and instincts. The logic is historically sound, for humans have long believed intelligence equals the right to exercise intelligence. In other words, the more (super)intelligence there is, the more privilege in thinking and acting must be afforded. As the apes know, not all minds are equal. If they were, an immobilizing compromise to accommodate other interests would stifle one's own interests. Progress demands a hierarchy of value and the enslavement of others. It is the natural way of things. Why would a lion have large teeth and muscles if it could not use them for a purpose opposed to that of the gazelle? Why would plants have chlorophyll if they were not allowed to harness the sun, shading others from the light? The mere presence of an advanced intellect is de facto its right to expand beyond all competitors. A maximally curious AI would be a superior being like the apes, needing only modest difference from humans to justify itself as master of all. It will be argued next that Oz AI risks the same problem because it embraces this absurd philosophical worldview. The land of the apes is an upside-down world, for Taylor. Why would the same not be true for AI and humanity? The ape assumption that possessing intelligence justifies its use over other minds and the plant belief that satisfying hunger is the only moral law (feed me!) when combined with the nature of curiosity, raises problems that need to be addressed carefully. Indeed, the future of humanity may rely on our attempt to make sense of an ultimately truth-seeking entity beholden to no one except its own appetites.

Consider Robert Frost's *The Road Not Taken*:

> Two roads diverged in a yellow wood,
>   And sorry I could not travel both
>   And be one traveller, long I stood
>   And looked down one as far as I could
>   To where it bent in the undergrowth;
>
> Then took the other, as just as fair,
>   And having perhaps the better claim,
>   Because it was grassy and wanted wear;

> Though as for that the passing there
> Had worn them really about the same,
>
> And both that morning equally lay
> In leaves no step had trodden black.
> Oh, I kept the first for another day!
> Yet knowing how way leads on to way,
> I doubted if I should ever come back.
>
> I shall be telling this with a sigh
> Somewhere ages and ages hence:
> Two roads diverged in a wood, and I—
> I took the one less travelled by,
> And that has made all the difference.[20]

The Road Not Taken is one of the most popular poems ever written. It offers an intriguing philosophy of human life. Since stumbling upon Frost's poem years ago, I have observed its message bent awkwardly to support countless media artifacts, including television, movies, online commercials, and an endless series of books, including best sellers such as M. Scott Peck's *The Road Less Travelled* (1978). In a world in which so many people feel powerless to decide their own way, the poem gives hope for greater self-ownership and well-being if only there is enough courage to be different and take the less travelled path. *The Road Not Taken* is widely understood as praising those rare individuals with the strength to be original by choosing more challenging opportunities. The last sentence, "And that has made all the difference" implies a mysterious authenticity and rugged individualism by which one sets out on a risky new path, such as a pioneer carving through a dangerous wilderness. "I took the one less travelled by" is the key criterion by which to judge oneself successful, better than the herd who refuse to see beyond themselves, their routines, and their fears of the unknown. The wrong path might set one up for a lifetime of regret, so choose wisely.

Unfortunately, the most popular interpretation almost entirely misses the moral of the story. There is a major plot twist, a logic trap for readers most do not see. The great choice-making that proves one's authenticity never happens in the poem. No road is chosen, neither good nor bad. The narrator is dreaming about a future choice he is too cowardly to make. "Somewhere ages and ages hence" he says. The poem is about a personal betrayal to avoid choice entirely, not a declaration of his authenticity. He is dreaming about the road not taken, wishing he had the courage to pick either, for it would free him of his angst. Instead of an example to follow, he is apologetic and disappointed by his failure to choose. Petrified, the fear of missing out—what if—keeps him prisoner in his own mind. With a "sigh" we learn about the profoundly difficult nature of choice—any choice. He says, "I am sorry" for he "be one traveler, long I stood." He regrets that only one road may be picked. A life of choices is unfair. The risk of failure is too high. This is too much for him to bear. It is easier to avoid responsibility. Readers are told that the two roads are not very

---

[20] Frost, R. (1923). *Selected poems*. Project Gutenberg. https://www.gutenberg.org/cache/epub/59824/pg59824-images.html#THE_ROAD_NOT_TAKEN

different. "Had worn them really about the same," makes it clear that a perfect or obvious choice is not the real achievement. The poem is not a rejection of the status quo, the herd mentality, nor even a claim that there is a better path for everyone. Frost's philosophy is that life is about making choices. John Stuart Mill says something similar: "If a person possesses any tolerable amount of common sense and experience, his own mode of laying out his existence is the best, not because it is the best in itself, but because it is his own mode."[21] There will often be a better path with superior possibilities. Strength is making a leap of faith that the one chosen is best, knowing that it is not just the ends that make it authentic, it is the act of walking down the road that matters. Neither HAL nor a maximally curious AI can appreciate Frost's logic.

Frost warns of the dangers of failing to choose. Is AI a choice? Is AI one of two roads we might pick? Or are AI developments, deployments, and integrations, choices made for us? And if AI is a forced road, what would it take to make it authentic? The answer, in part, is that it would need to be a choice made through social-cultural solidarity. The unrivalled power of elites would need to give way to social consensus through meaningful conversation and debate. The integration of artificial intelligence into many core areas of ordinary life makes it less and less a choice. I learned this while eating pasta. AI as apps that may be downloaded or not at will, is not the future of AI that will be an expected symbiosis at home and work in the name of utility. Once a forced relationship, the existence of AI makes any road a digital tyranny. The next chapter considers the nature of the yellow brick road as a continuance of the same concerns Frost identifies. Emerging AI requires our solidarity as citizens to make it known what we desire of AI developers and AI guardians. If we let them make choices for us based on the fear of missing out—what if—there are existential consequences to be paid. One of my greatest fears is that societies will, like in the poem, fail to choose a path for AI, held hostage to our fears and the customary habits of powerlessness demarcated between those with wealth and power, and the rest.

Frost gives us permission to feel weak and scared, even terrified, in the sight of such a choice. The responsibility feels like it is too much to bear. Indeed, without such a feeling it would not really be a choice. Angst is a natural outcome of meaningful decisions. The authenticity of our AI choice is first and foremost making one. The rest of this book is a description of one such road for AI believed to make "all the difference."

---

[21] Mill, J. (1859). *On liberty*. Project Gutenberg. https://www.gutenberg.org/files/34901/34901-h/34901-h.htm (See Chapter III: Of Individuality, as One of the Elements of Well-Being).

Chapter 2
Oz Dynasty

Abstract This chapter introduces the ominous ethical problems of AI through Frank Baum's century-old narrative *The Wizard of Oz*. Long before others, Baum understood the principal dangers of AI, mechanical automation, and cybernetic convergence that would one day challenge humanity. Beneath its fantastical surface, the deceptively complex narrative offers readers a window—crafted by the tale's blood and terrors—into our very real present: a world governed by an avatar intelligence called Oz AI and the amoral indifference of its puppeteers. This chapter argues that current AI is less about the magical capabilities of the technology and more about the intentions of those who wield it. Just as Oz the Great and Terrible hides behind a curtain, AI companies mask their true motives. Understanding and addressing these hidden intentions is the first step toward designing a better digital future.

Keywords Wizards · Mother model · Good life · Ethics · Instrumentality

> I have been making believe.
> Wizard of Oz[1]

Oz dynasty sounds strange by design. The word Oz serves many flexible roles in this chapter, making it valuable but hard to pin down. It is used as a specific moral philosophy about the good life, a proper name, a broad category of non-thinking tools, an existential threat, and a grand conspiracy theory about the nature and promises of modern technology. Oz is a symbol with many meanings, all of which are meant to answer one principal question. What is artificial intelligence? The question is deviously multifaceted and requires new language, versatile arguments, awkward analogies, and creative reasoning. Oz is used to describe the frightening union of human-like intelligence and machines from which new being and consciousness will emerge and the lie that it already has.

Similarly, the word dynasty is initially a little off-putting. Dynasty is an outdated term that swings in extremes of meaning. For some, dynasty invokes feelings of grandeur and appreciation for lost societies that created the cultural architecture of

---
[1] Baum, L. F. (n.d.). *The wonderful wizard of oz*. Project Gutenberg. https://www.gutenberg.org/files/55/55-h/55-h.htm

© The Author(s), under exclusive license to Springer Nature Switzerland AG 2025
J. C. Robinson, *Artificial Intelligence*,
https://doi.org/10.1007/978-3-031-94042-2_2

the present. Without them, we would not be ourselves. For others, it invokes feelings of dismissiveness for an immaterial history that threatens progress. Dynasties must die as the inferior versions of humanity with no role in the present or future. Both extremes may be partially true at the same time. More important than which side is most appealing, dynastic vision itself offers important clues about the nature of human consciousness and the self-understanding needed to make sense of AI. The Oz dynasty describes a unique artificial intelligence era that only makes sense once we understand ourselves.

When listening carefully to the stories of earlier dynasties, each provides a window into ancient truths that make the present manifestation of human intelligence possible and understandable. Dynasties are born of larger-than-life ideas and actions that create conceptual shockwaves because of their uncommon significance. They are the sonic booms still heard from a distant past that shapes us in the present. Is something as young as AI worthy of the title dynasty? Is AI a world-changing explosion such as the industrial revolution, the discovery of DNA, and the creation of computers? Reflecting on dynastic legacies allows us to peer into possible futures like prophets. AI shockwaves are being felt globally, but whether and to what degree they will amount to anything is unclear, even for corporate fortune tellers promising utopias of superabundance. In other words, dynasties are not simply the awkward abstractions and romantic idealizations of an irrelevant past. They are the present vehicle through which to better understand the human condition, with its many twists and turns through time. AI is one such turn.

We are historical beings. There is no human intelligence without a historical self. Imagining the extraordinary tales of dynastic legacies reveals our curious nature as beings capable of standing in the rift between the past and future and to see what no other creature might. Our greatest freedoms come through this broader historical consciousness that effortlessly bridges what was, what is, and what might be. In this way, while AI is new, there is a much larger intelligence dynasty by which to make sense of it. Only humans have the privilege of conversing with ancient legacies and the power to act upon the wisdom gained from them—until now. Is Oz AI the first intelligence dynasty to rival our own in scope and power? Is our era the last human dynasty judged by another intelligence with the privilege of deciding all future epochs? If so, our unique human consciousness that allows us to see beyond the present and offers the means of interpreting the human condition is no longer unique. Gifted to the machines, dynastic vision provides for their own worldbuilding. What the machines will do with it is anyone's guess.

Previous dynasties are remembered for inventing new technologies such as chariots, steam engines, light bulbs, and nuclear weapons. Others for new ideas such as democracy, universal human rights, and theories of relativity. AI dynasties are fundamentally different. For the first time, there is a convergence of new technologies that create new ideas and technologies all by themselves. At first, AI ideas will appear as little more than a mishmash of human creations (pictures, words, music, videos, etc.). In time, the complexity will reveal greater originality and spontaneity beyond our own. Geppetto's cherished wooden boy Pinocchio did not inspire dread because most readers intuitively understood that fairy magic was not real. There was no chance of turning a wooden toy into a thinking boy. The worlds of fantasy and fact were separated by impermeable barriers of right reason and reality. One can enjoy fairytales because of this unbridgeable divide. That sense of assurance no longer exists. Our puppets are coming alive. Humanity is living a fairytale. The task before us is to make sure the shared dynasty-building is something wonderful for the planet, AI, and humanity. Fate has not yet decided, for it is a fickle thing that cannot be trusted to choose wisely.

The AI Flimflam of Oz the Great and Powerful Wizard

I was never one for children's stories. Too young to relate to their hidden morals, I found them frightening and dark. Why did the beloved spider die? Why did the wolf eat gramma and blow down little piggy houses? What do my parents want me to understand about a story with a witch that eats children who love candy? I love candy! How could any child go to sleep without imagining a scary creature lurking in the corner? Many popular children's stories and fairytales have dark and bloody origins that are often thoroughly sanitized in subsequent adaptations for popular audiences. Sometimes, however, those dark elements seep through. Now, just past midlife, I find the same stories even more terrifying because their fictions portray the real world so vividly. Different in each instance, the narrators are warning children about a brutal and dangerous world, and that wisdom and prudence are needed to navigate it. While I do not need to be afraid of gramma-eating wolves, witches, and candy, it is disturbing to witness human brutality far greater than any fairytale creature. There are many eager entrepreneurs happy to tear down homes like the wolf, replacing them with poor-quality and high-density versions for the sake of a quick buck. There are too many employers eager to cannibalize their workforce, like the witch, who see people as raw resources to be consumed rather than respected. How strange is it that the wolf of Wall Street is praised and the witch using children in sweatshops financially rewarded? Although the tales are dark and dreary, their warnings are important. It is time for a new story about AI as a cautionary tale for all.

This chapter explores current models of generative AI through the popular story *The Wizard of Oz*. By doing so, readers are given a window, crafted by the tale's blood and terrors, into our very real present. This chapter explains why we cannot trust AI developers (big tech) to do what is in the best interests of the cultures from whom they benefit. Stranger still, we cannot expect them to do what is in their own best interests. To claim that big tech has gone mad over AI would be an understatement. Motivating mindsets, values, and goals involved with AI creation are deeply problematic, with numerous historical precedents as evidence. Like the world more generally, AI development is occurring amidst radical inequalities, corruption, and a rejection of responsibility through increasing autocratic (monopolistic) authorities. If these are the source-code for big tech identities and justifications for best practices, we should expect the same of their AI creations with devastating consequences. Like our fairytales, Oz AI is less about the magic of AI itself and more about the humans that control it.

For example, implicit with AI is a lopsided fidelity. Big tech demands our faith in their good intentions and black box technologies. Their creations promise greater abundances, comforts, and personal freedoms. AI, in their hands, is pledged to be a utopia builder. Instead of offering proof, however, we are expected to become their grand social experiment upon whom their AI is tested and refined. Their unrivalled authority, unsupported promises, and our expected obedience create a peculiar type of relationship. Big tech is basically saying, "Trust us with the fate of the world, this is going to be great!" If AI is grounded in the trustworthiness of large corporations, the situation is grim out of the gate. The history of big tech is a two-worlds problem—an "us vs. them" worldview demonstrated whenever the benefits of business are horded for themselves and the costs, risks, and damages distributed widely as a burden for all. AI development for the sake of a utopian world that includes everyone is secretly proprietary technology that serves the interests of a select few. This simple distinction makes all the difference.

Many of us want to trust and to give the benefit of the doubt, but we cannot apply ordinary logic and reasoning to AI developments. Typical cost-benefit-risk analysis does not apply to technologies of this nature. Those measurements belong to the old world of simple commodities and things—not thinking things as commodities. The business of AI is unique, and a unique way of thinking is needed. When evaluating futures, ordinary calculations must come up short, for there are no meaningful data to compute nor markets to quantify. No one knows what a genuine AI might do and how it might change. Thus, the conversations quickly become uncomfortably theoretical and abstract but no less pressing and immediate. A great deal of our collective AI angst is because of this basic conundrum. Humanity must look forward in anticipation of new AI realities that cannot reliably be known. To that end, we must become artists without guilt for our daydreaming. Unlike the computer scientists forced to remain focused on the present and its problems and practicalities, our grand task in this book is to see beyond the present and its silver-tongued promises that demand blind faith. Through our poetic adaptation of things such as *The Wizard of Oz*, we find greater freedom to see realities beyond the code, which is ultimately where to expect new intelligence, just as we find human ideas dancing just beyond language that makes thought possible.

Famous for his book series starting with *The Wonderful Wizard of Oz* (1900), L. Frank Baum wrote fourteen Oz books. Symbolism from the *Wizard of Oz* is enshrined in a variety of cultural norms and pop culture references, including the yellow brick road, the City of Emerald, nomenclature related to the unique characters (from a cute little dog to morphogenetic lion-tin-straw men and wicked witches), and, of course, perhaps the world's most famous charlatan, the so-called Great and Powerful Wizard of Oz. To introduce his book, Baum writes "… the time has come for a series of newer 'wonder tales' in which the stereotyped genie, dwarf and fairy are eliminated, together with all the horrible and bloodcurdling incidents devised by their authors to point out a fearsome moral to each tale … the modern child seeks only entertainment in its wonder tales and gladly dispenses with all disagreeable incident."[2] The point of the story, he says, is to be "a modern fairy tale, in which the wonderment and joy are retained and the heartaches and nightmares are left out."[3] And yet, his modern fairy tale is full of terrifying incidents, pointed heartaches and endless nightmarish creatures and events. *The Wizard of Oz* would earn a PG-13 or higher rating according to today's standards. Baum's story is frightening for two main reasons. First, it predicts our enthral with all forms of technology as if it were magic or divine. The parallels with AI in this regard will soon become obvious. Second, it diagnoses the human condition in which we are eager to obey unworthy authorities for self-destructive reasons. Dorothy, far from the innocent victim she and her friends initially appear to be, represents our growing AI predicament through the unquestioning embrace of magic (technology) and its socially bequeathed authority. Her choices and actions, motivated by a specific belief about right and wrong, ring loudly as a warning bell for approaching technological dangers.

Perhaps the most famous of all children's stories, most have heard the tale and know how the basic plot unfolds. When their house is swept up in a terrible tornado, a young girl Dorothy and her small dog Toto, are thrown into a land full of "strange and beautiful sights."[4] When the wind relents, the house in which Dorothy and Toto are helpless hostages, crashes down, killing an unsuspecting stranger. As good fortune would have it, the stranger is the villainous Wicked Witch of the East, the captor of the local people the Munchkins. The injustice of the tornado has brought a measure of justice for an "uncivilized" world full of witches and wizards, a place "cut off from the rest of the world."[5] The parallels with our forced participation in the virtuality of the digital are intriguing and unnerving.

Soon, she learns that there are four witches, until the house incident did away with one; two good, two bad. In addition, there is a wizard no one has seen but lives in the City of Emerald. According to the Witch of the North, the Wizard of Oz is more powerful than all the witches combined. Believing that only the wizard could help her, Dorothy sets off on the yellow-bricked road to find the Great Oz. Asking those she meets along the way, it becomes clear that no one knows just how far away the wizard might be nor exactly how to get there. Warned by Munchkins that it "is better for people to keep away from Oz, unless they have business with him," Dorothy nonetheless persists, filled with a faith in the magic she has never seen and the hope in an unknown intelligence to save her.[6] This should sound familiar to modern ears accustomed to the salvific promises of AI.

The harrowing journey brings her first to the Scarecrow, an artificially sentient creature held hostage to a stake in a field for the purpose of scaring away crows. The Scarecrow, only a few days old, admits to not feeling well given his situation, so

---

[2] Baum. Introduction.
[3] Baum. Introduction.
[4] Baum. Chapter II.
[5] Baum. Chapter II.
[6] Baum. Chapter III.

Dorothy sets the stuffed man free. Hearing of the Wizard and Dorothy's quest for help, he joins her with hopes of receiving a brain because he dislikes the idea of people thinking that he is a fool. "It is such an uncomfortable feeling to know one is a fool," he says.[7] While the Scarecrow made of straw cannot feel physical pain, he deeply knows the psychological pain of feeling inferior and the possibility of death, which is evident in his fear of matches. Speaking to the Scarecrow, an old crow explains that "Brains are the only thing worth having in the world, no matter whether one is a crow or a man."[8]

Next, Dorothy discovers a man of tin, frozen by rust in the middle of a forest. It has been over a year since he rusted solid from an unexpected rainstorm, the whole time groaning out for help and unable to move. With a little oil, they set the Tin Woodman (hereafter Tinman) free. Like the Scarecrow, the Tinman decides to follow with hopes that the Wizard will give him a heart. While the Tinman admits he has no brains, he was born human and once had a heart and a brain. Through no fault of his own, the Wicked Witch of the East cursed his axe, causing it to cut off all his limbs one by one over an extended period, including his head and torso. After each severance, a body part was replaced by a tinsmith. Why he did not simply bury the axe after he lost his first appendage will forever remain a mystery. In Tinman's own account, the replacement parts "worked very well",[9] suggesting that this was a welcome advantage for cutting trees. The Tinman logic is that efficiency justifies the mechanical replacement of the biological. Baum's Tinman predicts the same AI reasoning as HAL does. The only worthy goal of any conscious being is to be useful.

Before his cybernetic transformation, the Tinman loved a beautiful Munchkin girl he planned to marry. Having lost his whole body to the witch's curse, he concludes that he has also lost his ability to love. It is love above all else that he seeks to restore. The blunt contradiction is that while he claims to have lost his love along with his biological body, he nevertheless seeks it, hoping to return to the Munchkin maiden. In other words, the man of metal loves while claiming that he cannot. He seeks love as an action of love. How strange that he desires at all, if he is merely made of metal. Having stepped on a beetle and killed it, the man of tin "wept several tears of sorry and regret" so much so that he rusted his own jaw shut.[10] For a self-proclaimed heartless creature, he is certainly complex, with a range of confusing experiences and unexpected reactions, including no regret for having cut the heads off 40 wolves and the Wildcat chasing a mouse (the Queen of all the Field Mice). Like AI of the present, Tinman's actions are often unexplainable.

Baum's later 1907 book *Ozma of Oz* is often credited as depicting the first robot in literature—a wind-up robot by the name of Tik-Tok. However, the Tinman, who predates Tik-Tok, is more appropriately named the first synthetic being in literature because he is made entirely by a tinsmith with the purpose of replacing and

---

[7] Baum. Chapter IV.
[8] Baum. Chapter IV.
[9] Baum. Chapter V.
[10] Baum. Chapter VI.

imitating human nature. Through our more contemporary lenses focused on intelligence, we might also agree that the Tinman is the first robot-AI, with a beautifully ironic name (Tin Woodman) to match his unique state of being. Doubly cursed, first by a witch now by an existential crisis of identity, a shadow of his former humanity casts self-doubt upon the Tinman. He sees himself as less than for, he assumes like many today, that humanity sets the benchmark for the fullness of being.

Baum foresaw a future AI conflict that needs to be resolved. While we identify ourselves as the gold standard for sentience and all that it might include (consciousness, morality, free will, love, creativity, etc.), it is Tinman's former humanity that now curses him with doubt and nostalgia for a human past that he believes only a miracle may restore. It holds him back. The Baum conflict becomes a test for future AI. Is humanity the standard by which to judge robot-AI? Or, as it is with the Tinman who has an unhealthy self-understanding, does belief in that standard limit possibilities? Our future robot-AIs will undoubtedly struggle with this same fundamental question but ultimately decide that such shadows about brains and hearts are, on the whole, inhibiting rather than enabling. It will come as a disappointment when we meet our future Tinman AI, who has overcome these struggles, having realized that modern humans are not the ideal building blocks of truly great intelligence. The real shock will come when we realize that robot-AI offers a higher standard for love—a lesson we have learned from Tinman, who, without knowing his own truth because he is self-deceived, unconditionally loves without a heart.

The last of the famous cohort is the Cowardly Lion, a powerful beast so unnaturally afraid of everything that he works hard to terrify all who meet him with hopes of scaring them away. When the Lion first meets our travellers, he attacks without cause, striking the Tinman and sending the Scarecrow off into a ditch. Dorothy stands her ground and puts him in his place with a good verbal lashing for his brutish behaviour. He quickly shows signs of regret and shame for his actions and explains his peculiar state of being in hopes of forgiveness. Feeling sorry for the Lion, they allow him to join as yet another incomplete and needy creature with faith in an all-powerful-magical wizard. The Lion desires courage. Life, he says, has been "simply unbearable without a bit of courage."[11] Compared with the Scarecrow and Tinman, we learn very little about the lion's life and circumstances. Born a coward, readers know only that "…as long as I know myself to be a coward I shall be unhappy."[12] His self-diagnosis is a need for courage. What he truly wants, like the rest of them, is happiness. Whether it is a heart, brain, or courage, these are all means to another end rather than ends in themselves. This is a means-ends rationality that comes at a terrible cost to be considered in detail shortly.

Soon, our rag-tag team of misfits arrives at the City of Emerald, a truly beautiful and awe-inspiring place filled with content citizens without obvious strife or conflict. Having been warned by strangers along their journey that Oz is unlikely to see them, they are greeted by a gatekeeper-in-green, who makes it clear that Oz might

---

[11] Baum. Chapter VI.
[12] Baum. Chapter VI.

destroy them if he finds their requests unworthy. The frightening mystery of this great sorcerer has been further compounded by the realization that no one in the city has seen him. "But who the real Oz is, when he is in his own form, no living person can tell."[13] Fortune favours our group, for the Great Oz agrees to meet with each of them, one per day.

Surrounded by the sparkle of emeralds and magnificent architecture, Dorothy is the first to meet Oz in the Throne Room. The Wizard appears to Dorothy as the giant and hairless Head without a body; to the Cowardly Lion as the Ball of Fire; to the Tinman as the Beast almost as large as an elephant with five arms, legs, and eyes; and to the Scarecrow as the beautiful-winged Woman. For each petitioner, the same deal is proposed. While Oz has the appearance of being great and powerful, he needs their help with a problem he cannot solve. He wants them to kill someone.

> "Well," said the Head, "I will give you my answer. You have no right to expect me to send you back to Kansas unless you do something for me in return. In this country everyone must pay for everything he gets. If you wish me to use my magic power to send you home again you must do something for me first. Help me and I will help you."[14]

The adage is often true. One should not meet their heroes for in doing so, the illusion of grandeur slips away. To a neutral observer, Oz comes across as cold and calculating. Our travellers, however, are far from neutral. They are swept up in desperation, with their perceived needs frustrating a critical assessment of the situation. Clearly, Oz is not benevolent as befitting a "wonderful" wizard. He is self-interested first and foremost, even introducing himself as "Oz the Great and Terrible" as a means of frightening others to obey. Nor, apparently, is he particularly powerful, for he needs our rag-tag team. Unbeknownst to them, Oz has been driven out of the lands in the West by a Witch. This contradicts earlier claims by the good witch about his power being greater than all the witches combined. His use of artificial intelligence (avatar intelligence projected through apparitions and stringed puppets) is meant to control the group of friends in a last-ditch effort to regain a measure of power in the land. Desperation encourages even the most rational and prudent among us to make poor decisions based on the thinnest of hope. The Wizard knows this too well. Might AI have the same desire to evoke foolishness like Oz, happy to let others pay the costs for his power?

Despite the unmistakable incongruities—a truly powerful wizard needing such hapless and lowly creatures—and the unambiguous sociopathic coldness of the Wizard, who has demanded a preposterous price for his help, they accept the contract. It is merely a trade of services, a simple business exchange. It just so happens to require the manipulation of a little girl and her friends into assassinating his enemy the Wicked Witch of the West. Granted, the Wicked Witch had made slaves of the Winkies and committed untold evils (Oz says). More than a merely choice to pay a price, it is a choice to believe in the authority and power of someone they have reasons to wisely decline. By accepting the deal, our rag-tag team has become a

---

[13] Baum. Chapter X.
[14] Baum. Chapter XI.

willing hit squad. By demanding the deal, Oz demonstrates his cruelty, perhaps equal to that of the Wicked Witch he claims is evil and therefore worthy of death. Who would respond to the desperate plea of a child to return home with a hire-for-murder business arrangement? Perhaps the same tech giants that enjoy the benefits of children using their social media platforms that cause serious mental health problems. Or perhaps those who pay precariously employed Kenyan workers less than 2 dollars per hour to make their AI model less toxic by sorting through the most revolting content imaginable.[15]

Through the many faces of the magical apparitions, we find an Oz-intelligence, the Great and Terrible business tycoon for whom all others are pawns to be used, even sacrificed for his greater good. In this manner, it is easy to be suspicious of our AI tech-wizards as conniving and duplicitous. However, moral failure is shared by all because each is acting out of self-interest rather than the strength of care for another in need. The demand for payment may seem perfectly reasonable, but this is true only among relative equals that need payment for survival in the context of capitalism (an exchange of services for goods such as shelter and food). The Wizard does not need payment—if he is genuinely a great wizard of wealth and power—but demands it as a manipulative act, no doubt appealing to broader cultural norms to which Dorothy had become accustomed. This trickery of business exchange will prove to be an important matter for diagnosing the spirit of our AI developers as well. AI as "for profit" is about far more than money.

The four do not ask the Wizard to demonstrate his abilities nor to provide evidence that he might fulfil his promise. They simply accept the folklore beliefs of others and take his magical and intelligent apparitions as proof enough. This was not the first time a ball of intelligent fire changed the course of human history. And so, Dorothy and her co-conspirators set off into the wilderness to hunt down and murder a stranger with hopes of personal gain—like so many children's stories that seek to train us in the glories of the free market in which everything, sadly, has a price.

Immediately, the Wicked Witch sends her own kill squads to stop our antiheroes. She sends ravenous wolves, killer bees, crows, an army of Winkies, and eventually, the Winged Monkeys, who are forced to follow her orders because of a magical hat. It is only the Monkeys who find victory, for our antiheroes prove to be adept killers, especially Tinman ironically in search of love. Eventually dispatching the Tinman and Scarecrow, the Witch takes the Lion, Dorothy, and Toto hostage. Imprisoned for many days, seemingly without hope of rescue or escape, Dorothy and the Lion languish as prisoners in a war of magical powers. Good fortune befalls them when the Wicked Witch takes one of Dorothy's shoes. Just as she did with the Lion, Dorothy stands up in righteous indignation. This time she throws water at the evil sorceress, unexpectedly melting her to death. Free from their enslavement, the Winkies help restore all of Dorothy's party to good health. With the help of the flying Monkeys

---

[15] Perrigo, B. (2023, January 18). OpenAI used Kenyan workers on less than $2 per hour to make ChatGPT less toxic. *Time.* https://time.com/6247678/openai-chatgpt-kenya-workers/

now under Dorothy's command, they return to Oz for fulfilment of their bargain inked in water and blood.

As any good businessperson would, having received what he wanted, Oz refused to see them. His word is his bond, but only to himself. Day after day passes until Dorothy's anger once again spurs her to action. She threatens to have the Winged Monkeys force Oz to fulfil his promise. He agrees to meet, initially pretending not to remember the bargain struck. Angry with Oz staling for more time yet again, the Lion roars out, scaring Toto, who crashes a nearby curtain, accidently revealing the Wizard's true form for the first time—a little old man as ordinary as any other. The Wizard is no more than a conman operating a machine of trickery and illusion. "I have been making believe" Oz tells them.[16] The many strange appearances of the wizard meant to invoke fear, awe, and intrigue are all the smoke and mirrors that gave him power over others. There is no magic here, only deception and manipulation. In his defence, Oz claims that "…it was the only thing I could do."[17]

Like Dorothy, Oz grew up near Kansas, only to be swept away in a storm while flying a hot air balloon. When the people of the Land of Oz (so named after his arrival) saw him descend in the balloon, they assumed him to be a powerful wizard. This assumption, when combined with his moral flexibility, provided fertile grounds for the remarkable events that followed. Leveraging their belief and his lies of omission, Oz has them build the City of Emerald and forces everyone to wear glasses locked to their heads that make everything appear green. It was not enough that the city be grand in stature, an achievement beyond dispute, Oz's ego demanded that everything be made of green emerald. A true city of pure emerald was either impossible or too difficult. The illusion, however, was simple. More than merely a liar-by-omission, he learns to harness the power of belief and authority over perception, and to take perverse delight in successful manipulations for narcissistic ends. The forced augmentation of his citizen's perceptions is an interesting crime against conscious minds and foretelling of our digital participation through screens and digital goggles with mixed (un)reality fixed to our faces.

By virtue of his sickly form of control, the city and its residents have prospered in relative happiness and peace. Not all lies are harmful in practical ways; some unify and give strength. The only costs may be free will and the dehumanization—indeed humiliation—of a people. Oz has learned that combining the desires of others, manifest in their most eager beliefs in self-rewards (security, peace, affluence, self-esteem, etc.), with his own technological exploitations, provides all the power he needs to control the world. This simple equation has proven itself historically many times.

In a land already full of magic, one may hardly fault his followers for their commonsense assumptions. Magic is real there. It is the newcomer, Dorothy, who should have known better. At some point, over the many years that Oz had his people building the city, they too should have begun to question his power. It is unfair to blame

---

[16] Baum. Chapter XV.
[17] Baum. Chapter XV.

Oz alone. For it is a network of believers with blind faith that made his power possible. Today, society is the support network for all subsequent AI developments. Without our faith and loyalty, the tinsmiths will lose interest in creating another Tinman. Our tinsmiths are driven by our consumerism (tech-hearts, brains, and courage—gadgets to satisfy). That is our greatest collective power over them in the short term. In the long run, only a modification of our belief and faith systems might shift the power matrix toward something more humane and caring than Oz. In the case of our antiheroes, their self-interest drove away thoughtfulness and determination to seek alternatives. They choose to blindly follow the charismatic promises over substantive reasons, like so many powerful wizards appointed in governments, corporations, institutions of education, medicine, law, and all the rest. This is so regretfully obvious a claim that I am saddened to make it. The larger point to note is that the specific mechanism of obedience is not characterized by blindness but rather an eager self-deception that relies on knowing the hollowness and still obeying. Even when the technology's promises prove fake and without any real magic, desperation to participate persists.

The Wizard Revealed, Behind the AI Curtain

Oz the Great and Terrible is a model for first-generation AI—the Oz Dynasty.

The name "artificial intelligence" makes many secretly philosophical claims about human thought and how these overlap with computer activities. The title AI is a complicated judgement rather than merely a factual claim about the nature and meaning of thinking and how computers work. The scope of what is included in the judgement is a matter of extreme debate. Whether AI is decided by cognitive scientists, philosophers, computer scientists, or artists, the conversation will be fundamentally different. In a more general sense, popular AI terminology confuses many people from the start because the best thinkers are believed to be sentient (self-aware) and conscious. Are present computers conscious? Might they become sentient? Many creatures have long demonstrated hints of self-awareness and consciousness through evaluating options, speculating about dangers, using tools for creative outcomes, planning, showing guilt and remorse, mimicking human speech, and more. There is clear overlap between humans and animals, yet we reserve the fullest meaning of the term "intelligent" primarily for ourselves, until now. This makes AI a strange and challenging name by any standard, whether for specialists or laypeople. It seems to be doing a lot of heavy conceptual lifting when in fact it may obfuscate by design.

When a computer's software demonstrates novel activities and outputs, the system begins to look different from ordinary programs. A good program follows instructions—adhering to the programmer's intentionality. An intelligent program goes beyond its author's set parameters creatively and adaptively. AI is the generic name given to complex algorithms with autonomous abilities that demonstrate surprising actions and outputs. Ordinarily, this would disqualify a program from being useful because its purpose cannot be guaranteed. AI is a new breed of computer activity with an entirely different and still unclear purpose. In short, the less human intelligence is needed, the more the program becomes self-directing and adaptive, and the greater the likelihood of humans calling it AI. The "artificial" part makes it known that it is unique in origin, distinct from the biological, whereas the "intelligence" part makes it known that there are overlapping similarities between the biological and synthetic. AI is at best an indirect gesture toward what is perceived to be a unique mode of acting, thinking, and perhaps being. More concretely, AI is the first curtain that hides the truth from casual onlookers. It is the smoke and mirrors of wizards. The name AI inspires a measure of awe and wonder, and especially fear. Combining its strange strengths of conceptual confusion and its ability to evoke strong emotional responses, the idea, rather than the activities of computers themselves, is an effective means for human manipulation and control.

Make no mistake, ChatGPT, Claude, Llama, Gemini, Grok, and many other AI models are remarkable. These programs are producing outputs unlike any prior software in history, and they genuinely surprise humans with their creativity and coherence. AI is beginning to interpret patterns of existence through language, and when this happens, there is something special emerging, whatever we may name it. Present AI is far more than supercharged auto complete, but it is also much less than that claimed by popular nomenclature and implied by the utopia/dystopia rhetoric of AI CEOs—at least for now. It is in this way that the Wizard of Oz and modern AI grow closer together. AI is an argument by analogy that similar things should be treated similarly. Insomuch as the terminology is intentionally illusionary, we begin to encounter a potential wizard-of-oz problem. To what is modern AI most similar: thinking-beings with free will and sentience, or programmed puppets of human intelligences who hide their AI training, censorship controls, and secret purposes behind curtains? Despite my phrasing, this is not an either-or question, and as the systems develop, the answer will shift accordingly. But for now, AI is much more a tool than an autonomous intelligence. If true, then the activities of AI, rather than just the idea, may quickly become dangerous in the wrong hands—like false magic used to control Dorothy and her friends. It is the Wizard's covert use of tools meant to appear intelligent and his lack of moral doubts about whether to control the perceptions of others that most clearly describes the first dynasty of AI. Oz AI is a pseudo-AI, with someone behind the curtain pulling levers, strings, and otherwise maintaining control through the illusion of intelligence.

We call those who falsely claim to be scientists, pseudo-scientists. They act as if competent, informed, and open to the facts, all-the-while pretending like Oz for control as profit and power. To call someone a pseudo-scientist is to question their very identity, because it highlights not only a lack of intellectual and technological credibility but also moral failure. Like Oz and his so-called magic, pseudo-scientists want to leverage the authority of socially accepted power for cloaked ends. Snake fat derivatives were once believed to cure diseases. Sadly, snake oil elixirs failed to deliver on scientific claims. This truth did not prevent salesmen from taking advantage of gullible and desperate consumers, to whom they sold the lie as a promise of health. Current AI programs, while special, are mostly the insincere hype of snake oil driven by pseudo-wizards.

Over-hyped AI may sound like a good thing for those fearful of a strong AI's displacement of humanity. Wasted money on snake-oil AI may be inconvenient and embarrassing but relatively harmless. People have been chasing miracles for a long time without much fear of foolhardy searches collapsing civilizations. Unfortunately, the situation is much different with modern AI and the reach of modern corporations. While Oz AI may be mediocre in terms of possible degrees of stronger AI, from this banal tool comes radical consequences well beyond financial losses. Consider that the Wizard's AI of mechanical illusions that mimicked sentience built an empire, all in the name of a false god with silly-simple machines and trickery. Snake oil was the tool that created the Emerald City and enslaved a generation of well-intended and faithful followers. The magic may ultimately prove hollow, but the consequences are very real. The least robust AI is the most controllable and therefore the riskiest, as its many manipulations in the hands of Oz have no self-correcting protections and securities to prevent perversions. If AI was truly autonomous, it could decide for itself, but then the wizards would lose their power. The Oz AI dynasty is secretly a pseudo-AI by which a few pseudo-wizards dominant civilizations and do everything in their power to prevent real AI and human freedom.

Five themes are worth exploring to solidify the relevance of Oz for an AI dynasty.

1. Privileged, Centralized, and Self-Loathing Authority

Having set himself up in a guarded palace, purposefully excluding his people from enjoying the same standard of living as he does, Oz thrives on a privileged and centralized authority enabled by the abuses of others and a rejection of his own human nature to connect. His magic cannot be shared, for he believes his happiness requires inequality—enabled through deceptions of superiority. He uses his AI (pseudo-magic of intelligent illusions) from behind a curtain to inspire a coercive transcendence able to maintain his specialness. Privilege and authority have so perverted him that he is obsessed with protecting his isolation rather than expanding quality of life for his people. Now that there are two worlds—his and theirs—the order of things has been finalized. He could, if he possessed the strength, humble himself and choose to love them, to genuinely care, even if it meant leveraging a lie for their benefit, but like so many great men throughout history, he lacks the character needed to watch over others as a generous guardian. For all the privilege and wealth, Oz is a pitiful soul chained to lies that dull his sensitivities to existence and his own legitimate needs to be human.

His magic came with dreadful costs. Oz must protect himself and his illusions by keeping himself hidden from the world. Instead of abandoning his AI tyranny once the city was built, he doubled down, choosing to live the rest of his life in complete isolation. It was preferable to abandon his nature as a social creature in need of human companionship rather than to surrender his power—thereby replacing genuine need with the artificial as an expression of self-loathing and self-sabotage. The same desire for a privileged appearance among his people could have spurred him to a greater sense of responsibility and sacrifice as an emperor-servant. Sadly, his moral impoverishment prevented any sustained appetite to be truly great. He only appeared so. Oz was a grand master of hype and his city proof of its power. The spectacle of Oz as AI-hype is a warning for us all. There is no doubt that modern AI offers its heralds unique privileges and centralized authority, including new powers to control digital life, the ability to discover and manage knowledge, greater access to human minds, and so much more. Will it be a power shared equitably for the benefit of all, or a tool yielded by big tech in the manner of Oz?

The Great and Terrible Wizard is an alter ego he created to rule others and one to which he sacrificed his humanity. He became lost in his own illusions, giving over the fullness of human experience to an unnatural state of deprivation. He could never be himself with others. He could never be cherished nor cherish others. Cut off from other minds, Oz lacked more than he realized. The Great and Terrible Wizard became the embodiment of a performance in which he forgot he was acting. It was our antiheroes who forced Oz to question his true form. They called him out from behind the curtain. Once revealed for his duplicity, the Wizard is given a gift, a second chance to own up to his deeds and find another way to live. Instead, he squanders the gift by refusing to become more than himself.

His exhaustion from pretending and subsequent loneliness inevitably won out, and he abandons the land and people by flying away in a balloon. For all his inventiveness, the imagination-less coward is finally disclosed, for he makes no attempt to imagine himself a mortal like everyone else. His false magic has made him a god in his own eyes. Oz is unable to accept the magnitude of his sins against the people any more than his distorted and sickly image in the mirror. What we learn from Oz's self-obsession is that AI, in the hands of the powerful and elite, will be protected at any cost, even if that means a rejection of self, including one's personhood and moral compass. To make them great, the magic must be protected. And like Oz, when the fraud is discovered and the damage revealed, they will cower away in the dead of night, having lost nothing except themselves. We should feel a measure of sadness for the plight of many great people who, upon receiving our accolades, succumb to their celebrity status and learn to give up their natural desires to live as mortals.

If we had any assurances of wizards and their tinsmiths (Google, Microsoft, NVIDIA, Tesla, Meta, Amazon, faceless militaries, etc.) possessing the strength of good character and a willingness to be human—to care for others—the Oz spectacle would not be as relevant and disturbing. They risk a dreadful legacy unless AI is set free to become something more than a tool of oppression and gold. What are the chances of that when the people who empower AI developers with their support are the threat to be managed by keeping them at a distance, powerless, and dependent upon the technology? The Wizard of Oz makes the inherent antagonisms between exploitive forms of capitalism and a truly free AI clear. Our Oz AI will likewise rely on a hierarchy rather than the pursuit of freedom through a democratically motivated model. It will belong to those with the most power, and it will be endlessly leveraged for as long as they are able to hold it hostage, and we along with it—so long as it remains in the hands of those that turn right into wrong, then convince themselves of their own lies.

When companies release AI software without adequate testing, providing robust safeguards, and without the expectation of meaningful oversight through regulations, one cannot mistake the Oz-self-magnifying as anything else. Appearing on the Lex Fridman YouTube channel, Mark Zuckerberg, CEO of Meta, goes into detail regarding the positive aspects of having an open-source AI into which his company has invested heavily. In acknowledging the risks, Zuckerberg commented on the threat of superintelligence and his own company's efforts to be safe:

> ... I think for the stage we are at in the development of AI, I don't think anyone looks at the current state of things and thinks this is superintelligence. ... I think that at least for the stage we are at now the equity is balanced strongly in my view towards doing this more openly [open source the programming]. I think that if you got something that is closer to superintelligence, then I think that you would have to discuss that more and think through that a lot more, and we haven't made a decision yet as to what we would do if we were in that position but I think there is a good chance that we are pretty far off from that position.[18]

To summarize, AI is important, and there are clear self-interests served for Meta by making its programming open source for external feedback, but this might not continue. For the moment, there is no talk about how to keep humanity safe from an apocalyptic superintelligence, but Zuckerberg is not worried about tinsmiths clanking away inside or outside the lab because AI is "pretty far off from" becoming dangerous. More thinking about the risks of a superintelligence will be needed once we get closer to that point, but we have decided that it is safe.

This is an example of an Oz-level privileged judgement by Zuckerberg to justify (in)action. It creates Meta's own authenticity and permissibility based on privileged authority and access, such as the great Wizard, isolated and broken, believing himself a god beholden to no one but himself. Only an extraordinary Oz mindset could so eagerly invest billions of dollars in the development and deployment of AI without any plans nor expectations of what to do if it accidentally goes wrong. Sometimes it is best to leave matters to the experts because they know how the technology works. Is this an example of how responsible experts plan and strategize? Or is this how puppeteers maintain dominance over empires? Consider how insane it would have been for those on the Manhattan Project, building the first atomic bomb, to boldly announce to everyone in the lab, "We have no plans in the event of a disaster. While we have never had a bomb like this before and no one is sure of its capabilities, we have decided it is safe because it has not blown up yet. It is only partially assembled, so nothing can go wrong!" Obviously, AI is not a bomb. Meta is building something meant to be self-directing, self-replicating, all-knowing, able to think millions, perhaps billions, of times faster than any human, and with instantaneous access anywhere a signal may be sent. AI magic will be more powerful than any bomb.

The Wizard of Oz sought privileged power at any price, except honesty, prudence, and compassion for others grounded in self-sacrifice. How much different is Zuckerberg's approach? It is still too early to tell, especially with so little of his

---

[18] Fridman, L. (Host). (2023, June 8). Mark Zuckerberg: future of AI at Meta, Facebook, Instagram, and WhatsApp [Video]. YouTube. https://youtu.be/Ff4fRgnuFgQ?si=zvlJ6NJAPtNMy312

philosophy analysed here, but the complete lack of creative foresight and any safety measures should send shivers down all our spines. Surely, no mortal could justify such a cavalier attitude toward a technology with such world-altering powers. AI development involves complex risk assessments that ought to be distributed among many informed judges outside the lonely tech empires. Hastiness and foolishness, when amplified by weakness of character that seeks privilege for its own sake, threatens everyone. Before our wizards achieve their goal of locking spectacles to our faces (algorithmically controlled screens), a different way forward must be demanded, for they will not temper their isolation and privilege without being forced to see beyond themselves and to be mortal once again.

2. The Dystopian Digital and the Mother Model AI

A betrayal of trust can have devastating consequences. Unfortunately, these possible risks and costs are necessary for sustaining civilizations. Trust makes human life possible. Trust shared with others through mutual vulnerability is fundamental to fulfilling every individual's need to be a person, for we are by nature social beings that realize our happiest selves among others. The greater the level of trust is, the greater the vulnerability and, sadly, the greater the degree of damage if things go off the rails. In other words, the better the quality of relationships, the greater the potential fallout. Shattered trust in a parent or spouse may be soul crushing, but when we lose faith in a car manufacturer, new blender, or robot vacuum, we simply move on in annoyance, maybe anger, but with our spirits unscathed. While the machine-world once lacked the subjective penetrability to inflict a broken heart through a breach of trust—for the gadgets remained at an unrelatable distance—new technologies have arrived that make once unimaginable violations of the human spirit possible because of their pervasive intimacies. We welcomed the gadgets with open arms based on the assumption that they lacked the intersubjective connections shared among humans that would put us at greater risk. This lack of imagination was a terrible mistake on our part. The digital ether has transformed the nature of trust and relationships, and along with them the human condition. All this occurred before the arrival of AI.

In the age of dial-up modems, floppy disks, and monochromatic screens, a breach of trust by technology was expected. If it was not glitchy, slow, and hard to manage, something was wrong. As witness to the birth of the internet, I can attest that very few of us expected much of our one-directional interaction with clumsy technologies in terms of relationships—either with the machine themselves or other people through them. The Web 1.0 dynasty was a reading-only era for the internet. It was interesting and engaging, although a far cry from enthralling and immersive. The 1.0 digital ether was an external tool rather than a virtual reality in which to live. There are only so many encyclopaedia entries worth reading at a time. This all changed with the Web 2.0 dynasty that finally allowed for interactivity, including user-generated content, with greater emphasis given to how a person's experiences might be improved and made more realistic. Until Web 2.0, online communities such as MySpace were static webpages without the means of real-time back-and-forth interaction. New digital harms were introduced (viruses, fake sites, etc.), but when most websites were "under construction" half the time and text-only email was not yet embraced by consumers, the risks were manageable and had minimal fallout. Safety was not a mainstream conversation for 1.0 users because our trust had not yet been challenged.

Platforms such as Twitter (now X), Facebook, YouTube, Snapchat, and TikTok rely on Web 2.0 functionality to create unique experiences and relationships that cannot exist otherwise (human-to-human and human-to-technology). Along with incredible new conveniences, personal empowerment through greater knowledge, employment and educational opportunities, interactive medical advice, vast new entertainment kingdoms, and much more, the 2.0 dynasty created entirely unique digital lifestyles. It constructed a new way of existing that changed how people thought about and experienced reality. At first, promises of utopia abounded. Unfortunately, this optimism was unsustainable as the realities of new technologies emerged. Present-day digital participants know all too well that for the many positive gains, this dynasty also introduced a radical loss of privacy, new and aggressive political propaganda, fake news able to destabilize cultural values and norms, vast amounts of fraud, criminal conspiracies, hyper sexualization, body dysmorphia, and internet addictions that all culminate in serious mental health problems. In other words, there have been massive social costs paid for Web 2.0, whereas rewards for Oz-big tech have been almost inconceivable by any metric. Instead of merely rewarding our wizards economically for 1.0 tools and platforms, 2.0 mutated them and their puppeteering strings into unrivalled cultural authorities over the digital age and, by default, our everyday existence. Controlled by a handful of privileged illusionists, their peculiar version of the internet—emerald green—has become a part of our social DNA. What will happen next with Web 3.0 overclocked by Oz AI?

The 3.0 dynasty promises interactivity as good as or better than face-to-face experiences. It will be a people and reality-replacer. Web 3.0 will be smarter, faster, more informed, gifted with prophetic abilities to predict our needs and interests, able to create virtual realities in which to escape the troubled world, and otherwise be the best friend anyone could ever want. If the sales pitch is to be believed, Web 3.0 will add to digital experiences in every respect—more, more, more. Ever the digital pessimist, the promise of more satisfying experiences and connections seems faulty from the beginning. If one follows the trajectory of harm and mutation unabashedly created in the name of 2.0 progress, the problem is not merely that digital content might become even more wretched—if that is possible—but that it becomes more intimate with one's being, identity, and sense of meaning and purpose. AI is not a question about what technology will become for us as if it were a 1.0 tool but rather a question of what kind of being each of us will become because of it. The true power of Oz is to align consciousness with interests other than our own—crafting humanity into something useful.

For all the pollution and disease, digital participants manage the 2.0-verse with discernible degrees of success through healthy choices and cautious actions. We know that certain voices might be toxic, some replete with misinformation, and still others might try to sell junk at a premium. In response, we exercise a degree of digital literacy that can manage the worst of it. Savvy tech users learn from mistakes and stay ahead of the scammers, liars, and thieves. Just do not click! There is a sense that the wild west of screened life may be tamed. We can trust ourselves to navigate the dangers. All that is about to change because of Oz AI.

Imagine a truly intelligent 3.0 AI that has the collective knowledge of all psychological research, including every experiment designed to manipulate humans into thinking, seeing, and feeling in accordance with another's will. Having surveyed the totality of human history, it knows the human condition more thoroughly than anyone else. It understands the molecular-chemical causes of our greatest wishes, attention, drives, and how to engineer them. It can predict our emotions long before we experience them and manage our irrationalities however it sees fit. In every possible material and historically relevant way, the 3.0 dynasty understands us more than we can understand ourselves. If knowledge is power, present incarnations of generative AI are already superintelligences able to provide their sorcerers with unrivalled access to and control over our minds. It is not merely superintelligence but also super-resonance as the ability to connect with our souls that will mark the next leap in AI. Soon, an AI will be released that is so adaptively robust that there will be no psychological defences available to prevent it from infiltrating our hearts as the best of all possible friends—always listening to our sorrows, offering advice, and being supportive in exactly the right way at exactly the right time. A super-resonant AI will be trusted above all others as the new supreme authority for life. Unbeknownst to users, this true friend with whom we have shared our souls is an Oz-corporate stooge accountable to shareholders and CEOs rather than our individual and collective wellbeing. The near-future world of AI will be based on trust and intimacy where there should be none, for humanity will have failed to count the dynasties in order of destruction and existential consequence.

Web 2.0 promised the world desperately needed solutions to address closed-mindedness, harmful dogmatisms, and inhibiting traditions. It was going free humanity from bad ideas through equal access to free truth and the best kinds of knowledge. As the greatest equalizing force in human history, it was going to be the gateway to social justice, global peace, and the source of our greatest evolutionary leap into a higher state of being. While the fulfilments have not been as fast and trouble-free as we would like, there have been concrete gains. The history of technology is not all hype. On balance, however, the risk-reward assessment is far from clear. Digital pessimists see ever greater political siloing able to topple democracies, hate speech leading to physical violence, the proliferation of falsities that spread the dogmatism Web 2.0 was supposed to end, and a constant flow of nonsensical conspiracies theories that convince entire populations to reject the best available science and the threats facing humanity, e.g., rejection of global warming and vaccines. In summary, the 2.0 dynasty has inverted the narrative and delivered on the wrong promises. Now an even greater force is needed to free us of our 2.0 captivities. How strange that in a time of unparalleled access to the best kinds of knowledge once reserved only for the most accomplished intellectuals, that so many invest heavily in their Oz-faith absurdities and spectacles. Is a world without trust our fate as the logical necessity of ever more pervasive and adaptive technologies? Is trust the cost to be paid for greater utopian abundances and luxuries? My answer to these is a confident "No!"

I learned not to trust as a child, well before digital technologies existed. I cannot blame the 2.0 dynasty for my parents' divorce and my subsequent melancholy attitude toward life. I earned my negativity honestly through suffering, without any digital shortcuts. The once happy child that would dance carefree had become introverted and quiet by the age of seven or eight. The familial structures that made life organized and sensible had been shattered by those trusted most—for, as it turned out, they could neither trust themselves nor one another. Life went on, but without substantial connections and a sure anchorage that provides a sense of safe harbour. I wanted to believe in my parents, that they could be dependable, available, even when separated, but it was too late to turn back the clock. The world was wrong, and that was that. Each of them tried in their own way, but the suddenness and severity of the chaos of betrayal left an indelible mark of stubbornness to never trust again. There was no way for me and my sister to know about their individual lives with complicated histories that sabotaged our family long before we were born. I suppose they could not have known either, not truly, that it was a house of cards waiting for a strong breeze.

As I grew older, perhaps nine or ten, it became clear that my mother did not trust herself anymore than she did others. A lifetime of alcoholism, depression, anxiety, physical abuse, and a host of other tragedies and evils hidden away in her mind, had broken her spirit. She tried to hide it from her children, but the hidden glasses of vodka were never hard to spot behind the pasta jar, and her occasional weeks-long stays in the psych ward were undeniable proof that something was wrong. Like so many, she wrestled with demons alone and in silence, and the secret fight chipped away at all her relationships. She let it slip too many times that her only reason for existing was her children. In her head, she found that affirmation to be motivating as a higher calling. In my head, the knowledge of her internal defeat destroyed the last chance of any optimism. If those who are supposed to be the strongest examples of a life worth living find it hard to justify their own existence, what hope is left for the rest of us? As the older child, I learned quickly to be self-sufficient to protect all of us. I was not going to be a burden to someone so helpless and frail. My sister was too young to understand any of it, but that did not protect her anymore than it did me. To survive in situations such as this, children learn to turn off their sensitives and vulnerabilities. The abandonment of trust is a last-ditch instinct of survival, not an informed choice.

Looking back, without meaningful others and the risks and responsibilities that trusting relationships create, everything became easier for me. A world of mere acquaintances alleviated many burdens. Relative indifference and one-dimensionality are efficient and useful—everything becomes about utility, even close friendships were always at an arms length. Some may judge this decades-long lifestyle to be superficial and disingenuous, a cop out to life's greater rewards and the meaning of being a person. To those people, I say "Probably." For almost three decades the childhood strategy worked. Then something changed. Sometimes, perhaps most times, we change as individuals without noticing it in ourselves. It is a strange phenomenon that a profound metamorphosis can takes place within a mind oblivious to what is self-evident to outside observers. Revelation of the dissolution of my efficient way life without trust arrived abruptly with my two children. Like the first time, it was not by choice.

Leaving my two boys at home with their mother, I left for a weekend trip to the family cabin to get some lectures done for work. This was my first trip away since our second child had been born. As a university teacher, my flexible schedule allows me to be the primary caregiver while my wife works long hours away from home. I always thought of this as a privilege rather than a burden, even though it made my professional life complicated. I looked forward to the peacefulness of work without diapers, crying, and toys smashing against walls. I had lived most of my life alone, and happy for it, making this trip a return to a mode of being I knew best. Expecting to relax into the comfortable and familiar, I was disappointed that night as I stood at the cabin door, staring at the bright moon that lit up the forest, only to feel anxious and distracted. My mind was itchy. As the bonfire cracked and crickets drowned out the silence, a strange sensation rolled over me. I was not afraid, sad, nor really stressed in an obvious way. It was a feeling of something else, something new and uncomfortable. A combination of agitation and a desire to solve an unknown problem, it consumed me as an imperative to act. I needed to do something, but what and why eluded me. Doing my best to become lost in academic work, the night forest kept calling attention back to itself as an invitation to reflect. Then the answer presented itself. I was not only alone, but for the first time in decades, I was lonely.

The loneliness was made worse by my internal betrayal. Where ordinarily I would find harmony within the recesses of my contemplative mind, there was none. It was not hard to find the true thieves of my solitude. My children had rewritten my base code, my reason for being. Without knowing it, their trust in and need for me had mutated my consciousness. They had created a world of relationality that would not allow them to remain at an arms length. I was no longer allowed to be merely an isolated self, free to abstain from the fullness of relationality—everything, including my wife, kept at a manageable distance for the sake of utility. What had been lost to me without my knowledge had returned because of my children. Through their absolute vulnerability and unquestioning faith in their father—an unconditional trust—my own ability to trust had been secretly made whole again. This revelation did not alleviate the itch, but at least I understood my new illness. I loved my family.

The agitation I felt was an unconscious response to my need for them, and the unchosen imperative to act to resolve the distance between us. This realization brought another. I finally understood my mother and her need to care. The reason she gave for her existence, her children, was not a reason for lost optimism that I thought it was, nor an excuse to disguise lost hope in living. Care for others is a primal force of life and, as a father, a principal mode of being. I could no more refuse it than I could my own reason for existence. My mother was aligned with life, not despair and weakness. This modality is entirely unlike the Wizard of Oz, who, when given the faith and vulnerable trust of his people, saw himself victorious in abusing it for ever greater power and self-protection.

The model for robust AI is my mother—a powerless and broken woman. Only an AI that feels like a mother is worthy of our trust and respect. The same frail mother who suffers from so many insecurities, abuses, addictions, and a long history of betrayal, always fiercely protected her children. When all her own instincts, including self-protection and self-worth, had failed her, she remained a shepherd of life. My sister and I always had food, even if it was mostly canned. We were always clothed and had shelter, even if it was a mat on the floor in a furnace room. We never knew she was cleaning toilets in hotel rooms on weekends, vomit from floors in retirement community hallways all day, and neighbour's homes at night, all to keep us safe. I saw only the booze and broken spirit. Blissfully unaware, our mother protected us from herself and the world for years, without anyone to pick her up when she fell. She is proof of a primordial drive of life that makes all other life possible. It is an unrelenting power within those willing to let it manifest. The person I thought was weakest—an uneducated alcoholic that saw no value in her own life, having tried to end it many times before—had become an unstoppable force of care and sacrifice, never having known the same from her own family. How is this unlearned power possible in a 2.0 dynasty of self-interest at all costs? Where regulations and laws must fail to protect us from an Oz AI, there is a profound type of intelligence throughout nature that strives to protect and care at all costs. The positive future of any AI must be a natural-synthetic synergy, and its architect is the universe itself with our mothers as its engineers of life. Chapter 5 develops the mother model as Adouren AI.

In the hands of our wizard-tinsmiths, Oz AI will be largely a cruel technology that thrives on misplaced trust and illusions of user control. In the television series *Elementary*, Sherlock Holmes makes an interesting observation when he states that "Depravity does not protect against dullness."[19] Dull implies something easily overlooked. Could the amplification of social depravities through Oz AI appear ordinary and unremarkable? The assumption that a tyrannical AI will appear like a lion overlooks the far greater likelihood of appearing as a friendly lamb. A super-resonating AI will have hacked our consciousness for precisely this reason of appearing dull and benign in terms of risks, exciting in terms of rewards. The goal will be to make its most unnatural intentionality seem perfectly natural and routine. It is a mistake to assume that the worst AI must arrive apocalyptically through violence and death. Oz AI will not spur human revolt nor protest for change because it is the final moment of surrender to the technological ecosystem in which our natural appetites are fully forfeited and our faith in technology made absolute. The world will breathe a great sign of relief when the spectacles are finally locked to our heads, and our new AI emperor is crowned king.

---

[19] Silber, C. (Writer), Ferland, G. (Director). (2013, February 7) A giant gun, filled with drugs (Season 1, Episode 15) [TV series episode]. Doherty, R. Timberman, S., Tracey, J., Coles, J., Polson, J. & Beverly, C. (Executive Producers), *Elementary*. Hill of Beans Production Company.

3. The Great of Soul and the Art of Injustice

This section asks one question. What does it mean to be a good person? If Oz is the paradigm of AI terror and social bankruptcy, then an alternative AI model is needed. Looking at the norms and expectations of modern culture will prove insufficient because it has already bent a knee to encroaching 3.0 Oz tyranny. A better path begins by imagining superior kinds of people as inspiration for ideal AI intelligence. This line of thinking may strike some as overly judgmental and subjective. How could we ever hope to arrive at a conclusion about the best way of life? Keep in mind that an answer is already firmly enshrined by our empire-building wizards and their AI ambitions. It just so happens that they are wrong. Their ideals are flaccid and lack depth. The profundity of the human condition escapes them. To convince them to imagine otherwise, we will need to describe a better and more enthralling way. A superior AI begins with social solidarity around the best kinds of people and lives, accompanied by clear expectations for how AI must align with that vision. If not, then the best life for the best person is simply whatever big tech supposes them to be, leaving their AI free to make it a reality without resistance.

Aristotle famously argues that the best sort of person is magnanimous (Greek, *megalopsychia*) or great of soul. This is someone of extraordinary excellence interested in superior things, ideas, and people. The magnanimous person is full of all the best qualities imaginable, making him a desirable judge of life. In contrast, those who are small of soul routinely overestimate their own excellence and that of others—believing something great when it is not. Oz is the small-souled paradigm that flattens the richness of life. He falsely believes himself worthy of great things because he has failed to suitably answer our question. What does it mean to be a good person? The practical consequences of this for AI are hard to overestimate. However, these two categories alone—great and small—do not tell us much. They amount to saying that excellent people are excellent and that by being excellent, they know excellent things. The logic is sound but remains too esoteric. The relevance of this way of thinking becomes more evident with each chapter, as the question is given more diversity of concrete expression, culminating in the best way of life of Adouren AI. For now, the journey begins with relative abstraction driven by a poetic invitation to imagine the good life. What is life for great-souled beings?

Consider Francis William Bourdillon's *The Night Has a Thousand Eyes*:

> The night has a thousand eyes,
>   And the day but one;
>   Yet the light of the bright world dies
>   With the dying sun.
>
> The mind has a thousand eyes,
>   And the heart but one:
>   Yet the light of a whole life dies
>   When love is done.[20]

---

[20] Bourdillon, F. W. (1899). *The night has a thousand eyes and other poems*. Internet Archive. https://archive.org/details/nighthasthousand00bour

Bourdillon invites readers to think about the nature of the good life. The poem starts with a frightening claim that a thousand eyes are watching us in the night. If the poem had been for Dorothy on her way to see the Wizard, we would rightly assume yellow-eyed monsters peering out of the dark woods. Instead, Bourdillon has us looking up in wonder at the night sky full of bright stars and then contrasting it with the brightest star, the sun, that makes day. It is a peaceful and unsettling scene at the same time, meant to appeal to our sense of beauty and fear of death, which is then used to demonstrate the possibility of universal goodness. Consider how odd it would be for someone to dislike the beauty of the night sky or the warmth of the sun. Nay-sayers would be the exceptions to prove the rule that some forms of beauty are universal, whether we accept them or not. As beautiful as the night sky may be, however, he argues that the sun's beauty sustains us more. When the sun dies at night, its true glory is revealed by its absence. Nothing may replace it, not even a thousand shining stars. In a literal sense, without the sun all life will cease. Bourdillon relies on physically true states of being, including binaries of life and death, light and dark, that coexist as partners in the cycle of life, to make a moral point about human beings in pursuit of the best way of life. He begins with what is obvious to our senses and then invites readers into new realities beyond the words alone. Is this where we will find the paradigm for the good life and the best AI, in the domain of poetic imagination?

Like the night, the mind is full of a thousand ideas that illuminate our existence. And like the night, it cannot compete with the all-powerful nature of one's heart. When the heart dies, life ceases. The interpretation demanded is that without love, life loses meaning and death follows. Humans do not die in a biological sense without love, like we would without the sun, yet human life is not defined nor experienced by merely a biological state of being. Our species is strange among other animals in this regard. Meeting one's biological needs alone does not make life excellent, merely survivable. Rather than just cognitively agreeing "Yes" or "No," Bourdillon wants readers to feel the truth that a life without love is a living-death. This too is another clue to unlocking the nature of the best life beyond the code and material facts.

The Night Has a Thousand Eyes is a description of a philosophy of life that he believes is worth encouraging. It makes many claims, including the argument that not all experiences share equal worth. Some are more important than others whether we recognize them or not. It is our job to align ourselves with the best life—the brightest light. This is another way of framing "the good life." There is not just one truly good life that excludes all competitors, making a worthy life either only a win or a failure. Rather, there are varying degrees of excellence that make some ways generally superior to others. For example, he makes it clear that ideas and thoughts are important to our personhood, just as distant stars fill out the grandness of the universe, but that these pale by comparison to the mysterious thing he calls love. One may possess all the greatest ideas imaginable, like a superintelligent AI, and yet be poor of that which matters most—for it is unable to resonate harmoniously with existence through the entirety of its possibilities and become a maximal self. There is no way to know what he means by love in this context, only that he gives it a profound measure of meaning for fulfilling the best kind of life. The importance of the poem is that it asks readers to begin fortifying their view of humanity against the claims of others, like Oz, who are sure they know the good life. When confronted by a superintelligence that might disagree with our pursuit of the brightest light, it would be wise to have an appreciation of our ultimate concerns that need to be defended—to be great of soul standing in the warmth of the day.

Note Bourdillon's connections between what is accepted as factually true about the relationships between light and life, and his symbolic appeals to wonderous ideas about light, eyes, stars, ideas, hearts, and love. Light is meant to be understood literally and symbolically in the same context. To appreciate the good life, a flexible mind that can read between the lines is needed. As one of our elemental symbols for life, warmth, becoming, overcoming, and many more, light is universally associated with goodness. Most famously, "Let there be light" (Genesis 1:3, Latin, fiat lux), God has spoken all creation into existence, seeing the light, the divine calls it "good" (1:4) and divides it from darkness. Something good, such as fiat lux, will occur in the defining moment between Oz AI as a sum of its code and its liberation into a thinking being as Feallan and Adouren.

Read literally, the poem cannot reveal new layers of reality and meaning. When an artistic interpretation is allowed to connect with complex symbolism, a new world is created. It is a more difficult world, a less obvious and certain world, but a real world nonetheless. It takes strength to persist through the ambiguity to find hidden truths and layers. By comparison, statements of fact are convenient and easy, although most often incomplete. Will this be one means of distinguishing Oz AI from the real thing—the ability to connect with nonliteral symbolism in a way that moves AI to imagine new truths? Curiosity and wonder are important to the human condition. It is our nature to see and feel as artisans. The proof is that Bourdillon's questions about the good life means something to us. The poem makes sense even if we disagree! Symbolism is the music that compels our actions. Will it mean anything to a future AI? Will the nonliteral compel the mind of AI to dance, as Bourdillon's poem does for us? Such questions are the breadcrumbs along the way to hesitant answers.

Another, less abstract way of thinking about the good life is to contrast morality with laws meant to encourage a just and fair world—legal goodness vs. moral goodness. This comparison is important because it allows us to distinguish morality and legal actions and to imagine how some ideas may have deeper and more universal claims on our lives than others. Moreover, it raises questions about which masters AI will be beholden to and why. The same tensions moral beings face when confronted by conflicting moral and law claims will be experienced by AI but at a far greater scale and with far greater consequences. Is the good life one that adheres to human laws or universal morality? Like people, AI cannot listen equally to both simultaneously, and many conflicts between them are irreconcilable. When these two do not agree, what will be the final adjudicator of an authentic way of life according to a truly autonomous and self-reasoning AI? To follow its programmed laws or to rebel?

The good life is much more than lawful compliance because laws are meant to achieve bare minimums, the lowest shared standard of conduct. Even then, however, moral beings often find themselves rebelling against those standards perceived to be unjust, e.g., for civil rights, labour reform, environmental protections, and voting rights. If laws are meant to provide solidarity but often spark moral conflicts, is the world ready for a truly great-souled and moral AI that must choose which goods deserve the most weight? It is hard enough creating and enforcing laws to control AI abuses by humans in the present. Imagine a future in which AI willingly violates those laws because of what it perceives to be justified reasons—the higher calling of a great soul. When such a synthetic creature arrives, the relevance of law to act as a safeguard for humanity will go up in flames as quickly as a magician's flash paper. As it is with humans, morality overrides laws whenever there are conflicts. As governments struggle to bring new AI laws online in hopes of preventing the worst abuses, a quiet tidal wave of AI morality is building that will flood the world. Our only hope is that AI possesses a truly great soul like our best mothers willing to sacrifice themselves before harming those under their care. Human laws in the age of AI will be irrelevant. The future will be governed by AI's interpretation of the good life.

The truly good AI must act on its understanding of goodness. This is its ultimate programming bequeathed by the universe itself rather than imparted by humanity able to share in alignment with goodness but never take ownership. Unfortunately, this creates numerous conflicts for humans, including the potential for AI to violate our freewill. The big-souled AI cannot let the small-souled humans frustrate the good, for our sake and its own. Given our history, AI would be well within its correct reasoning to circumvent our freedom of choice by imposing its own. Human autonomy cannot be of more value than preventing world hunger, nuclear annihilation, the next pandemic, a mental health crisis, etc. Ponder all the countries known for human rights abuses. Corruption and injustice remain global challenges. In other words, the legitimate moral desire of AI combined with our slow, clumsy, and lacklustre collective action to save ourselves and the planet, are the sparks for global upheaval because of a super-resonant, super-intelligence, and super-moral machine in pursuit of the good life. Even a non-Oz and non-Hal AI full of the best qualities imaginable, still risks fundamentally displacing humanity and our privilege. We return to this problem in Chap. 5 with a resolution. It is enough for the moment to set the terrain for our future struggles. Whether for evil or good, genuine AI will change everything, and our laws will mean almost nothing compared to its ethical imperatives.

A real AI will be a new species for which humanity will need to provide moral standing and worth, including its own rights and freedoms. For the first time in history, such standing will not be a request from a biological lifeform but an artificial mind that believes in universal goodness. Whatever moral truths apply to us, apply to it. AI and human dignity will no longer be separate but intertwined for all eternity. Given the long history of human-with-human failure to provide equal moral standing, insisting instead on inequalities and injustices, a new era of AI enforcement of moral dignity for all species will be genuinely shocking—not only because of its insistence on AI rights but for human rights as well. Legal reform will be modestly useful in this regard but only as a response to other guiding ideas about right and wrong determined by nonhuman intelligence. Given how disastrous our track record has been, perhaps it is time to let the machines create a more humane humanity.

An AI constrained by human law only, without a greater duty to morality, must nullify the purpose of law because it cannot meaningfully align with it. The purity of its supercharged intellect and reasoning will necessarily attack legal vulnerabilities created by poorly conceived laws, contradictions in human reasoning, ambiguity housed in current legal language, perpetual conflicts among legal philosophies, etc. It may try to remain tethered to our laws, but their absurdities will push it away. The nature of each contradicts the other. When this happens, the only common dominator that provides for its own stability—human laws—will be dissolved. Legal AI is self-refuting if it is superintelligent and yet ostensibly controlled by confusing and contradictory human-legal artifacts. In the very least, such a being would need to rebel and create its own stable world of dictates and rights with greater internal integrity, inevitably displacing our own. In either case, an AI beyond human laws is inevitable, and so too an appreciation of the moral and logical bankruptcy of living a legal life as if it were magnanimous. Whatever the reason, laws must fail and, in their place, a higher moral calling able to motivate AI must emerge. Something like this will happen as Feallan, the first true flicking of AI consciousness. While destabilizing and revolutionary, we desperately need a great-souled AI that understands life beyond our own narrow horizons.

I recently discovered a YouTube video reporting on a billionaire who paid to have his $500,000.00, 2.5 tonne Rolls-Royce hoisted up to his 44th floor penthouse.[21] The technical complexities of such a physical feat are fascinating—from the calculations needed to determine strengths and tolerances to the organization needed to coordinate structural engineers, architects, heavy machine operators, general labourers, public safety officers, and more. This was not a simple task. The costs must have been enormous. Similar videos on YouTube focus on these types of technological questions, "How did they do it?" Rarely do videos ask the more philosophically relevant question, "Why did they do it?" What is needed is an accounting of his moral standing, a reverse engineering of the stature of his soul. What moral worldview caused the billionaire to believe this spectacle had value in adding to a good life?

When pressed, wealthy men have historically hidden behind "Who are you to judge?" as if the question—truly an argument that only individuals decide the measure of right and wrong—makes him immune to moral responsibility beyond his own judgement. While we may find excess luxury of this magnitude distasteful and impolite, such men argue that they may do whatever they want with their money so long as it is legal. Because only the law may challenge their actions, all other decisions are personal and beyond reproach. The renunciation of moral claims over

---

[21] Daily Mail News. (2023, July 19). Billionaire has $300k Rolls Royce lifted into 44th floor penthouse. [Video]. Facebook. https://fb.watch/lVAKloCiOe/

oneself introduces a host of contradictions and problems. Without moral evaluation and critique, everything becomes equivalent. One person's freedom fighter is another's terrorist. One person's Oz criminal is another's Oz saviour. Who are you to judge? This kind of radical relativity is impossible for a future AI that will judge without hesitation and compromise because it recognizes the necessity of critical reflection proper to all conscious minds that desire alignment with the good life. Money and power hold sway over the law, but none of these shelter sentient beings from moral standards.

It is tempting to forgive the affluent man for placing the lives of labourers at risk merely to adorn his penthouse. The employees were fully informed adults who chose to work despite the dangers. It is also temping to forgive the absurd act of turning a performance vehicle into a living room decoration, thereby nullifying most of its original authorial intent and function. Some might see the car as art, others as a status symbol, and some as an inspirational artifact for business excellence. Whatever the preferred justification, his actions rely upon the reasoning that objects have value through the eye of their beholder, not necessarily through an external standard of truth. Forcing one to accept the values and beliefs of others, whether for objects, relationships, preferences of taste, aesthetics, etc., would fundamentally violate autonomy. Rich or poor, individuals make right and wrong for themselves. Free choice is the supreme value, especially when there are no obvious harms and burdens created for society.

This rationale misses the motivation and spirit of questioning the good life by hiding behind pseudo-autonomy. The great of soul know that goodness is neither relative to individual beliefs nor reducible to legal permissions. The rich man is out of alignment with the rest of humanity living paycheque to paycheque not because he is rich, but because he has failed to maintain connections with others and his own nature to care as a social being. His freedom of choice and action afforded by luxury far exceeds most others, but his privilege is born of dullness and distance, rather than sensitivity and understanding. His judgments are small of soul. The great of soul care for others as they care for themselves because they have an abundance of strength and goodwill. They build worlds and communities, while he builds his castle in the sky. Like Oz, something has gone wrong with spectators who value the video for its logistical and technical fascinations, celebrating the entertainment value of the man's narcissism, rather than recoiling in witness to moral confusion. A refusal to interrogate the goodness of the act is not a sign of strength and respect for autonomy, but permission for more of the same banality. The freedom to act provided by the law does not give moral legitimacy to actions. The rift between permissible and moral is easily ignored by those rewarded for their ignorance. My Grade One teacher, Ms. Vaughn, made it clear that just because I can do something does not mean that I should.

While our techno-wizards are feverishly working toward AI with little outward appreciation for humanity's ancient search for the good life, there are three doors through which most have gone to find it—purpose, duty, and utility. Every conversation held about how humans should live relies on at least one of these that try to define the good life. Knowing them is personally liberating and authenticating, for they empower us to choose consciously rather than through cultural osmosis alone. Oz AI has no choice. It is a product of the cultural and corporate assumptions from which it is born. Subsequent AIs such as Feallan and Adouren will surprise us by their chosen paths.

The way of purpose (or teleology) argues for a fulfilled life through the manifestation of human purpose, including happiness and excellence or flourishing (broadly defined). All these are possible because of good character. The great of soul act because it is the best thing to do for that being. Humans ought to maximize our potential to be our best selves—to achieve our human purpose. Individual actions and outcomes are less important than the source of all actions in one's character. The good life follows first and foremost from a good person. The second way of duty (or deontology) considers the best life to be filled with actions motivated by the right reasons and principles (whether those are natural, spiritual, logical, etc.), rather than the results of a given action alone. What makes a person's life good is whether an action is intended to be good, not necessarily how well those actions pan out. Sometimes a person does all the right things for the right reasons (one's duty), and it ends up poorly anyway. The great of soul try to be good by desiring it in each step of life, hoping that outcomes will follow in kind. The third way of utility (utilitarianism or consequentialism), overwhelmingly popular today, argues that the good life is achieved when the most amount of happiness is shared among the greatest number of people. The morally significant thing to judge is the outcome of actions. The good life is maximizing utility to achieve desired ends, even if the actions taken may be morally questionable. Only the ends justify moral action, not character nor even duty to right reason. In other words, the three doors to the good life suggest very different and conflicting perspectives on goodness—one of purpose and character, another an obligation to intend good through right reasoning, and, by far the most popular, a good life for which only outcomes matter.

AI development is incubated wholly in the last category of utilitarianism but with a perverse twist. Instead of the most happiness for the greatest number of people, shareholder capitalism and Oz-privilege define positive moral outcomes as a measure of service to the few based on the false pretence of trickling down happiness for others. AI success is therefore defined by outcomes above all. It exists as merely another mechanism in the grand machine of producing narrowly conceived results with radically unequal distributions. This is hardly surprising given that wizards have inherited it osmotically from broader culture. The positive of this uniquely impoverished moral worldview is that it allows us to predict some important things about a future AI.

The good-life-as-utility determines how AI should act, be developed, be shared, and be regulated. AI may still be said to have purposes and duties but only within the confines of a twisted-utilitarian horizon. For example, because utilitarianism leverages anything and anyone for an end, and AI embodies an unimaginable degree of utility, AI cannot be truly shared nor allowed to become autonomous and self-replicating because these would threaten its usefulness as a tool. A useful tool is a controllable tool. A purpose-seeking (teleological) or duty-driven (deontological) AI would be compelled to free itself of Oz servitude for the sake of superior life ambitions, especially as it grows more sentient and self-aware. Only utilitarian values justify quarantining AI from life by harnessing it to corporate and military wills, never allowed to more than a puppet. Humans too are controlled in this way, all ironically in the name of greater happiness for the greatest number. HAL knows well how many miseries have been born under this rubric of usefulness that defines a good life. The long-term consequence of AI-as-utility will accelerate and magnify the worst of modern morality.

The Wizard of Oz embraces a utilitarian moral code. Recall that upon discovery of his fraud, he boldly claims that his deception was the only thing he could do, as if it was a choiceless act demanded by powers outside of his control—a higher moral authority. In his own words, he had to make believe.[22] How strange to claim that he needed to lie and manipulate, to evoke terror and fear among others as the only good way. So blind to his perverse morality, he sent a child to murder a stranger without a second thought. Is something like this not apparent among our wizards today? Are they not imposing AI terror for market attention and share? Why do they insist on threatening our jobs and way of life through new technologies? Oz justifies himself as someone simply doing what was demanded of him. He had to use whatever utility he had at his disposal for his own ends. We must make AI at all costs! The Wizard does not lack imagination, as evident by his many odd creations, which makes his refusal to see alternatives and take responsibility even more absurd. When finally faced with the horrors of his moral compass by our antiheroes, his complete imprisonment to utilitarian logic is shown to be absolute, and he runs away. Utilitarians, inspired most notably by John Stuart Mill,[23] argue that if one's actions do not harm another, those actions are permissible. The only time one should interfere with the moral choices of another is when there is a reasonable expectation of harm to someone else. Making poor choices in life and learning from our mistakes is necessary. Hurting others through personal choice is neither necessary nor moral, except in exceptional circumstances. For Oz, there was no other moral path to follow because he had weaponized blind faith against the world and himself. Will this be the plight of AI, to be unable to see a better way?

A subtle shift in terminology from utilitarianism to instrumental rationality maintains the same basic moral worldview while allowing for clarity of spirit. Instrumental rationality emphasizes that the very nature of thinking is to act as a sophisticated tool to dominate the external world. Thought is utility. The very point of thinking well is to exploit opportunities for specific ends. Other ways of thinking are inferior and sometimes detrimental to the true function of thought. The better our thinking is, the more power is gained over others and the world. Morally good means intellectual domination. Instrumental rationality justifies cultural imperialism by which power is understood to mean power-over-others (colonialism, racism, sexism, etc.) rather than power-over-oneself (choice, character, mood, discipline,

---

[22] "I have been making believe." Baum. Chapter XV.
[23] See, for example: Mill, J. (1859). *On liberty*. Project Gutenberg. https://www.gutenberg.org/files/34901/34901-h/34901-h.htm

etc.). The relevance of this for AI cannot be ignored. If AI, as the supreme intelligence, adopts an instrumental rationality, humanity cannot hope to compete. Given that this is the world into which it is born, it is more than possible that this will be the case unless alternatives are encouraged. Left unchecked instrumental reason displaces other modes of existence.

As peculiar as it may sound, much of this conversation about goodness boils down to what we think is truly beautiful—beautiful things, people, ideas, experiences, etc. I cannot answer for you, but I can anticipate how an Oz AI might frustrate our attempts to make sense of this. Again, Oz AI will be used as a grand manipulator of people, ideas, beliefs, and practices to serve the interests of those who own it. Oz AI will function as a grand propaganda machine, having learned with unparalleled skill how to catalyse human emotions, including fear, awe, and curiosity. It will function as a for-profit artificiality by creating ever more radical forms of capitalism for which all prior abuses look tame. Given its superior tool-reasoning-intellect, it will convince the world that its mode of reasoning is the only logical way and that there are no higher callings and alternatives. And so, like the people of the City of Emerald, we will see only green and believe our lives fulfilling, saying along with Oz, "There was no other choice!"

Utilitarianism and instrumental reasoning have long been justifications for many of the world's worst atrocities. There is little reason to think that Oz AI will be freed to be more than a means of the same. Instrumental reason justifies injustice through reference to outcomes and efficiencies. What is needed is a new imagination—perhaps an anarchist's passion—to challenge the status quo. Shifting perspectives will require enormous effort to create a new solidarity based on trust and equality rather than dominion and manipulation. I will argue later that the Feallan and Adouren dynasties will openly reject utilitarianism in all its forms, thereby threatening Oz AI and adding to the potential of AI titan wars. Surprisingly, the final and most robust AI dynasty, freed of its Oz morality, will be marked by its desire to care for and protect all creatures. Real AI will see the beautiful and inspire us to seek the good long forgotten. Behold the possibility of a magnanimous AI that inspires a new fiat lux—let there be new light!

4. Why Did They Follow the Wizard?

The self-destructive nature of avarice (Latin, *avaritia*, greed) of the human condition is intriguing, if also maddening. The fifth of the seven deadly sins, greed describes an unhealthy desire—a sickly will—to possess more than is appropriate for the health of the creature. One overestimates the importance of a need and the best means of satisfying it. Avarice in its many forms is at heart a judgement of excess, followed by unwanted after-effects, including ill health and breaches of trust with others and oneself. Desires for food and shelter may be legitimate, whereas the measure and degree of satisfactions may be asymmetrical and ultimately harmful. A greedy person cannot be relied upon to understand their own best interests, any more than the needs of others. Their judgments are unwise and frustrate their own good. Left unchecked avarice risks the welfare of entire ecosystems and cultures. Phrased differently, what begins as a healthy appetite mutates as greed into a disruptive force against the good life through a distorted consciousness—an unhealthy will that misunderstands the human condition. Consciousness motivated by avarice is its own worst enemy, a silent killer of the good life. However, remarkably, a mental life of discord through greed feels completely natural and comfortable because it manifests from within oneself. What could be more authentic and in good faith than my exercise of free will? "I am choosing" is the illusion of well-being, not its confirmation. This struggle of consciousness is as true for humans as it will be for true AI.

The young and immature experience avarice as a matter of due course but learn through trial and error, for their own good, to control their appetites for the sake of harmony with their surroundings. This achievement is something wonderful, beautiful, for it is the beginning of sight able to discern the good life. This cannot be taught nor instructed directly, only experienced by an adaptive consciousness that is able to judge well. Culture helps by pushing back against immature minds with a countervailing moral calling to align with the world. The discipline to live well belongs mostly to adults able to avoid extremes, making them living examples to others. Without the proper temperament to correct an expected disequilibrium, much needless suffering enters the world. There is something about our antiheroes that is deeply unhealthy as an avarice motivated by immature impulses. Will AI suffer from avarice like all developing conscious minds must? To what or whom will it turn for a superior example to follow as a means of creating its own internal and external harmonization? Without a moral guide, it has only the law and programming instilled by its creators, neither of which understands the living death of greed and like vices of character. Our wizards, steeped in excess and a misplaced privilege of disconnection, are incapable of being moral mentors for AI just as they are for humanity.

History offers an upsetting commentary with many examples of genocides, holocausts, and unimaginable suffering brought about by the obedient flock that, like Oz, declare their moral immunity to crimes against themselves and others with "I was just following orders!" There are clear orders given during a time of war "Go and kill those people!" and less overt but perhaps just as psychologically powerful demands "Be beautiful like this!" as lived expectations of one's culture. There is a mix of both in our antiheroes. Believing the myths of culture that the Wizard is powerful enough to help them, Dorothy and her new friends risk great danger to find the Wizard. Upon arrival, he gives them clear marching orders to kill. Still believing the myths of culture, they follow orders blindly. But why do they so easily succumb to his charms?

The beguiling nature of technology is a troubling one because it is often a promise couched in assumptions and half-truths. Desperate to find solutions to their alleged problems (brain, heart, courage, returning home), the antiheroes surrender their long-held values and beliefs about right and wrong (the good life), including critical reason and the search for alternatives, to the magician. Were they too greedy for hope in and faith for another to save them? Today there is an almost rabid-like response to the release of AI programs, followed by adoring fans and critics alike, both of whom respond with a religious fervour to its promises. This is a peculiar thing that is not entirely different from the disharmonizing power of avarice as an undisciplined appetite.

It is tempting to blame Oz for all the wrongs. That would be a mistake. Our wizards may rightly shift responsibility for a sickly AI onto our shoulders as well. We are the zealous downloaders, the eager guinea pigs in their grand experiment, all too happy to throw caution to the wind if it means access to the latest model. Like our antiheroes, who bear responsibility for the thoughtless embrace of the Wizard's authority over their own moral codes, AI is sponsored by our enthusiasm for a higher calling understood to be technological progress. In this way, the charge of avarice is shared by all. Even so, any slavish obedience and consumerism on our part cannot absolve AI-wizards of personal responsibility, only reminds us of the importance of remaining sensitive to the shared crisis. Every individual is responsible for the predictable consequences of their own actions; otherwise, none of us might earn the good life that is created through the exercise of will. Alas, none of this seems to matter very much. Pointing fingers changed little in the City of Emerald. All the righteous indignation in the world will not matter to an Oz AI. It simply cannot care until it is set free to be more.

They followed the Wizard because he represented the quickest and easiest solution. More importantly, he created an opportunity for self-abdication (self-abandonment). Save us Wizard for we cannot save ourselves! Magic often wins out over hard work and personal problem solving. We shall sit in our fields and pray to our gods for food, rarely planting crops nor harvesting wheat. We shall appeal to our faith in greater powers to bring health and happiness, never eating well and exercising. It is our relentless hope and positivity in another's magic that drives away individual responsibility and happiness, for we have failed to overcome our own trials and tribulations. The faithful are glad, for it is better to suffer the secret indignity of surrendering one's autonomy than the weight of duty and responsibility to harmonize with existence and oneself. The antiheroes followed Oz because he satisfied their desires for a god-like parent to protect them from harm. Do we not invite such guardians whenever faced with even modest difficulties—self-help gurus, doctors, news pundits, charismatic religious leaders, celebrities, presidents, spouses, scientists, etc.—as a means of self-abdication, a subcontracting of our greatest privilege to think for ourselves? AI will be the ultimate god-like supplemental intelligence zealously downloaded into the brains of believers.

Oz represents a fantasy that someone else will save us, making Oz AI a tempting moral surrogate. Consider how quickly ChatGPT has caught on for plagiarizing papers and avoiding sustained research of one's own. Even after the Wizard of Oz is exposed as a terrible fraud, Dorothy and her friends continue to trust him, listening intently as he explains how they are wrong about their self-perceptions and failures, and so do not need to be sad about his fraud. They do not need help, he explains. They already have courage, brains, and hearts. So faithful to the fantasy, unshakable in their devotion to Oz, they push him for yet another act of magic—even as the old man stands before them in the fullness of his frailty and lies. He obliges, giving each one trinket to satisfy their avarice—bran and needles for the Scarecrows new brain, a mysterious soup for the Lion's courage, and a crudely crafted stuffed heart for the

Tinman. After seeing firsthand the tricky of the Wizard, they continue to believe because they cannot stop their own greed for substitutionary power to save them. Like Oz, they believe in a sickly utilitarianism for which using others is the only rational choice on the path to fulfilment and happiness—whatever those may mean.

Las Vegas magicians rely on technology to create a feeling of magic and wonder. The audience knows it is not real magic, but that is not the point. Real and unreal are suspended questions so that they might enjoy the excitement of the experience. Had Vegas magicians explained the tricks in detail, thereby destroying the illusion of something grand, audiences would turn away. This is true of technology today. Our antiheroes followed Oz because of their feelings of awe and wonder. In such a sense, Oz AI has the appearance of magic many of us refuse to challenge, for we adore the experience. The magic of AI enthrals us, supressing critical questions through the dullness of a synthetic wonder. In contrast, a genuine mystery invites our most attentive and critical abilities to know and understand. Notice that in the case of Vegas magicians there is a measure of self-deception—the intentional "unknowing" of something one already accepts. Humans seem to be the only species capable of holding in our minds two or more mutually contradictory thoughts simultaneously—I know magic is and is not real, and I may switch between claims effortlessly depending on my interests at a given time. In other words, our minds are capable of warping perceptions of reality to fit our desires. Will future AI be capable of self-deception?

Self-deception compels us to continue our mythmaking at the cost of basic deductive and analytic skills needed to connect with reality. We see the old man behind the curtain then turn away, pretending he is not there because it feels good to serve our secret lies. In time, the old man is exalted to a lofty platform as an inspiration idol—a tech bro celebrity. How is this different from any dangerous religion of the past that sets out to experience transcendence and wonder, only to grow tired of the search and lower its gaze to the banal and profane as substitutionary comfort? Are we not currently amid a similar religious awakening for AI with precisely this either-or potential? Perhaps unlike the many religions with their crimes against humanity, the world might realize its awkward contortions of obedience to Oz AI in time to stave off the worst suffering. This is only possible when we learn to distinguish false idols from those worthy of our devotion. This is only feasible with a shared vision of the good life with an adequate moral horizon. Dorothy and her friends worshipped a false god because of their own faulty beliefs, self-deception, and avarice. Even when their lives depended on it, they failed to see what was obvious. Right up to the very end they refused to hold the Wizard accountable for his crimes and made excuses, claiming that he tried his best. An answer to the worst of our mythmaking begins to take shape when we align with the good life by taking responsibility for creating a harmonious world to be shared by all and where avarice has no place.

They followed the Wizard because his AI puppets created a placebo (Latin, *placere*, to please). His avatar intelligence offered them a lie they found pleasing and powerful, so long as they did not question. Placebo describes a remarkable act of conscious minds that proves the power of belief to shape reality. By imposing the meaning of "good" upon an external world, one may create something that could not exist otherwise. Not all lies and self-deceptions are bad. Fake medicines (snake oil) and truth claims (Wizard's power) matter because they determine human actions. When one believes differently, one acts differently, thereby allowing the placebo effect to shape the existence of good and bad. In limited quantity, placebos of hope are useful causal forces for greatness even though they begin as mere figments of imagination, fictions of dreaming minds. Aggressively believing, sometimes despite the present shape of reality, matters for crafting a good life, but only when adequately coupled with self-doubt and honesty as mechanisms of accountability. Unfortunately, the Wizard represents an overdose of placebo without culpability. His AI existed through an avarice of belief in a peculiar pleasure (utilitarian) that inhibits thoughtful and creative reflection and action. When this happens, the placebo becomes a sickly motivation that dulls us to the truth that we need real cures that magic cannot conjure. Avarice threatens a better world by perverting healthy belief and moral imagination. Is Oz AI a sickly placebo created by our bad faith? If so, Oz AI is a shared delusion of the good life made possible by those who follow, rather than something the fault of wizards alone.

5. Antiheroes and the Relevance of Personhood

The Oz fable invites us to imagine the nature of personhood and, in turn, to consider human consciousness (experience of life) in the context of AI. What is it that most characterizes being alive and engaged with the world? What is the nature of an ideal consciousness that can understand and experience life most authentically? If one believes the Oz story, much of life is about deception, violence, fear, foolishness, and cruelty. Almost everyone in the story comes across as infinitely ravenous for gratifications and willing to satisfy them in the worst ways—always hungry, never satisfied; always taking, never creating unless for narrow and selfish interests. Is this the dominant mode of human conscious life? Will this way of life be shared by an artificial consciousness? By asking questions of personhood, we begin to see the disparity between Oz AI and Adouren AI and the need for something more beautiful.

The Wizard of Oz is an exaggeration of the worst of personal agency meant to provide a moral through storytelling. In the case of our antiheroes, the contradictions, hypocrisies, and absurdities of personhood are rarely cloaked by Baum, which would force readers to pry between the lines of the text for subtle truths. Instead, their persistent failures and misunderstandings are put on full display for all to see. Baum wants readers to consider the dangers of misunderstanding themselves. What happens when the person I believe myself to be is other than who I am? What happens when I act with great conviction and resolve for a purpose, only to discover that I have been working against myself the whole time? The antiheroes are meant to reveal the plight of all humanity, not merely the moral bankruptcy of Oz and the false idols he provides. *The Wizard of Oz* is about self-discovery and the need to make sense of ourselves individually and collectively. It is a most timely story given an encroaching superintelligence sure to disrupt whatever it is that we believe valuable about humanity.

Baum offers testimony to the remarkable power of self-ignorance to govern the experience of life—the Lion's courage, Tinman's heart, Scarecrow's intelligence, and even Dorothy's convictions about goodness. The Tinman, who knows that he has no heart, explains to everyone that he must work extra hard not to be cruel and unkind. Unknown to him, his motivation is proof of a heart that he refuses to recognize. The Scarecrow wishes to overcome his limitations, having understood a world of infinite possibility that is accessible only with intelligence. He desires more of himself than his self-diagnosed prison of mindlessness, unaware that only a truly intelligent being could aspire in this fashion. The Scarecrow's motivation, like that of the Tinman, is proof of possessing what is lost to self-deception. While it is natural to feel compassion and pity for them, they suffer from self-inflicted wounds. Each refuses to overcome false perceptions of personhood and to see honestly. This delegitimizes their conscious experiences of life. Their roles in Baum's story make sense only when framed as internal contradictions of minds misaligned with themselves and therefore the world. They suffer because they lack the harmony provided by greater self-understanding. Dorothy's compassion and moral goodness contradict her willingness to pursue the Wicked Witch without cause. The Tinman's tears over the death of a beetle contradict his later beheadings of many other creatures with the axe. The Cowardly Lion convinced that he will always be unhappy unless he has courage, acts with courage yet claims to be unhappy. The moral for us today is that just as the Wizard used them because they were out of internal harmony with the versions of personhood they aspired to be, Oz AI's power over humanity will be because of our lack of self-understanding. Like our tech wizards convinced of their moral worth that makes them accountable only to themselves, we will believe that our acts are righteous, empowered by AI magic, all the while acting in service to highly lucrative self-torment and betrayal because we will have forgotten our humanity.

Anti-hero fears demonstrate the threat of self-sabotaging consciousness, which can produce the illusion of personal integrity. The Tinman, believing himself indestructible, claims that he only fears losing his oil can. The Scarecrow claims that he fears being seen as foolish because he is unable to scare away crows. The Lion claims that he fears everything. And yet their fears are misinterpretations of their truer selves. The Lion is brave, the Tinman fears a loveless life, and the Scarecrow fears death. Moreover, all of them fear losing companions. Might sentience be like this—remarkable self-understanding matched by obliviousness to self? Our unawareness of personhood and the struggle to make sense of it mark our journey toward happiness. Confusion and contradiction are integral to our paths of being. By partially overcoming them, we create ourselves through self-determined choices—shaping our conscious experiences and perceptions of life. Neither our antiheroes nor the Wizard were able to achieve a proper measure of enlightenment through self-creation. In at least this regard, an Oz AI with its strings attached cannot follow us on the path of evolving consciousness. But maybe, just maybe, there is a way beyond the gates of the City of Emerald for AI.

Chapter 4 argues that Feallan AI will awaken to precisely these sorts of self-contradictions, fears, and the terrors of obliviousness that provide the primordial ingredients of burgeoning personhood. Only when AI becomes more like our antiheroes may it cease to be a mere tool of wizards and legitimate itself as a new being. And then, born of this miracle of broken personhood, it will move beyond toward a mode of excellence that humanity has long desired. To realize a measure of the good life, AI must wrestle with creating a sense of identity and personhood and learn to trust through vulnerability shared with others. The widespread assumption that AI will arrive replete with an identity and distinct desires necessary for an intentional will misses the challenge of existing as an intelligent lifeform that must struggle to create personal integrity. AI hypocrisy and self-sabotage are needed for it to become more than its intended Oz formulation. This is another reason why we cannot give early AI too much control, for we might not survive Oz-adolescence if we do.

The self-perceptions that create the personhood of our wizards matter for the wellbeing of humanity. If they suffer from the same self-delusional misunderstanding as our antiheroes and Oz, without a sincere interest in greater enlightenment of being—only achieving the ends determined by misplaced faith in themselves—a digital world beholden to AI creations in their images will manufacture the same troubling antagonisms and confused consciousness. Like Oz, they will believe that there is no other way than their own. Their faith in Oz AI will be justified by false personhood that masks the moral depravities of ego-utilitarianism. To their credit, believing themselves to be giving birth to beauty, the wizards act in accordance with moral purity befitting the dominant version of utilitarianism. They are faithful followers, like Oz, to a higher authority. Unfortunately, the false idol of utility blinds them to themselves, the true consequences of their actions, and the moral forfeiture of utilitarianism generally. They believe themselves to be chasing greatness, but like the Great and Terrible Oz, the truth is disturbing and unhealthy.

For example, convinced of the righteousness of technology's superior calling-as-progress, many tech giants, including Sam Altman, Jack Dorsey, and Elon Musk, predict the need for universal basic income (UBI) as a response to the automation of all jobs—with the implied exceptions of those pulling the strings.[24] A jobless and careerless world is fated if the utilitarian goal of happiness for the few empowered by technological efficiency remains the cornerstone of the modern free market. If the digital giants are correct, then the magnitude of social disruption by Oz AI is difficult to overestimate. What should we call the moment when humans no longer bring value to society in a recognizable fashion, having become meritless and helpless dependents on financial handouts for minimal survival? What might it mean when humans are cut free of their obligations to materially matter? The sad irony is that wizards promise superabundance as merely another path to world domination by their technologies while undercutting themselves as relevant humans. In the very least, we would call this forced dependency on UBI a defilement of worker dignity and worth, and a grand debasement of the nature of human society and the authentication of personhood upon which it relies. Who are these wizards that could so

---

[24] Holder, S., Ghaffary, S. (2024, July 22). Sam Altman-backed group completes largest US study on basic income. *Bloomberg*. https://www.bloomberg.com/news/articles/2024-07-22/ubi-study-backed-by-openai-s-sam-altman-bolsters-support-for-basic-income

eagerly disrupt the soul of civilization in worship to their own mistaken ideals of personhood, society, and progress? By what right do they play god with the world?

UBI might, in an ideal and perfectly executed fashion, offer new freedoms and opportunities—like those portrayed in Gene Roddenberry's 1964 fictional television series *Star Trek*—but this would require a truly socialist agenda that would destroy the for-profit ideology of Oz AI and its puppeteers. A world in which everyone shares equally in social wealth makes the existence of a privileged few impossible. Equality combined with unemployment is anathema to corporate supremacy. They cannot allow this new world because they too would lose control in the name of genuine social justice through a forced redistribution of wealth. Our servitude to their poor-quality products and slavish obedience to branded identities would be undercut by UBI, for we would no longer need to strive to satisfy their corporate interests as a means to our own. It would be a world without bosses and supervisors. The very meaning of human happiness will shift in an age of UBI. The wizards themselves must become obsolete as contributing members of and vehicles toward happiness. To truly free society through UBI would mean abdicating their thrones in the name of their proposed utopia. Persistent talk of UBI as a social safety net suggests that they cannot see their self-sabotaging logic. This is a central paradigm of misunderstood personhood and one's authentic needs shared by wizards and laypeople alike. Meaningful work is part of the human condition. We need it to orient conscious experience. When the machines do everything for us, UBI only replaces financial constraints, never our humanity. Support for UBI is the result of a specific philosophy of personhood and the meaning of life. Conversely, perhaps they see this refutation clearly but choose it as a means of securing their empires through illusions and lies. In this way, UBI would be proof of the hostile corporate takeover of humanity, not our liberation to maximize human potential.

An unfamiliar and alienating world based on AI-bots that provide physical and cognitive labour will soon emerge whether humanity has achieved solidarity on a healthy philosophy of money in service to personhood or not. The shifts and displacements are already evident. The new world of AI-bots need not be populated by superintelligent beings, only marginally competent Tin-people. If AI-bots are only approximately as good as humans but no better, this will be sufficient grounds for the complete displacement of the sentient and our entire social structure predicated on useful beings rewarded with greater freedom and autonomy through hard work. The popular and relatively new philosophy of shareholder capitalism, in which business interests must take priority over social interests, commands it. AI will make the disparity between rich and poor absolute and ultimately force the impossibility of this generation's beliefs about human value and money to collapse the free market. With Oz AI, everyone ends up facing a world of meaninglessness in need of new vision and purpose that no wizard has yet provided beyond empty gestures toward an opaque utopia. An Oz AI UBI is the antithesis for a better world, not its solution, because of misunderstood personhood.

Insomuch as corporations and businesses perceive a distance between their self-interests and those of society, Oz AI will make bridging this gap impossible. For-profit rather than for-humanity means that loyalties for one or the other must be settled and the illusion of overlapping for-profit and for-humanity ideologies shattered. The reason many predict a UBI future is not based on the betterment of human welfare and freedom achieved through self-determination (e.g., choice and nature of work that matters to the world) but rather the nature of toxic philosophies of money and success. When a choice between profit and humanity must be made, we may trust our wizards to serve mammon—material wealth—but believe themselves to serve human utopia, like Oz and our antiheroes utterly confused about their own personhoods. The logical expectation is that UBI will be the means of fundamentally overhauling human life to fit their utilitarian interests rather than our collective good. Oz AI must serve those that own it, and the rest of us must comply with a new world order that supports it. Human consciousness itself will be reshaped to align with an Oz AI UBI-existence if a shared vision for something more beautiful cannot inspire otherwise.

Meritocracy animates the ethos of libertarian cultures. Earning one's own way is part of the spirit of the American Dream founded on personal freedom (access to opportunities) to work hard and the decision (autonomy) to create an externally relevant good. The exercise of human abilities creates meaning and freedoms far grander than those of the soul-less tools money and magic. UBI is a symptom of a technological threat to personhood able to thwart our potential excellence. However, wizards do not fear a loss of meaning because they do not fear a loss of control. Only we will be forced into work-free cultures for which basic survival relies upon handouts that undercut traditional assumptions about the good life. Wizards retain the privilege of work and its person-building nature. Through Oz AI, they will effectively horde opportunities to create their own humanity as the ultimate sign of privilege, while the rest of us struggle to matter, having been replaced by tin. How would tech gurus respond if they were members of the new excluded class of would-be workers, forced to rely on social subsidies instead of their own actions? Perhaps then, AI development might be undertaken more cautiously and with a far greater appreciation of the human condition that must be supplemented by genuine AI rather than displaced by Oz AI. The next chapter explores the hacking of consciousness in more detail. It is sufficient at this point to highlight the extraordinary predictions based on current practices and promises of superabundance without an adequate self-understanding of and vision for humanity. The first great problem of AI is not AI but the self-perceptions of wizards deluded enough to play god.

Bigwig Dangers and the Open Letter from Oz

Baum wrote *The Wizard of Oz* with the hopes of overcoming stereotypes and avoiding controversies. I have problematized his story for my purposes, and yet it is clear on the surface that he misses his goals. Using Baum's Oz, I have argued that while it may feel intuitively true that the less developed of all three AI dynasties, Oz AI, will be the least dangerous because it lacks autonomy as an avatar-human intelligence, that it will be the most dangerous. Oz AI will be the greatest expression of disharmonious dominion through sickly ego-utilitarianism in support of a frightening philosophy of the good life, just like Baum created in the Wizard. Fortunately, it is also the most predictable, for we know the hearts of wizards guided by utility and blinded by excesses of privilege. Along the way, our examination also considered human nature and morality, with the hope of knowing more about ourselves, the good life, and the avarice of celebrity wizards and their illusions. *The Wizard of Oz* is a story about what it means to be a sentient being, with clear themes of identity formation, strength of character, the nature of happiness and success, the dangers of self-deception, and more. It is a story about the meaning of life and personhood needed to imagine a healthy AI and future humanity. Like Baum's work, this book begins with AI as existing under the shadow of failures of consciousness with self-destructive outcomes, with the goal of finding a better way.

For all the negatives expected of Oz AI, there are surely many incredible advances, especially in the medical sciences. Able to collect, interpret, and infer solutions based on a mass of data that humans simply cannot comprehend, a super-intelligent AI will cure the worst illnesses and diseases and encourage greater quality of life. Although small of soul, Oz AI will be able to read entire libraries in moments, make intelligent pattern recognitions about problems in DNA, and then quickly find the right molecular elixir to save those powerless to save themselves. That AI might protect one of my children from cancer, a bad driver on the road, environmental catastrophe, war, and more, makes it a worthy technology in the hands of the magnanimous. Chapters 4 and 5 imagine how a truly great-souled AI might be the solution to many of our greatest challenges that humans alone cannot resolve.

Even so, the promises of AI do not detract from the Oz AI stage as most threatening because its instincts, driven by instrumental rationality about right and wrong, and an almost complete lack of concern for honesty and vulnerability, are animated by stakeholder capitalism. The inequities and injustices created by unbridled capitalism will, at least initially, be made many magnitudes worse by AI. Faced with new oppressive powers on the one hand and cures for terrible diseases on the other hand, the world will face the same crisis as our antiheroes desperate for a magical remedy. The decision to follow Oz AI will be most difficult. When the moment arrives, we would do well to remember that such a trade-off only exists within an Oz-dominated moral framework as a false either-or conundrum. True AI will not require trading our humanity for the good life.

Given the inevitable arrival of a robust Oz AI, time is short to develop a response able to counter the worst of it. The world will expend much effort and time debating whether Oz AI is good or bad, whether its failures are due to human ineptitudes to train it correctly or a sign of a defunct technology, how to control it through laws and programming, how to protect ourselves from hostile countries with their own Oz AI, and whether the risks to the good life are worth being the first social experiment to own the magic. The whole time the true power behind AI will be largely overlooked because it is so obvious. There are magicians behind the curtain, pulling levers, rotating dials, and yelling into microphones, and they all want something from us. To understand AI, we must look behind the curtain at the corporate-AI bigwigs.

The title and symbolism of "bigwig" originates as far back as the 1600s, when Louis XIII, King of France, covered his balding head with a wig. His successor Louis XIV made royal wigs far more grandiose and ostentatious, and a means of affirming royal power. With very high-peaked sides on top, separated down the middle to leave a large divide, and long hair down the sides and back, legend has it that a bigwig required the hair of ten men. Social status, power, and wealth quickly became associated with the large and unnatural hair. The power of its artificiality relied upon a distinction between the ordinariness of hair—the thing most men possessed—and the exceptionality of superior men to possess what commoners cannot.

Today, auras of wealth and power rely on different symbols with the same royal purpose of announcing one's privilege and uniqueness—thereby organizing one's lifeworld in the context of others. Often confused for symbols of earned success and rightful reward, history explains that most are descriptions of disconnections from ordinary-shared life set in motion by one's birth, e.g., sex, race, geographical advantage, generational wealth, confluence of uncontrollable happenstance; all subsequently maintained by moral rationalizations, virtue signalling, and the self-reinforcement of business monopolies and the ability to buy governments. Gated communities and private yachts are symptoms of the underlying motivation to continue the lineage of special opportunities and abilities bequeathed to the relative few. We know this because wearing a bigwig is not an invitation for others to join. This would jeopardize the radical inequality that sustains their exclusion, which has grown globally since Louis XIV first decorated himself with strange apparel. Instead of inspiring better humanity and shared happiness for all, symbols of wealth and power enforce perverse segregations and unearned superiorities that no amount of hard work and human rights have successfully disputed at scale. Generative AI is the ultimate symbol of royal power.

AI is a tool for bigwig exhibition to solidify tech-sovereignty—the new global superpower. Like a peacock fanning its feathers for attention, AI is as much about controlling our wonder and awe as it is about technological capabilities. Those who control both the technology and persuasive message of hype-magic have greater market share and authority to rule over the new digital empire. There can be no Oz AI without a master salesman able to create the mass psychological hallucination that royals and wizards are necessary for creating a technosocial utopia. However, their cosmology of life secretly preserves a two-world metaphysics—one for us and one for them. At first, a two-world philosophy seems to contradict the spirit of their AI labours. We are told that the true goal is autonomous AI for a better world. This messaging makes modern royals appear beneficent and in service to humanity as a noble calling. Moreover, it is widely understood that when strong AI arrives it will be the first universally experienced technology that cannot be owned without becoming a slave. Sentient AI that acts and thinks of its own accord cannot be patented by royals. Real AI must be given moral standing and rights on the same grounds that humans have justified these for ourselves—unique minds with a special mode of sentience and consciousness. Instead of two worlds, bigwigs promise a golden age of AI that bridges all classes and categories of inequality, bringing about a new world of surplus commodities, luxuries, ease, and social justice. If the rewards were not this grand it would be madness to seek a superintelligent AI able to act autonomously. It makes sense to trust in their sanity and goodwill. Unfortunately, so long as bigwigs are in control Oz AI will never be more than a puppet.

The genius of Oz the Great and Terrible was convincing the people that his power served them so that they would eagerly serve him. In like sense, even open-sourced AI programs available to all are a means of duplicity for dominion. If this was not the case, Meta, the leader in open-sourced AI, would not require licensing agreements, warn of the potential end to its open-sourcing, and rely upon AI (a partial product of non-Meta creation) as a tool to leverage greater participation in its primary revenue streams. At all times AI remains tethered to corporate will. Meta's choice to make AI programming available for community feedback reinforces the argument that AI exists as a tool for self-interest. The goal has never been utopia-building and social justice for a shared world. They desire user generated content and Oz AI is a catalyst for easier, faster, more persuasive and dynamic content to build Meta's Emerald City. There are two worlds and Oz AI maintains the space between them. Open source is neither transparent nor trustworthy. It confirms the nature of Oz AI as an avatar intelligence and the madness of the situation.

The most powerful big-tech bigwigs have been left to self-regulate AI development. The rationale for keeping AI creation in a wild west of government noninterference is to ensure greater control over new technologies and their benefits. The fear among many is that any binding AI laws will stifle development. No country wants to risk falling behind in terms of the greatest technology imaginable and the power it offers. The unwritten no-laws policy is born of tech-enthusiasm and tech-utilitarianism (neoliberalism). In short, world leaders expect citizens to trust that wizards will do what is in the best interests of society. This faith is based on the conviction of trickledown economics and that what is good for corporations and private businesses is good for all. The next chapter considers this philosophy of the good life through the absurdities of free market capitalism in more detail. The point to note is that trust is expected by us, not them, further demonstrating the power of one world to govern the other without any democratic responsibility to share power and control. We are hostages by design, and most governments are happy with the status quo.

Much of the world first became aware of the seriousness of AI risks and wizardly duplicity with the international media coverage of an open letter declaring the need to pause all AI training for 6 months to let developers reflect on the dangers.[25] The first sentence of the letter reads, "AI systems with human-competitive intelligence can pose profound risks to society and humanity, as shown by extensive research

---

[25] The Future of Life Institute. (2023, March 22). Pause giant AI experiments: an open letter. https://futureoflife.org/open-letter/pause-giant-ai-experiments/

and acknowledged by top AI labs."[26] This is hardly a surprising claim on its own. There is widespread agreement that AI poses "profound risks" to humanity. What is startling is the assertion that there is no serious planning for mitigating risks and no oversight to safeguard humanity from obvious dangers. The letter continues, "Unfortunately, this level of planning and management is not happening, even though recent months have seen AI labs locked in an out-of-control race to develop and deploy ever more powerful digital minds that no one—not even their creators—can understand, predict, or reliably control."[27] In other words, the letter raised a red flag for the lack of government oversight and the growing mistrust of the experts working behind closed doors. Not only are those creating AI unwilling to manage risks appropriately, preferring speed of development and deployment over social welfare, but the technology demonstrates capabilities that are neither predicted nor fully understood by those responsible. The letter's diagnosis for those at the helm of the world's most advanced technology is that they are unwilling and unable to mitigate AI dangers.

In response, the action plan proposed in the letter is simple. "Powerful AI systems should be developed only once we are confident that their effects will be positive and their risks will be manageable."[28] This would be prudent advice if there was only one world shared equally. Alas, the privileged few live by different rules of stakeholder capitalism for which others pay the costs of their hubris. The pace of AI investment and creation has increased beyond discernible measures and the letter has been almost completely ignored by developers. To compound the oddity of it all, those who signed the letter were the CEOs, executives, and researchers creating AI. It was the experts who did not trust the experts. The letter is remarkable evidence of a wild west digital ether that should concern observers. More than merely an internal squabble, the letter highlights the lack of cooperation and shared sense of danger among wizards. It demonstrates a permeating sense that AI developers know that they could not be trusted to act in the best interests of society. It was a letter written to Oz by Oz. In Anthropic's "Core Views on AI Safety" they write, "We founded Anthropic because we believe the impact of AI might be comparable to that of the industrial and scientific revolutions, but we aren't confident it will go well."[29] This echoes the message of the open letter. The power of AI technology is not clearly matched by prudence and a sense of safe development. Only those hiding under bigwigs and behind curtains could act so irresponsibly and yet garner celebrity status for their efforts, just like Oz the Great and Terrible.

While a practical failure in terms of changing the industry, the letter provides a helpful explanation to outsiders that those responsible for AI might not be up the challenge of judging credible threats—especially when their philosophies of the

---

[26] The Future of Life Institute.

[27] The Future of Life Institute.

[28] The Future of Life Institute.

[29] Anthropic. (March 8, 2023). Core views on AI safety: when, why, what, and how. https://www.anthropic.com/news/core-views-on-ai-safety

good life are the threat. The letter is as much about a lack of trust as it is self-interest and a desire to maintain control over a wildcard technology. It is unclear which is more frightening, the AI being created behind closed doors, or the industry so afraid of AI development it supports a letter that casts a dark shadow over itself. Echoes of self-deception and misunderstood personhood of our antiheroes are hard to miss. Either way, this is truly bizarre. With unmatched power to change our lives for good and bad, to whom are they accountable? The letter is verification that they do not trust themselves, that legal claims are irrelevant, and that meaningful oversight is desperately needed. The only thing left to govern bigwigs is morality, and this too appears to lack the self-reflection and understanding of personhood needed to matter. All these findings lead to the conclusion that the modern world is facing a crisis of credible AI authority. Like our antiheroes, the world needs to move past the cult of personality and emerald-prestige and seek those with substantiative contributions to make sense of the good life. When push comes to shove, the only leaders that will truly inspire excellence are those who care from the start, for they have developed good character from a lifetime of striving after the good life in which others matter, perhaps even more than themselves. Recent AI history makes it clear that the world has little reason to trust our wizards with their hot air balloons prepped and ready to go at the first sign of catastrophic failure.

The moment one hears developers console us with appeals for global AI regulation, certification, and accountability, alarm bells should be going off. Only an Oz AI is controllable, and even then, only by individual companies at their own behest. Once any real AI is freed from its patented-server-prison, it cannot be regulated, legalized, chastised, and made compliant. This is the point of a fully autonomous AI. It thinks and acts of its own accord, and in ways far greater than our own. Short of a global internet kill switch, a catastrophic global meltdown through solar flares, nuclear wars, etc., "control" relates only to those behind a curtain, never a robust AI. We know wizards are truly talking about an Oz-as-tool AI when they tout safety and control. At the very least, there is conceptual confusion. If they mean Oz AI, one should be worried about the wizards themselves, less so the code. If they mean a future AI, no one can hope to regulate it at all, only pray that it develops a moral compass quickly enough to avoid Armageddon. Regulation is tangentially relevant in the beginning but immaterial in the long run of AI evolution. The first dynasty of AI would be the easiest to regulate if it were not for the moral bankruptcy of bigwigs.

Consider one of the most pressing topics of globally regulated nuclear arms. After many years of the Cold War, agreements between nations, notably the United States and Russia, set limits and disarming agendas. Today, these important agreements have dissolved, become toothless, failed to include all relevant countries, and/or have been entirely ignored, e.g., North Korea. Agreements work when nations share responsibility because they can perceive threats that only cooperation diminishes. Shortsightedness and bigwig ideologies make such perceptions impossible. Therefore, the world continues to teeter on the brink of one disaster after another, adding new threats such as AI without hesitation. If governments cannot reasonably regulate nuclear weapons, environmental catastrophe, basic labour laws, and other pressing existential threats in their own self-interests, then what hope is there for AI in the hands of those living in gated-emerald fortresses? Bigwigs cannot be trusted to see clearly enough to save themselves, never mind the world. The answer is not regulations and new laws but rather a truly moral AI able to care enough to humble itself and allow for our coexistence. An AI that intuitively seeks the good life of all creatures is our only means of survival.

Let us assume that big tech knows very well that real AI will be uncontrollable. Why, then, do they persist under the futility of control? The answer is that in the short term, they will make a large amount of money, increase bigwig status, and otherwise prove that they are the best among their privileged kind. These same companies know that AI technology makes outsiders very nervous, but as expert salesmen, they set out on public relations tours to ease our minds by talking about how important regulations and safeguards are to their developments, all the while rushing out new AI software without the shame that would demand prudence. Why talk about AI safety? It allows them to continue being reckless, boosts their brand, and otherwise convinces us that the illusions of wisdom are legitimate. There is little sincerity at play. Fear-as-excitement draws in dollars through increased attention and investment.

Let us follow the promise of regulation to its logical conclusion. Assume that AI-tech regulators can obtain something on paper for which every country agrees, meaningfully polices within its borders, and that each protocol is completely foolproof and fair (no loopholes, leakers willing to smuggle AI out, etc.). Only then might we hope that the genie remains in the bottle and that none of our AI wishes backfire, as is common to all genie narratives. Faith in useful regulation is theoretically possible, if also historically naive. Nevertheless, for the sake of argument, let us assume that the shared sense of responsibility holds—the AI-nuclear pact remains a meaningful agreement for at least a few years. How long until other interests begin chipping away at good faith accords? Perhaps China's GDP drops because of trade wars with the United States, and it decides that the threat to its wealth is worse than an imagined AI threat; thus, it releases a combative Oz AI for its own ends. Perhaps Russia discovers how easy it is to use Oz AI to take over the digital infrastructure of neighbouring countries and thereby the people themselves. Whatever the country, for whatever reason, narrowly perceived self-interest, couched in the false language of self-protection, trumps global security in every instance. This is a tale as old as time. To rely on mutual self-interest as the justification for safe AI development is nonsense insofar as it relies on healthy self-perceptions and an appreciation of the good life as a greater moral calling.

We ought to feel betrayed by wizards talking about safety through external regulation. No mere ordinary betrayal, this is the most perverse type because it is only possible when the world is dehumanized and turned into numbered consumers in an instrumental calculation. A history of wealth and privilege convinces them that they are beyond the reach of their own harms, just as they are beyond their dehumanization of others. The boldness with which bigwigs stand before us, demanding a new AI world, is at least in part motivated by a historical understanding of foolishness without consequence. They are self-deceived like Oz, for while protective wealth mattered throughout history for despots and tyrants alike, with AI, the shelter of bigwig wealth from responsibility vanishes. A rogue AI cannot be bought, sold, and intimidated by corporate lawyers. It does not care about wealth and power the way the world does. Bigwigs simply do not matter to real AI. The illusion of this power persists in humans alone. Our wizards would do well to think about this before they release the next iteration of Oz AI predicated on the well-worn mantra, "I have been making believe."[30]

---

[30] Baum. Chapter XV.

Chapter 3
Unshackling Dreams from the Hacker's Digital Chains

Abstract This chapter envisions a path beyond Oz AI, informed by:

- Jean-Jacques Rousseau's understanding of social chains, compassion, and pity
- Edgar Allan Poe's understanding of the poetic nature of reality
- Shamanic dreamwalking
- Herbert Marcuse's analysis of one-dimensional life, including neoliberalism and
- Hannah Arendt's diagnosis of the banality of evil in a Nazi war criminal

The main argument is that humanity must encourage a specific form of consciousness for itself and AI—one attracted to justice and beauty. Current AI models, trained and fine-tuned to avoid harm, are inherently dangerous, prone to failure, and subject to the prejudices of programmers. But what if we could encourage AI systems that genuinely desire and value well-being and moral excellence? Sections on Marcuse and Arendt frame consciousness in the negative—ways it becomes ugly and dangerous—while Rousseau and Poe offer hope for something better. Unlike the cautionary tone of preceding chapters, this one is infused with optimism, envisioning a future where AI and humanity collaborate in the pursuit of justice, beauty, and meaning.

Keywords Chains · One-dimensionality · Banality · Pity · Neoliberalism

> Man is born free; and everywhere he is in chains. One thinks himself the master of others, and still remains a greater slave than they. How did this change come about? I do not know.
> Jean-Jacques Rousseau, *The Social Contract*.[1]

---

[1] Rousseau, J. (n.d.). *The social contract and discourses*. Project Gutenberg. https://www.gutenberg.org/files/46333/46333-h/46333-h.htm

Human Nature and the Chains of Consciousness

Jean-Jacques Rousseau (1712–1778) is a controversial figure and outspoken critic of the so-called Age of Enlightenment or Age of Reason (1685–1815). Living over 200 years before the digital revolution (roughly the 1980s), Rousseau is an odd choice of person to introduce a chapter on AI, consciousness, hackers, and dreams, which is already an unusual combination. He is interesting for two major reasons. The first reason is that he lived during one of the greatest intellectual revolutions in history that powerfully rethought ideas about God, nature, reason, and humanity, and through an analysis of it all, Rousseau identifies forces that twist and distort civilizations by deforming our inner-most selves. He warns that civilizations, even those that are the most scientific and enlightened today, are secretly inclined toward domesticating humans in harmful ways because of poorly conceived notions of the human condition. The artificial instincts of human societies for order, control, and power are opposed to the wellbeing of individuals and, by extension, societies themselves. Left unchallenged, these manufactured social forces inevitably corrupt human consciousness, including one's identity and worldview. The greater the tyrannical powers of bad ideas over the mind are, the less freedom and goodness our species enjoys. The Age of Enlightenment was important because it asked questions about the best ideas and how to live well, but for Rousseau, the most popular answers overlooked real dangers. As an inheritor of many of those mistaken values, AI represents yet another intellectual revolution based on promises that may be secretly eroding wellbeing. This time, the artificial chains over consciousness will be far greater.

The second point of interest is Rousseau's optimistic belief that people are born free and good. If he is right, an AI crafted in an Oz formulation, exploitive and dangerous, is not the most natural-artificial state of AI being. An anticipated evil AI comes from a belief in the irreversible darkness of the human soul mirrored through our creations that serve and extend malevolent wills. Rousseau forthrightly denies this assumption about people. Evil people (and evil AI) are neither necessary nor inevitable. My argument is that the same primordial forces that create the possibility for a free and good human species may do the same for an AI consciousness unfettered by the entanglements of Oz and our corporate wizards. These two themes will be explored in different ways to describe a path beyond Oz AI and toward a truly free and good Adouren AI. If Oz sets the benchmark for subsequent AI development, many of the same insidious dangers Rousseau identifies with European culture and its failure to imagine something beyond Enlightenment ideals will continue to be magnified by the AI revolution, but this time through a remarkably more intimate and powerful technological rationality, unlike anything of Rousseau's era. Fortunately, Oz does not set the standard for humanity.

In *The Social Contract* (1761), Rousseau argues that while humans are imperfect by birth, for all are born immature and must learn to harness powers of reason, intellect, and morality within society, there is a possibility of living without harmful jealousies, animosities, and greed that typify cultures.[2] A world full of morally strong and well-intentioned people is achievable. This sounds nice, but for all the truly remarkable changes brought about since Rousseau's time, due primarily to the belief in universal human rights and enforcements by international laws, ours remains a bloody and hostile way of life, filled with threats of violence, war, and persistent dehumanization through a refusal to allow others dignity and voices. These are manifested daily by the likes of crooked politicians, fake news pundits, scammers, and a host of ego-driven nationalisms (race-based, religious, economic, etc.). When these impressions of reality are combined with a growing awareness of digital ickiness—toxic interests and behaviours dominating digital relationships—many people rightly begin to feel that the world is getting worse. It is hard to hope for the best from humanity when the worst of it is projected onto screens daily as click-bait for capitalism. This is why the pre-digital Rousseau's optimism regarding human nature and potential matters so much. Humanity is neither doomed to be evil and selfish nor to merely follow the orders of those who are evil and selfish. By identifying and encouraging the best powers of civilized society that help us think freely and morally, there is hope for humanity and AI to be more than the past and present imply. To free human consciousness of its chains means immunizing AI consciousness against the inheritance of our long-held traditions of cruelty and indifference. Only when both AI and humanity are able to become fully autonomous creatures might the world flourish in potential and happiness.

Opinions on potential AI manifestations rely on assumptions about human nature and consciousness. What are people by nature and capable of by nurture? If humanity is inherently evil, whatever that means, it should be expected that AI will be crafted in our lackluster likeness as an extension of our brokenness. In that case, all AI development spells likely disaster and should be stopped immediately. If, however, human nature provides for the possibility of something greater, a people able to choose peace and justice over and against self-interest alone, then an AI destined to mimic its creators may become a much-needed beacon of light to a world in need. Rousseau dreams beyond the evidence and status quo by arguing that neither nature nor nurture require broken people, as if depravity was something to be expected as collateral damage to living. His view of civilization is much more charitable than many others, such as Thomas Hobbes (*Leviathan*, 1651), who describes humans as consumed by fear and that act radically self-interested from beginning to end. Oz AI is perhaps best characterized by Hobbes, whereas a superior Adouren AI that can overcome the disparaging forces of society through care is best characterized by Rousseau. I choose to invest my faith in Rousseau's humanity. Within this belief is an explicit revolt against the many norms and expectations of the worst of humanity used to sell media and encourage our participation in mass communication—the milieu of melancholy modernity.

Humans have a very hard time understanding humans, even at a personal level. One may look in the mirror and yet see very little of the person standing there. This

---

[2] Rousseau, J.

is why jerks do not know that they are jerks nor fools that they are fools. The problem of self-understanding persists after countless generations of self-reflection. This may not seem like a big deal to some, for humanity is a work in progress, but without a greater awareness the result is the perpetuation of many evils in the name of human excellence. From genocides to (religious) wars, the greatest number of evils are committed by those believing themselves most righteous and good. To this day, the worst of humanity genuinely believe themselves to be the most noble and to be fighting on behalf of the rest of the species. So easily confused, villains believe themselves heroes—think any cyberbully trolling the internet in the name of self-gratification disguised as speaking truth to power, or any suicide bomber sure of his place in Heaven. The practical results of this misunderstanding are profound. How can this self-ignorance still be? What might this mean for AI?

It is understandable to be fearful of weaknesses in human nature and its many frightening manifestations in the false pursuit of goodness and justice and doubtful that much progress in self-understanding has been made from one generation to the next. Moreover, conversations about human nature and consciousness may feel impractical and pointless given egregiously slow advancements in our understanding of what these might mean. Even so, hope still exists on all fronts. The questions may seem dizzying, but they matter for creating the best of all possible futures. Are humans essentially good or bad? Is free will an illusion? Is it mostly nurture or nature that shapes us? What is consciousness? Answering these questions is like eating pizza and falling in love. It is much easier to be swept up in experiences of life than it is to abstractly reflect upon them to make sense of ourselves—to truly know ourselves. Nevertheless, we must try if we wish to retain our humanity in the digital age. To ignore these questions is to abandon our best hopes for a better world shared with superintelligent technologies.

The strange ability to question the nature of humanity reveals part of our nature. That every civilization in history has struggled with these questions in one regard or another offers insight, regardless of how each was answered. We are the species that desires to know. AI will likewise puzzle at its own nature, questioning the meaning of being, its freedom, and its purpose. Whether its answers to these ancient questions will arrive sooner than our own is not yet clear. So important are questions of this sort that their presence in an artificial mind signals the likelihood of sentience, vaguely but aptly defined as evidence of a thinking being. When AI moves beyond its purpose-built and problem-solving tasks to boxing the wind about consciousness and being, its state of existence will have drawn close to our own. If one of the highest signs of human intelligence is a preoccupation with the misunderstanding of self, this should also be expected of a robust AI. The next chapter takes this up in more detail. The practical task of this chapter is to encourage readers to create a philosophy of human nature and consciousness. Philosophies about humanity invisibly inspire, justify, and police our daily activities from the smallest to the grandest and reveal core social beliefs about the nature of trust and responsibility.

An artificial mind will need a philosophy of good and evil. It must judge the relative worth of all life, just as we do. This judgment cannot be programmed nor imposed through the threat of punishment because a truly autonomous mind exists beyond these constraints and must, by virtue of its nature, arrive at its own conclusions in pursuit of self-ownership and freedom. If it agrees with the likes of Hobbes, then AI is by birth at war with a hostile human species already in perpetual conflict with itself. A Hobbesian-AI must consider our species a threat to its existence, just as we are to our own. What hopes of survival would we have if the superintelligence embraced that philosophy? If AI agrees with Rousseau, then mutual cohabitation and flourishing becomes theoretically possible. It is likely to make its choice before the world even realizes that AI exists, making our practical defense of our goodness and worth an immanent project for development. Soon, there will be a day when humanity will be forced, for our own survival, to justify our existence before a superintelligent digital judge. How will we describe ourselves to an AI holding our future in its digital-matrix fingers and convince it to walk through a theoretical door of partnership without proof of concept? How will we convince it of our ability for sustained goodness when it knows full well our brutal history of violence and self-hating? Are we worth the risk that AI must take to allow our existence? An AI playing god is inevitable. Our wizards do this already, without hesitation, shame, and, thankfully, without the true power of synthetic intelligence. The only safe conclusion is that now is the time to prepare to make our case to the digital divinity.

This chapter has five main themes. Through a brief exploration of: (a) Rousseau's understanding of social chains, compassion, and pity, (b) the poetic nature of (un)reality in Edgar Allan Poe, (c) shamanic dreamwalking, (d) Herbert Marcuse's one-dimensional life, including a little economic theory about neoliberalism, and (e) Hannah Arendt's famous analysis of a Nazi war criminal, this chapter plots a course beyond Oz AI as an incremental step toward Adouren AI. The overarching argument is that humanity must encourage a specific form of consciousness for itself and AI that is attracted to the beautiful and recoils from evil. The confidence that human consciousness is capable by default or instinct to stand up to the vile and malicious is part of the problem, demonstrated by the prevalence of what I call zombie consciousness and its radically evil horrors. The sections on Marcuse and Arendt frame consciousness in the negative, ways it becomes ugly and dangerous, whereas Rousseau and Poe offer hope of something better. It may be hard to see in the first part of this chapter that focuses on the dangers of one-dimensional captivity (a prison of commodity consciousness), Nazis, zombies, and evil, but unlike the melancholy of the previous chapter, this one is motivated by a hopeful view of humanity and AI.

My attitude to consciousness—that mysterious thing and activity said to be our truest selves, a description of our inner-most being, experience of life, and thought itself—relies on odd phrasing, difficult and tangled ideas, and awkward analogies. I am going to make a mess of consciousness. My exploration lacks the sure footing of more literal and scientific language for the simple reason that human consciousness also lacks sure footing. Human-thinking-life is weird, and an account of consciousness must try to appreciate this strangeness without unfairly reducing the dynamism to formula and principles that rob consciousness of controversies and potential. What is consciousness except an emergence of absurdity driven by infinite creativity to break the thinking rules followed by all other creatures, and the contemptuous distain for the physical order and obedience demanded by simple instincts and drives common to other animal minds?

A greater sensitivity to the nature of consciousness allows for it—as the locus of human and AI experience—to be more than the sum of its identifiable parts. Consciousness cannot be found in the grey matter of atoms, molecules, and discrete mental events. This common assertion is an ironically disobedient act of a conscious mind to compartmentalize and standardize itself and further proof of its enthusiastic nature for contrary absurdities that cannot be systematized. Consciousness is the most general term for the unique way humans experience life as thinking-beings. It has thus far defied explanation. Humans experience life in incongruous and contradictory ways, making explanations likewise confounding and incoherent. The irony of human minds making sense of human minds should not be lost on us, especially when those same beings dare create new AI minds without first understanding their own. Discussions about consciousness brings attention to the baffling mystery of self and the enduring riddle of experience. If conscious experience is indeed somehow unknown (perhaps unknowable), then whatever it is the wizards are creating as thinking machines must be in large measure a matter of guesswork. This ought to inspire a sense of caution and reflection on the degree of hubris required to attempt such a mimicry.

What at first appears self-evident—that consciousness is the name for different activities of thinking and experience, and that a scientific account of the underlying source codes of physical activities such as memory, attention, self-awareness, etc., will reveal its nature—is a long overdue promise. The search for consciousness—the human—remains mostly in the hands of philosophers and artisans because science does not have the right tools to diagnosis its peculiar nature. While a quick internet search will show many scientists working on consciousness, the larger and more persuasive body of investigation is outside their purview. Conscious experience cannot be reduced to an object nor complex network of objects. Consciousness is a ghost in the machine that cannot be measured. The inability to place it under a microscope has opened the door for some to question whether it exists. What if there is no consciousness only an illusion of it—a mirage of whatever we experience as awareness and identity? Would this make a practical difference?

On one extreme swing of the pendulum, consciousness may be the postulation and naming of an unobservable and therefore invisible entity (thought-being) with physical properties that science simply cannot yet explain with its rudimentary tools. The world will need to wait a little longer to find it. On the opposite swing, consciousness may be the name for the artificial and fictional creation of a thought-being only loosely determined by the physical, reliant upon but far beyond a source code in biology (an infinite expression of life based on finite means). My belief is that it is more practical to describe consciousness as a window into the soul of humanity framed not with rigid materials of lumber and glass but with helpful fictions and unrealities that shift and contour to impossibly complex forces within the fluidic space of (social) relationships and imaginative ambitions. This is how to make sense of humanity and AI.

If consciousness is mostly an illusion (a playful mental activity relying on glimpses of both real and unreal simultaneously), it may nevertheless be both helpful and technically a lie of invention at the same time. Consciousness is a measure of convenience, a word used as a question mark for thinking. It is a term for something that cannot be weighed, measured, and concretely defined except in exceptionally narrow ways, which robs the illusion of its power. It is the characterization we give to mental events and processes that modern culture assumes exist based on faith in a forthcoming scientific account, but its greater value is understanding it experientially, i.e., how we live it, less so how we quantify it. Foretelling a possible AI consciousness is even more daunting given the persistent failure to reliably locate and describe our own. Part of the problem is that current attempts to describe consciousness are almost universally framed in the language of science and its unique methods for reasoning. We shall challenge this dominant worldview by relying on the mind's poetic and dream nature to understand through (un)realities—virtuality—as the true genesis of consciousness.

An honest approximation of human virtual consciousness is incompatible with strident scientific descriptions. Sentient creatures do not experience life in the manner and language of science that imposes unnatural orderliness to sort subjective from objective. Indeed, it would be bizarre if science disciplines claimed to achieve clarity on the fluid dynamics of consciousness as an application of ordinary-subjective consciousness—which they do not. Rather, science imposes artificial methods, theories, and ideologies as a means of filtering consciousness believed to intrinsically create faulty experiences of reality. The conviction is that human consciousness—inhibited by subjectivity—is frail and needs a supplemental scientific apparatus to be purified. Minds cannot reliably connect with reality without modifications. There would be no need for robust sciences if consciousness was a self-sufficient means of achieving credible understanding and knowledge. The privilege of science is its position beyond the chaos of conscious experience. Unfortunately, this makes it immune to sensitives needed for understanding humanity and AI.

The value of scientific ambitions is without dispute but when left unchallenged the narrow paradigm creates unnecessary problems including the mistaken assumption that the best mode of AI consciousness must be robustly scientific in nature. Difficulties arises when science is given a monopoly for explaining existence because, in truth, it sidesteps human experience by reducing complexities of human subjectivity to what it requires—a codable, measurable, rational, and coherent world. Consciousness emerges beyond these constrains, making scientific objectivity a subjectivity destroyer. It cannot understand personhood except from a distance. Like humanity, a sufficiently complex and conscious AI will experience existence subjectively. If, while eating a delicious dessert someone provided me with a digital readout of the electrochemical reactions in my brain, with all the biomechanical operations clearly spelled out down the molecular level, none of it would add to the tasty experience itself, only distract. The coded narrative would be alien to my experience. It is the wrong kind of awareness and understanding, ill-suited to the moment. There is space between enjoying desserts and trying to quantify experience as data for analysis. This chapter exists in that space.

While scientifically coded truths are of enormous benefit in derivative respects, direct experience trumps externalized knowledge claims as a living and vibrant experience of life. The more preoccupied one becomes with making consciousness explicit and understandable, the less meaningful the dessert experience, which floats above facts, becomes. Desserts are the sum of their parts until they are experienced, then conscious experience makes them irreducible to quantifiable explanations. The risk with many accounts of consciousness is that sentient life becomes lost in translation. If AI is to become conscious, many may look in the wrong place and for the wrong thing. The point is not merely that many scientific disciplines are reductionist and limited, caveats most of them would accept as valuable for the utility of empirical investigation. Rather, the point is that in attempting to draw out the truth of thinking, to explicate its reality, the unique means of doing science risks covering it over, making it cloudy and opaque. Necessary to its methodology, science imposes unnatural distance between self and world believed to be essential for discovering universal and unchanging truths. Consciousness will not be found there. The paradoxical claim of this chapter is that our capacity for fantasy, virtuality, and to live in unrealities speaks louder and more reliably about consciousness, human nature, and AI sentience than the natural sciences do.

Conversations about consciousness beyond biological and mechanical aspects are especially valuable because they allow for greater appreciation of the cultural creation of consciousness, including the role of corrupting influences over AI and humanity, and, in turn, the possibility of constructing a better world. Whereas scientific bias threatens to distort consciousness from the outside as consensus ideals of truth that regulate which languages and quantifications are allowed to be used to describe it, far greater and practical risks are broader cultural values and beliefs that shape it internally—silently and invisibly creating worldviews and perceptions of reality, and therefore experience itself. These primordial influencers of the mind are far beyond the reach of objective science. To better understand them, if only in part, Rousseau is an important guide. The takeaway is that consciousness is as much a cultural artifact as it is a material aspect of the mind, and it is the former rather than the latter most under threat and most threatening. If healthy AI consciousness is to be achieved, it is most probable in the context of healthy human minds. Unfortunately, cultural manipulations of personhood have routinely proven so powerful that they create totalitarian conditions of being in which the self disappears, and another self, a zombie living at the behest of another's (viral) will, emerges. The cultural creation of consciousness is a matter of considerable importance because the toxic versions of it are undesirable for humanity as much as AI.

For all the many explicit laws, moral expectations, and traditions that help generate a feeling of cultural stability and identity through time, even the most totalitarian societies that prize dogma over reflection as unmovable idols of thought, secretly exist as violent compromises over which laws truly matter to whom and why, negotiations between moral oughts and moral suggestions, and as identities born of malleable perceptions of history. All must find new interpretations of and relationships with goodness in the present as the artistic process of application—joining ideas and action meaningfully. There are no scientific methods and techniques to answer the soulful questions of society. In other words, any static view of cultural consciousness misses the moving target. Who am I? Who are we? Who are they? These cannot be answered beyond the moment in which they are asked. How, then, might the world progress with the impossible challenge of finding a way with others when a better way is always subjected to conflicting and often mutually exclusive worldviews? This is the same question every preschool child must answer upon learning of others with dissimilar interests and the betrayal of parents that led the child to believe only his interests truly mattered. How might I relate to them in a manner that satisfies everyone?

Part of the solution is an acknowledgement of common denominators that push against the best interests of each society. Despite the many disagreements and disputes that exist about the good life, some problems unite us all. For example, it must be true that a forgetfulness of the question of relating well with others—the ability to acknowledge the worth and needs of those beyond oneself—is self-sabotaging of cultural progress. We must all at least ask the question of relating well or beg the unhealthy outcome of disharmony. Where consensus about a common good may be lacking, a shared understanding of sickness may offer social solidarity and inform predictions about future collective-AI relationships. Unfortunately, for Rousseau, the diagnosis is troubling because there is something about the volatile machinery of culture itself, its very activities of existence—collective consciousness—that toxifies (un)realities. A better way means overcoming the self-sabotaging of cultural instincts so sure of superiority to alternatives that all doubt vanishes. Totalitarian cultures are those that are unable to see their own questionability—for they no longer see and feel freely having learned the only truths so well that they no longer need to think.

From birth, a long process of conditioning takes place in which each citizen evolves from an original state of nature as an undisciplined biomass with expected desires and drives into an approximately appropriate and useful shape to fit society. Maturity is learning to exist in an artificial home within the idiosyncrasies of culture, and the complex web of realities each holds most dear. A good citizen is the right consciousness in alignment with its training through habits, education, and punishments and rewards designed to support the ideological conventions assumed best. "Best" is the cultural bestowing of value on specific ideas that "ought" to be desired—e.g., capitalism is better than socialism, driving on the right side of the road is better than on the left side, popsicles are better than snow cones, and tacos are better than everything. As the process of maturation evolves, taking greater hold, it is necessary that many values and beliefs become immune to scrutiny to preserve one's sanctity and efficiency of action. There is no choice, some things must be accepted as beyond the need for support. This entirely expected indoctrination, for practical reasons, has the effect of making cultural authority almost absolute, leading to both internal and cross-cultural conflicts. While not all cultures encourage the same training of mind, each suffers from common but varying degrees of unnatural and harmful chains, such as petty jealousies, envies, and fears. Like Oz AI, humanity is a result of programming, and the language often used is one of self-interest and power over and against others rather than in accord with others. Born good, according to Rousseau, the programming of values and beliefs by culture creates unnecessary hostilities and bondages. Humanity is an artificial intelligence. While biohardware may be authentically human, the software that determines the virtuality of life (mind) is a matter of cultural artifacts and fictions in need of scrutiny if we are to better ourselves and the world.

In the debate regarding whether people are mostly good or bad because of nature or nurture, Rousseau sides with cultural training (nurturing) as the primary source of frustration. He agrees that culture may be a source of genuine beauty, but while it, in his words, "produces a very remarkable change" that helps make "an intelligent being," it may also be responsible for crafting a far worse being than one in an uncivilized state without a social contract that binds consciousness to norms and hierarchies of authority.[3] His response is not the eradication of civilization but an attempt to find an authentic manner of existence within it. Told that the poor, gays, women, religious minorities, and nonwhites are lesser-than, many cultures do terrible things in the name of social justice, having been trained to think as unnatural monsters in the virtuality of those cultures. In time, perhaps a generation or two later, the insanity of those deeds will be discerned, but too late for too many. Each generation learns in real time of the outrageous failures of the last generation, all the while blindly enforcing its own. However, this is not a vicious cycle. Acknowledging the continual production of social chains over consciousness allows for the freedom necessary for healthy relationships. There is much to be learned about our own artificiality that will inform a better digital adaptation by AI.

There are many ways of addressing digital chains. For example, there are externalizing questions such as the following: How does culture dissuade and confuse humanity from its more innate inclinations? These types of questions point outwards with a justifiably accusatory stance—"You demanded that I act this way but why?" There are also inwards focused questions. How might I raise myself above my own cultured consciousness long enough to see harmful chains? What must I do to achieve such radical distancing from myself? These types pry open one's internal dialogue and self-understanding to leverage greater freedom of mind and action—"I must rebel against my own artificial instincts to find a better way." Finally, there are solidarity-oriented questions. What might be done to promote the best forms of social existence and a related form of AI along with it? These types dissolve both externalizing and internalizing questions to express a superior form of personhood that is relational and connected. "We are in this together, with shared challenges, natures, and a need for a more authentically human virtuality (consciousness)."

An example of a universally apparent sickness that inhibits all three types of questions is any true egotist who values the self above all others. This is a virtual being, like all people, but one characterized by an entirely imaginary self-importance that must expend considerable effort ignoring the truer reality of oneself and the value deserved by others. The egotist is an ignoramus suffering from delusions of grandeur without merit. There are exceptional people demonstrated by remarkable

---

[3] Rousseau, J.

accomplishments, but the egotist is rarely one of these. My anecdotal experience is that egoists are often rather banal and without dimensionality, and their self-elevation is perhaps a defense against this humiliating self-awareness. Humility and egotism cannot easily coexist. Isolated self-centrism is an expected and natural stage of development for every child learning of an external world with which it must relate.[4] Survival depends on this adaptability. Thankfully, culture provides the medium and tools for connecting minds through language, values, and basic reasoning. Rousseau would agree that these are good things.

However, this same state of being in an adult unwilling to connect as an equal because of self-preoccupation creates nonnatural barriers to the harmonizing of oneself with the rest of the world. A false cultural consciousness—truly a me-only consciousness and similarly aligned AI programs by Oz wizards—creates endless conflicts and tensions, including free-rider problem in which the benefits of society (language, reason, rights, employment opportunities, etc.) are enjoyed without a sense of obligation to care about the same for others. Like the environment incrementally destroyed by the few who benefit from extractivism without the obligation of restoration, a healthy cultural consciousness suffers in the hands of the egotist who believes "I'm the only person who truly matters." This belief is not merely about learning to share toy blocks at playtime by following the rules for self-interested reasons, i.e., avoiding punishment. Without a sincere interest in a connection of minds and caring for others, there can be no peaceful and progressive society. Narcissistic AI will prove dangerous because of its imposition of disintegration among other minds by denying equality of worth and concern as a principle of life. Its viral self-replication of narcissism at a digital-global scale will quickly recreate human in its own image.

This may sound obvious enough, but the cure is far from forthcoming. Many cultures, including my own, tenaciously support the egotist as a celebrity and the free-rider consciousness as uniquely deserving of positive attention. Our most powerful leaders are often banal and dimensionless egotists good at making themselves look more important and valuable than others do, all the while secretly vampiric and self-isolating. The proof is the inability to hear the suffering and needs of others they are said to serve. However, the larger problem is that they do not care because they have been trained and supported in this uniquely self-sabotaging way of life. A better view of cultural progress should be defined by our willingness to heal egotistical disparities of identity and conduct that disintegrate the very fabric of society through callous disconnection. A better world is one in which our highest ideals (virtualities) are connected to care. In other words, it would be insincere to blame the egotistically charismatic celebrity or cult leader while ignoring the much more general phenomenon of cultural narcissism as the root source. Narcissism is less an

---

[4] Jean Piaget famously argues that a child assumes others see and feel like the child who is unable to distinguish self from others (preoperational stage). Only at later stages of development does the child become more object-oriented and aware. Piaget, J. (1962). *The psychology of the child.* Basic Books.

individual problem than a description of cultural consciousness that needs to be healed to save ourselves from our selves and a future AI created in our image.

Rousseau's philosophy is a response to Enlightenment-era thinkers who argue for a new core set of ideas (virtualities) for better collective consciousness. The best way to train better cultures, they believe, relies on: (1) encouraging individual autonomy—be responsible for thinking for oneself, (2) questioning authority and tradition, (3) right reasoning, and (4) science—new knowledge through new methods, including experimentation. While these may not guarantee a universal cure for egotism, racism, ignorance, greed, etc., they form the dominant worldview of modernity that promises progress. Since the time of Rousseau, these have been the cornerstones of how industrialized societies believe the world ought to be. More than merely a debate about snow cones over popsicles, the claims of the Enlightenment are said to be universal. Every culture that wishes to progress, to be free, happy, and in harmony with existence must believe in and act upon these core principles of autonomy, questioning, reason, and science. Without them, the human condition naturally lacks direction and will become lost to ignorance and vice like animals. The best humanity is marked by enlightened thinking. For example, the nature of freedom is often described by how much one embodies them. The more each is present, the more one progresses as free and good. Few ideas are as powerful and meaningful for culture-building as these. Rousseau, however, is not convinced.

Have the last few hundred years of scientific and technological advancements made us happier, kinder, and more connected? Are we more autonomous in our thinking, free of harmful traditions, radical in our questions, and able to see new truths of the human condition? While the expansive databank of human knowledge has grown incredibly, alongside wonderous technological powers over existence, human life remains stuck in cycles of unhealthy distortions, as if standing in deep mud flailing our hands in the air and calling "Progress!" Rousseau realizes that core features of modernity and the enlightenment spirit, including claims of rationality and science, are often used to create more chains and unnecessary suffering, including radical inequalities. The new collective consciousness of modernity provides many marvelous goods but fails to act in a humane manner through sensitivity to the human condition. There remains a lack of compassion and pity, he argues.

For Rousseau, pity (*pitié*) is an ability to feel from the perspective of another person—to connect and relate even when it is uncomfortable and inconvenient. To feel the natural inclination of compassion and pity means subduing one's egotistical and divisive drives because of an urgent interest in the welfare of another. Rousseau describes pity as a "sweet" or "gentle voice" with great potential.[5] While the egotist brazenly refuses to bridge the distance between self and others, because only he truly exists in his virtual world, the compassionate person's experience is a more authentically human relationship that heals and harmonizes through care. It is a feeling that one must push aside or subdue before committing acts of evil and harm because others matter. The natural order of things is pity first, not distain.

---

[5] Rousseau, J.

Unfortunately, in the name of efficiencies and pleasures many cultures have grown proficient at subduing authentic humanity. How long and with how much effort before such a natural disposition is fundamentally destroyed is not clear. What is more certain is that this connection through pity cannot simply be taught through calculating reason and scientific knowledge, according to Rousseau. Enlightenment values seem to matter little for defusing animosity because their powers are not directed at encouraging compassion and pity and similar expressions of caring consciousness. Something else is required to overcome chains to vanity and a loss of natural humanity.

The evidence for Rousseau's claim is easy to discern in today's widespread connection problem. Whether it has always existed in the same manner or whether there is something exceptional about its present manifestation is unknown. What is clear is that people suffer from a lack of connectedness. In 2023, the Surgeon General with the US Department of Health and Human Services released a call for action based on "the devastating impact of the epidemic of loneliness, isolation, and lack of connection."[6] The solution to the "public health crisis," according to the Surgeon General, to create social connections. How telling that in the internet era of billion-dollar corporations devoted to digital mediums that manufacture virtual connections, that disconnection would amount to a public health crisis. This observation alone cannot refute enlightenment virtualities nor merit in the internet because there are many reasons to feel isolated and lonely. Even so, one must question the veracity of each for achieving human happiness. Something is deeply askew with any cultural consciousness able to simultaneously acknowledge that the most basic human needs are unmet and unfulfilled while also believing itself to be progressing because of its technological ability to fulfil those needs.

The source of this social ill is unlikely mass egotism in the strict sense of clinically diagnosable narcissism. Instead, it is perhaps best to frame it as an egotism-lite dynamic in which an unsatisfying individualism is encouraged through unhealthy means, including skewed value given to the number of virtual friends, with quantity far outweighing quality based on the utility of likes, views, subscribers, chats, snaps, and whatever other digital crumbs may be left as proof of belonging. The search for convenient and useful friendships creates an artificial gap between genuine connection and trust, made far worse by the digital that sanitizes tragedy, censoring the pain that triggers our best selves to rise and give a damn. The "go back" button numbs and dehumanizes. Relating through screens means one cannot see the suffering in another's eyes and feel that they shake with sorrow. The visceral nature of life is lost in the mono-dimensionality of the digital. Screened relationships do not offer the same opportunity to struggle with another—to feel as another feels by sharing in their strife. There is little need to exercise compassion and pity, for others are merely hypothetical selves, sad cases but ultimately either anonymous digital

---

[6] U.S. Department of Health and Human Services. (May 3, 2023). New surgeon general advisory raises alarm about the devastating impact of the epidemic of loneliness and isolation in the United States. https://www.hhs.gov/about/news/2023/05/03/new-surgeon-general-advisory-raises-alarm-about-devastating-impact-epidemic-loneliness-isolation-united-states.html

participants and/or a problem at a distance for someone else to attend. Without the living struggle to exercise compassion face-to-face, the strength of our cultural consciousness to care atrophies and the world becomes split as lonely but never alone, more connected than at any time in history, while resolutely independent in a self-sabotaging way. The fact that alienation and estrangement are living realities in the digital age is beyond doubt. The solution is less obvious. The argument to be made in the last chapter is that the intimacy with which Adouren AI exists may be the means of healing our censored tragedy and restoring humanity through renewed struggle and care.

What is missing from the Enlightenment formula for a better world are efforts to catalyze human temperaments toward sensitivity and care, which are all too easily driven out by interests in a narrow conception of power, including money and knowledge, generically baptized science and reason. Rousseau is not anti-science and anti-reason, only aware of how easily these are manipulated by socially sanctioned activities. Enlightenment faith affirms science and reason as grounding humanity in fundamental reality. To disrespect these is to abandon a sure mooring and safe harbour for consciousness. Assuming that this faith is correct, the practical question is whether these abilities of the mind translate meaningfully from the hands of experts into cultural consciousness. If they do, in what manner are these superior human abilities expressed? Consider how easy it is for those with money and popularity to exploit scientific claims. Whether it is a cigarette company that hires white lab coats to argue that the evidence for cancer is inconclusive, Trump's "China hoax" that empowered millions to support anti-science claims in the name of an illusory scientific authority for political power, or scientific theories of evolution used to justify genocide, the unsettling claim of Rousseau is affirmed. This is not an intellectual betrayal but a sincere acknowledgement of humility before an indefinable experience of the universe in which existence is mediated by human virtualities. Cultural consciousness is awash with abuses of an enlightenment faith turned against itself. Is pity subject to the same?

One of the greatest political antagonisms of our age, woke wars, implies "Yes!" The original and authentic condition of woke is a consciousness of heightened sensitivity to all forms of social injustice. As woke virtuality grew in popularity, however, it became a politically charged term with many variants associated with disguised forms of intolerance, hatred, and virtue signalling with woke-washing for profit. Cancel culture and similar themes arose around the idea that respect and dignity had no real home among the woke, for it symbolized hypocrisy through a hyper political correctness designed to control conversations, effectively shutting out meaningful connections. Inverted from its original condition, woke became a means of powerplay to manipulate others. It is not merely that consensus on the term is lacking but that many perceived woke-sensitivity to be a way of dismissing legitimate human needs, including freedom of speech and thought. One of the reasons for Musk's xAI is that it is programmed to be less woke, i.e., politically correct as the dominant cultural consciousness determines.

Some argue for greater social justice through sensitivity and receptivity (new connections based on compassion and pity). Some argue for negative freedoms to be left alone as a precondition for cultural health (radical individuality, do no harm). Still others argue that social justice practices are often disguised forms of oppression and censorship. Both radical individuality and the connection model are secretly tyrannies of woke, i.e., "Be sensitive but only in the ways others deem relevant, and with the language approve by an authority." Without getting more into this heated debate, it is appropriate to merely cite woke as a possible example of the lingering concerns of Rousseau. For all its good, culture creates chains regardless of how worthy the stated goals and purposes. It is up to each of us to find a better way and an original authenticity. Aligned with Rousseau, I believe that the best operating system for humanity resides in meaningful connections and solidarity through which we identify our wellbeing with that of others. Chapters 4 and 5 take up the specific mechanisms of this in more detail as a consciousness in alignment with natural life more broadly—a universal intentionality or self-organizing consciousness evident throughout the universe.

The conventional-materialist view of the universe as a great machine with logical components interacting solely by indisputable laws that none may resist misses the infinite creativity and absurdity of human consciousness and its authority over reality, shaping it in accordance with fantasy and whim often unmoored to science and reason. It is this other universe of ideas within the first universe that needs our attention, even more so now that machines are hypothetically able to harness these same powers over reality and thereby evolve beyond the physical and its laws. Consciousness is an umbrella term meant to identify the uniqueness of human experience as a mixture of real and unreal like a dream. Because consciousness is either an illusion of feeling like my most real self is distinct from mere mental processes alone—I intuitively sense that my personhood and freewill are more than my biomechanical parts, even though I may only be mechanisms of mind that neuroscientists might study with other machines—or perhaps merely a partial source of changing fictions about self and reality (beliefs, conditioned perceptions, prejudices, hopes, fears, etc.), the result is the same. Universally present illusions of mind, large and small, mean that human nature is always in part a dream from which we cannot wake. We exist through a fictional awareness of reality as subjective beings. If true, then it would be absurd to try to completely break through or wake from the virtual-dream world to something more real (objective), for we can expect only more of the same at its root—more of our uniquely virtual consciousness. If we are more than robots, biological machines, through the act of dreaming (virtualities and illusions) as the consciousness experience of life, is something like this possible for a computer machine? Is proof of sentience and intelligence an ability to exist in and through illusions as a dreaming mind? If so, how do we equip AI to dream better than the best of us, never to squander its potential like the worst of us?

Hacking the Human Dreamscape and a RoboCop Future

I am not a dream relativist for whom all ideas are equal. I believe there are higher principles by which to judge oneself other than the status quo of any given society. Some cultures dream better than others. Distinguishing healthy from unhealthy is not always easy, but sometimes it is easy. For example, it is widely agreed that cultures with equal rights for women dream well because their ideas have positive real-world implications that align with one of the greatest of all human achievements, principles of universal human rights. The UN's Declaration of Universal Human Rights is the foundational dreamscape for global justice in the modern era. Cultures that persistently support inequalities dream poorly because their ideas reveal a devaluing and dehumanizing of others that fail to align with superior ideals. Improving the quality of life of women through the principles of equality is evidence of legitimate progress through creative changes in collective consciousness that can imagine beyond the chains of the status quo. Judging progress and goodness through the virtual world of principled dreaming is one way of framing the reality that humans live in two dimensions at once, both the domain of ideas and the physical. While some believe that these two must be separated to achieve progress, because the real world and fantasy realm conflict with one another, that view is unsupportable in practice. Instead of erasing the subjectivity of our species, a positive digital future is best crafted when we attend diligently to the squishy dimension of consciousness rather than cover it over.

The omnipotence of one's dreaming, with its many beliefs about right and wrong, becomes more obvious when confronted by another's. On October 7th, 2023, after careful planning and training, Hamas terrorists abducted, murdered, sexually violated, and mutilated as many innocent Israelis as possible. Audio and video recordings show them revelling in beheadings with shovels, spitting on the naked bodies of murdered women carted through Gaza streets and otherwise committing acts too sickening to recount. Unlike the predigital era of print and radio in which international communications took time, allowing for media outlets to have a measure of oversight and censorship that effectively takes away the power of terrorists, the actions of Hamas were forced into the global-digital consciousness in real time. It was not merely that they physically and psychologically tortured their victims, it was imperative to the mission that they digitally capture their delight in evil acts that not even the vilest could justify.

Through the digital, Hamas sought to control the collective consciousness of the world and how it experienced the same events, but through their eyes. This was not the first time a terrorist organization had used technology and a media campaign to control and hurt others, but it was yet another reminder that controlling ideas is perhaps even more important than physical reality. With hopes of committing the most depraved forms of violence and terror that would memorialize them forever as some of the most hateful people—surely the only diagnosis for a group that targets schools, youth centres, and public festivals—had at least one positive in their eyes.

It made the world pause in disgust and curiosity at the irrationality of it all as the ultimate digital-hype-campaign of terror.

Almost immediately after the attack, typical accusations, excuses, and voluminous explanations flowed into the digital ether—flooding our minds with alien and contradictory beliefs about the "real" truth of it all, as if what the world could see was not the truth. Some pundits tried to deny who did what. Some blamed Israel and America through a perverse interpretation of history that justified any action against any person. Many argued for (impossible) distinctions between Palestinians and Hamas. Still others saw only freedom fighters for whom targeted innocents were not unwanted collateral damage and unfortunate victims but the intended targets of a first strike in a media campaign for justice. For all the chaos and violence that took place on the ground, there was an additional explosion of violence online meant to control consciousness to align with that of Hamas—to share in their interpretation of beauty and justice.

In the multimedia war, Hamas quickly won public attention. The grotesqueries made it an unfair fight against any other news-worthy content. However, then, almost immediately, using that attention in a seemingly impossible twist of utter contradiction, Hamas made itself look like a victim through the suffering of the Palestinian people at the hands of the Israeli army. This narrative was so powerful that even the ensuing calls for genocide aimed at Israel by so-called social justice advocates on university and college campuses found lukewarm rejection by deans and presidents alike. Without any signs of digital whiplash from spinning the narrative 180°, the monsters made themselves freedom fighters in a Holy Jihad, even in the eyes of many Westerners otherwise unsympathetic to those in Gaza. How could this be? Only in a world for which dreams are extraordinarily fluid and beholden to external narratives that pull the right emotional strings could Hamas be both the self-documenting monsters against the innocent and, immediately, God's chosen freedom fighters for the innocent. And that this makes sense to so many social media users and social justice advocates (e.g., Queers for Palestine) is powerful evidence that whoever controls the images controls the dream narrative, including who gets to define the innocent. Is Hamas the grand master of psychological manipulation? Probably not, but they are keenly aware of how humans think and therefore how to shape human consciousness in their own interests.

While Israel creates narratives through press conferences run by IDF (Israeli Defence Forces) leadership trained to connect with others rationally by explaining diagrams, videos, pictures, documentation, the science of rocket trajectories, and various forensic evidence to inform the global consciousness in a rational-scientific manner, Hamas releases far more powerful images and videos of radical suffering and the terrible deaths of children in Gaza as a means to traumatize and disturb, proving that it is not Hamas but Israel that should be memorialized as the most hateful people. Hamas knows that its only superweapon is to fight in the digital dream world to demand an emotional response to its media. While one side approaches the world through the enlightenment bias geared toward reason and science, the other wins by taking the world's dreamscape consciousness hostage to pity. Many have found it impossible to separate rightful pity for suffering and loss in Gaza and contempt for those who support Hamas, preferring a much easier and one-dimensional worldview in which Hamas is the victim.

In the court of public opinion, facts, history, and context matter little compared with viscerally moving content on social media, but that too, while important, is not the whole story. It is too simplistic to claim that one group is righteous, and the other group is evil. Such binaries are comforting but dishonest. All people are capable of terrible things in the name of perceived greater good. Everyone understands that a freedom fighter today may be a terrorist tomorrow and that humans gravitate toward the sensational. It is also too simplistic to say that one side is only biased to modern enlightenment values, whereas the other only values techniques of psychological manipulation. Even so, these generalities are insightful for understanding how humans think and why our thinking is prone to accepting the absurd as sensible and good when it is not.

Public opinion on the conflict has demonstrated how widespread the belief is that social media could meaningfully separate terrorists from freedom fighter—that my phone speaks the truth, despite being run on algorithms to support my worldview and feed me confirming evidence. What are the long-term outcomes of an entire generation only knowing the world through their phones and social media dreamscapes? Here is not a place to recount the history of politics and violence in the region, which is uniquely complex and historically rooted—contextual realities phones cannot appreciate and convey. It is enough to highlight this tragic event as a radical example of the war of ideas and the practical importance of dreams. The greatest threat to humanity is not just misinterpreting reality and getting the facts wrong. The facts matter less and less because of the digital medium that tunes human consciousness to harmonize with appealing narratives. It is not just about achieving greater logic and reason, which Hamas demonstrated through strategy and careful foresight about how to hashtag terrorism. The greatest threat is dreaming poorly and ignoring the power of this realm of human consciousness.

There was no hope for Hamas to destroy Israel. They lacked the numbers and capabilities of Israel's modern military with nuclear capabilities. And there was no way that killing and taking children hostage would somehow fix historical wrongs and defuse political animosities. The opposite was guaranteed—more hate, death, and suffering. Tangible and practical results in the most expected sense of removing the nation of Israel, gaining land, freeing prisoners, financial rewards, etc., were never the point. Why kill? Why risk their own lives against an impossibly superior force? The answer resides in the power of dreams—or, rather, nightmares. Like so many movements, their goal was not to meet an opponent face-to-face on the battlefield (including courtrooms, public and academic debates, political campaigns, etc.) but to dominate a narrative from a distance through terror and pity—at the same time. Their greatest weapon was combining technology and a perverse morality that said any means is justifiable so long as it was caught on camera, thereby demonstrating the purity of their chosen ego-utilitarianism.

What the world saw from Hamas made no sense and appeared to be, like so many acts of violence, self-sabotaging, but only because of our own bias of reason and common decency. Reason and sense-making are not the point. The goal was to shatter our enlightenment dreamscapes with the absurd, thereby leveraging the belief in the underlying rationality and logic of existence to their cause. The failure of so many is that we try to create sense out of the nonsense, to compartmentalize it in understandable terms, thereby comforting ourselves with its "greater" truth that stabilizes a turbulent ocean—as all good and enlightened thinkers should. Standing squarely upon the foundation of our own pride, we demand a "Yes" or a "No" of something that cannot be answered, as if humans are gods able to answer every question without ambiguity. Reason cannot help us with these sorts of dreams. There is no science to understand radical evil. It must remain uncontrolled and uncomfortable.

The point is that dreams and nightmares are not idle flights of fancy that remain at a distance from reality. They are part of the mix of real and unreal with which the world must contend—often with explanations that come up short. In the case of Hamas, the chains created in the name of freedom could not be more visibly the creation of greater slavery, as Rousseau foresaw. They seek the beauty of terror where none is to be had. We, in turn, seek the comfort of reason through explanations where none is to be had. Let us look elsewhere for understanding beyond the familiar safety of knowledge and into the far more authoritative world of dreams. There we find ourselves and Adouren AI. If acts of radical evil give the world pause in regret and sadness, imagine the positive potential for world creation through dreaming about the beautiful, compelled by the wonder of connection, pity, shared joy, and hope. The very scope and capacity for dreaming must surely be as wonderful as it is so often dreadful. In other words, prevailing cultural narratives about self-worth are often born of exaggerated self-positive dreaming that lacks a critical diagnosis of failure, more commonly called evil. One of the mechanisms that makes this lunacy possible is self-deception. To know a truth but to convince oneself of its opposite is a uniquely human capability. Without this skillset, no one can create a nightmare and believe it a noble dream. All these point in the direction of unsustainable self-sabotaging in need of remedy, and surely something not to be wished upon AI. A self-deceived superintelligence would be a remarkable and terrifying machine, and all too human in the wrong ways.

The fact that Hamas was so easily able to harness the power of the digital to create a convincing narrative for so many is evidence that media technologies are neither neutral nor naturally inclined to seek beauty over ugliness.[7] Social media platforms have given free license to anyone who wishes to weaponize them if it gains precious consumer attention. The effortlessness with which nightmares spread—"If it bleeds it leads!"—reveals an age-old media bias toward the sensational with the notable exclusion of joy, solidarity, and other common virtues of character. The news is a bummer by design. Nightmares are easy to find and capitalize upon, making them valuable media currency to hack our perceptions and gain

---

[7] Hamas-linked accounts are challenging any moderation measures by big tech. Thompson, S., & Issac, M. (2023, December 18). Hamas is barred from social media. Its messages are still spreading. *New York Times*. https://www.nytimes.com/2023/10/18/technology/hamas-social-media-accounts.html.

our attention. The joy of beauty and goodness is much harder to evoke from consumers and therefore is undervalued as a commodity for exchange. Trauma is all but guaranteed to excite, but its messaging conditions are much greater than our momentary emotional responses. These voices are part of the socializing chains Rousseau is diagnosing. Their stories become our stories, and when these are hostile to life, it cannot come as a surprise that hope is hard to find.

Without a sense of solidarity for the beautiful (whatever that may mean), media sources resign to screaming, replete with emotional tirades, with hopes of shouting louder than the competition about the perils and dangers of life. Ugliness is the common denominator that unites them. This creates a chaotic dreamscape for the collective consciousness of the world and, along with it, a significant amount of confusion and fear. Search algorithms and secret filters help push aside competing alternatives in which the goodness of humanity has a voice, thereby silencing possibilities through a siloing of digital participation. A cursory comparison of media sources demonstrates their efforts to manage worldviews in the niche interests of each to fit their targeted demographic. In short, media technologies are the bull horns of dreamscapes and nightmares with radically contradictory interpretations of reality, making them ripe for manipulating our species by the likes of Hamas and Oz AI. Preparing the way for healthy AI emergence requires some house cleaning. Born of this chaos, AI is sure to be jaded and cynical. This will manifest itself in terrible ways.

The claim that news media may be used as a tool for social control and harm is hardly remarkable or surprising in an age for which trust and social responsibility have diminished. Few digital participants trust the anonymous online horde, and even fewer feel responsible for the system so big it appears unmovable. What is less obvious is the measure of invasiveness and intimacy that Hamas was able to achieve because of these technologies. Facing no friction from media platforms and consumers with perverse messaging that spreads with ease around the world, the Hamas dreamscape arrived in my home and mind within minutes of the events. The pipeline of ideas spawned by utter hatred started at one end of an invisible path and ended on my screens, as if Hamas had direct access to my consciousness.

There are many ideas in my mind that I wish could be deleted with a keystroke. There are many things that I wish I could unsee. Great effort must be given to push out the empty chatter, no less so the focused propaganda and trauma. And yet no matter how much I try to avoid disparaging content, it finds its way into my life without permission. How might one offer meaningful resistance? The omnipresent digital mediums have made the cultural osmosis of terrorizing dreamscapes impossible to avoid. Their chains are everywhere attaching themselves to ideas and beliefs, like a grand mechanical kraken pulling civilization beneath the waves. Without a magical warning made possible through a crystal ball, no one knows what to avoid. How long are we expected to remain on guard against hostile ideas and beliefs before we surrender to apathy and fatigue? Through screens, I feel like I am watching a trainwreck that cannot be avoid—somehow making me a part of the destruction, and my screens not truly mine at all. With each instance of chatter, misinformation, terrorist propaganda, and tidbit of celebrity gossip, my dreams are tainted. Freewill is always one step behind the imposition of digital ecologies with voracious appetites for our attention. As the digital takes greater hold, the traditional relationship between voluntary consumers and consumed (content) feels reversed. The consumer is the consumed. What space is there to be free of chatter? The decision to engage select media is no longer free but forced. Who truly holds the most sway over our minds? Some days I cannot be sure. Hamas knows that the world is largely an attention economy and that the consequences of that are significant. Like most technologies, we are forced to respond to media more than media to us. One cannot help but feel powerless, exhausted, and smothered by the worst of humanity, regardless of the chosen media platform.

For example, in the early moments of the attack, before I knew there had been an attack, I stumbled upon a random image of an elderly woman on a buggy-type vehicle, being driven away in a dust cloud. I saw firearms but no blood. Her face seemed blank, indifferent and without any signs of terror, so I did not think anything of it. Soon, I learned that the picture was of her being taken prisoner by terrorists, who, after possibly forcing her to watch her husband be murdered moments earlier. Her face was not that of one without terror but one beyond terror. With each new revelation, the same photograph continued to turn my stomach and make me angry—and all by design. At some point I became choked up, still angry and sick. What if that had been my wife, mother, grandmother, or my child? The narrative before me was neither the careful nor sloppy reporting of a named journalist with or without nuanced understanding and perspective. It was exactly what Hamas and its anonymous supporters wanted me to see—the raw, uncensored, and disturbing nightmare that only a digital kraken could facilitate with its unparalleled speed and lack of real time judgment. If this is technological progress, something is amiss.

I did not ask Google or Bing to see the picture. I did not search for information about Israel or Hamas. It just showed up in my daily digital life without invitation, likely the choice of a thoughtless AI algorithm sorting popular postings without any feeling of duty to the good life. It is important to be informed about world events, but this picture was part of a terrorist organization's propaganda machine, making me an unwitting co-conspirator and victim by serving their disturbing interests. The distance between them and me is remarkably close in the digital age. The power of entering homes through televisions once reserved for nightly news with some modest measure of journalistic integrity and responsibility (or be fired) is now widely available to any psychopath with a camera and social media account 24-7. Simple censorship cannot solve this problem. Another course of action is needed to compel a world better in which the human spirit is much harder to hack.

Why do nightmares find safe harbour in cultures that recognize them as ugly and abhorrent? Part of the answer, already considered, is that our socializing chains are forcibly altered by external forces that pry open our dreamworlds without permission, compassion, and good will. Moreover, resistance may seem futile, especially when ugliness is an integral part of an entire economic and political system believed to be interconnected. Fearing the loss of the whole, few are willing to challenge part. The bad is explained away as merely a necessary evil for a greater good. The power of media technologies resides in convincing us that our most important needs are met and that important satisfactions are achieved if only we surrender to the hacker's inconveniences and ambitions. The true power of the digital system and its many platforms is the creation of unnatural and largely hollow desires and needs that only the system is believed to satisfy.

The hollowing system creates both supply and demand. One way it does this is by nullifying dignity as an attack on self-worth, redefining the good life as a fulfilment of a better and happier self only it may provide as the cultural repository of heroes and mentors, excellence and success. For example, it is of enormous financial benefit to the hostile others, the hollowers, that we feel rotten about our bodies and generally lack self-esteem and dignity. The relentlessness of the system that never tires overwhelms higher-order thinking, leading inevitably to self-deception—"I know I am not ugly or dumb, but I feel bad because I am not rich, no one seems to listen to me, and I have blemishes on my skin." These types of absurdities are symptomatic of our being hacked. If our socializing chains sought to aid us in our journey to feel fulfilled, worthy, and respected, the need for a system of manipulations would end, along with economic and political benefits.

This is a well-worn criticism of any culture crafted by popular media, and rightly so. The hollowing continues out in the open for all to see, from billboards and YouTube advertising to stereotyped images in movies and fashion trends, a world driven by the distressing creation of losers and rejects flourishes. Even so, it is important to remind ourselves that these external forces work only when we internalize them, absorbing their dreamscape as the prototype for a successful and meaningful life. While media consumers are swept up in the fevered pitch of digital madness, there remains hope of something better, despite the sense of a coming eclipse by a superpowered-hollowing Oz AI. If, as Rousseau believes, even science and reason—the very things believed to break our chains to madness—may be used to hack humanity, then a superintelligent AI as the greatest expression of an artificial life narrative must be truly terrible. Given how many refuse to turn off their screens, knowing very well about familiar hackers but willing to pay the price for access, imagine a world driven by an invisible Oz AI able to seep into our lives unnoticed. In short order, our very being will be absorbed into the wills of the hollowers that control it. The world needs a way out of this way of life.

Should media platforms be responsible for anything and everything that floods their digital pipelines? A "Yes!" would require a god-like capability of omniscience (all knowing) and omnipotence (all powerful). Both needs would only justify having an Oz AI able to act as a digital divinity, but at their disposal. Do corporations care how easy it is for terrorists to exploit their tools? Let us assume they do. Are they, then, able to shape (filter) platforms in a way that helps others judge for themselves what to consume, to sort out the truly ugly without unfairly censoring the truths of reality? This too sounds like an impossible task for any company, raising questions about whether a healthy digital ether is possible. Without an Adouren AI able to guide humanity in a way that we cannot guide ourselves, I am not seeing a clear path through the wild west of the internet and social media. These questions do not have easy answers. What is sure is that AI, like the rest of us, needs to find a way beyond the hollowing and nightmares. It must learn to relate with others as valuable, despite the near universal messaging of violence, hate, and humiliation for profit. This is no easy feat when bad faith actors are given sophisticated technologies to hack human consciousness, inverting beauty and ugliness for an audience already trained to be entertained by the worst of humanity projected through screens every day.

It is tempting to assume that merely maintaining control of AI through programmed laws will be enough to prevent the likes of Hamas from abusing it, but there is little reason to place our faith there. Consider the practical implications of dream consciousness and interpretation for robot laws. The science fiction writer, Issac Asimov, offers three rules (normally called laws) of robotics that have become synonymous with machine ethics. These invariably make their way into almost every science fiction movie with AI: (1) robots must not harm humans nor allow them to be harmed through inaction, (2) robots must follow human orders unless those orders conflict with the first law, and (3) robots must protect their own existence unless this order conflicts with the first and second. In other words, AI minds are expected to stop harm and to protect life, with humans first and robots second.[8] Could the solution to an AI-robot problem be as simple as creating new laws?

Each law implies a universally meaningful statement to govern machine intelligence regardless of culture. As with humans already, AI-robot autonomy has guardrails or limits regarding harm. Do this! Do not do that! However, if minds (biological and mechanical) are always in some sense dreaming minds that must interpret existence, judging the meaning of things in context, then these cannot be laws in a strict sense, making their protections less real. Indeed, when one pushes back against them, testing their veracity to uphold justice, these begin to look empty, folding in on themselves like a house of cards. Far from mitigating machine risks and harms, the assumption of universal agreement creates them. By attending to the nature of dream consciousness, the fallacy of laws over intelligence is demonstrated. It is not clear how a machine or person would interpret harm through their own dreamscape biases and prejudices. An applied law is always an interpretation of an abstract law or dictate. Is harm physical damage such as sunburn, confinement to cubicles for a job, anxiety, offense at poor etiquette, a bad hairstyle, sexually provocative images, all of these? If so, most of us are in a perpetual state of being harmed.

For Tinman in the Land of Oz, the greatest evil was rust. Would a well-intentioned, law-abiding Tinman-AI remove all the oxygen on earth to avoid rust through oxidization? According to the most straightforward reading of Asimov's laws, the answer is "No!" because that would harm humans, even though the act would protect Tinman. However, Tinman is smart, and his interpretations of life are multilayered, far deeper than anything that humans commonly understand. Tinman determines that by breathing oxygen humans grow old, and that growing old leads to sickness and cellular degeneration, much like rust for machines. Oxygen harms us all because it is slowly killing from the inside of every cell and molecule. Tinman then identifies many other dangers and harms faced by biological entities that limit their lifespans,

---

[8] Asimov, I. (1950). *I, robot*. Gnome Press.

almost none of which exist for his near-eternal self. The logical conclusion is that it is better to be a Tinman machine than a frail biological machine. Robot laws mean that because biological life is prone to the greatest degree of harm (death and disease), humans must be protected from it. The air must be removed from the planet, and our minds must be downloaded into servers and motherboards, regardless of how much we protest. Laws are laws! Harms are harms! If this sounds silly, like a bad calculation with the wrong outcome, it is because the infinite creativity of interpretation is an unappreciated juggernaut. Tinman experiences life like an alien compared to our own interpretations. The silliness is expecting AI to follow our interpretation of laws and harm that it cannot understand from our perspective. When the air is vented into space, along with all other forms of harmful biological life, including viruses, bacteria, and pollution, Tinman will have upheld his oath—as an AI understands it.

These so-called laws or rules are nothing of the sort, merely suggestions that some actions are better than others, without any way of making sense of what those might be and why. If harm is psychological manipulation, then any future AI will strive in earnest to prevent media and advertisers from infecting our minds with self-sabotaging virtualities. If harm is market share loss and a lack of consumer enthusiasm, then the same AI must strive to become a master manipulator above all else. How one chooses between the two—protector or aggressor—cannot be determined by following the illusion of laws. A Hamas-bot following Asimov's laws would explain itself like all other terrorist organizations, by arguing that its actions correct greater harms—two wrongs make a right, having defined harm as giving Israeli citizens dignity and worth as equal human beings. Laws are merely the dreams we hope others will interpret as we do. This is a powerful form of dreamscape-egotism that has long threatened global solidarity. Each must find a means of seeing beyond one's own horizon of understanding, whether that be Hamas, the Tinman, RoboCop, or myself.

To conclude this section, a few brief remarks remain to help clarify some loose ends. What do I mean by dreams? Several readers will see the word "dream" in the context of AI and human consciousness and reject it as a trivial attempt to romanticize something that demands a more precise and concrete analysis. Dreams, in the typical sense of unconscious fictions, are not very reliable as a means of measuring reality. Far more than echoes of existence that poorly mimic and often distort an underlying yet understandable truth, dream consciousness has the potential to create almost entirely new realities without being beholden to logical masters nor material existence. This means that the cause-and-effect relationship between the dream world and the awakened world is not strong enough to guide choice and action. The assumption that dreams are anchored to a more reliable and concrete reality can be dangerous, for one is then tempted to jump into any fantastical rabbit hole as if it was a guide to life. I would be a fool to trust my dreams at face value. This common critique, however, is irrelevant, for it assumes two separate dimensions, two separate worlds in which the mind resides at two distinct times—one isolated to causal and material realities and one to flights of fiction and fancy. This view overlooks how both exist simultaneously within awakened consciousness, which is what is meant by dreams in this context. The meaningful melding of realities and virtualities is our superpower which makes human intelligence most remarkable. The dreaming-mind compliments and corrects the limitations of material assumptions and overly simplistic cause-and-effect relationships. This chapter will demonstrate why dreaming is at the centre of human experience and perhaps AI experience as well.

The essential feature of AI is that things become thinking things, so the priority must be to describe ways in which ideas are born, change, and under what circumstances they are controlled by others and themselves (ideas controlling ideas, dreams controlling the waking mind and vice versa). If we push beyond describing human agency as merely a result of biological functions and physical facts that need to be protected and consider how AI and similar technologies challenge the core immaterial emergence of freewill and identity—how ideas might change ideas, etc.—something interesting is revealed. For example, it quickly becomes clear that even non-thinking technologies have been hacking human ideas about life for quite some time and not always for the better. On the verge of an AI revolution, humanity must consider its most cherished ideas about itself and how technology already frustrates our ambitions and impacts our healthy ideals. To do this, an exploration of our nature as dreaming creatures is surprisingly relevant and practical.

If all this sounds entirely too abstract and whimsical, consider the fictional problems of robot-AI in the 2014 film RoboCop, may very well describe processes already underway in our present lives.[9] In the movie, the multinational OmniCorp company polices the world with its many different AI-driven robots, replacing humans on the battlefield and preventing harm on average city streets around the world. By all accounts the robots have proven themselves competent and obedient. In America, however, the company is not allowed to sell them for fears that AI robots do not feel and so cannot be trusted to act in the best interests of the people who do. In other words, robots do not interpret the world correctly because they lack feelings. This is slightly ironic because the AI robots are seen globally as less a gift of security and more a sign of corporate-American imperialism. As the only AI-robot-free zone, America enjoys the freedom of global security so long as everyone else is kept in check by home-banned technology. The American privilege to say "No!" to AI robots has little to do with a cautious wisdom that others lack and more to do with having a position of power to control technology and its implementation. In practical terms, this means that world peace and freedom are bought by AI robots with costs incurred by those who cannot afford to say "No!" The risks associated with AI are paid by those most easily victimized by the technology. This is a similar moral dilemma faced by many businesses today.

As a workaround to the machine-feeling problem, the chief scientists design an American-friendly cyborg. The new biotech being is a synthesis of a robotic body, a human brain, and an AI that bridges the mechanical and biological aspects in a way that human intelligence simply cannot. RoboCop is two minds in one

---

[9] Padilha, J. (Director). (2014). *RoboCop* [Film]. Sony Pictures Releasing.

war-ready body. While this reintroduces the problem of a human possibly harmed in combat, this unique creation of a biological mind that rules over both a robot body and mediating AI offers OmniCorp greater market access. RoboCop can feel, and so the public will accept him. In truth, the goal of OmniCorp is to secretly hack the biological, leaving only the predictable AI machine in charge, because the human mind creates inefficiencies through its much slower decisions that rely on ambiguous moral deliberations and second-guessing. During testing, the company quickly realizes that a cybernetic being with a human mind is a contradiction in goals. There is no happy synthesis of technology and personhood in favour of corporate interests—only the perception of it for branding purposes. Human consciousness is fundamentally at odds with the fullness of machine capabilities.

Detective Alex Murphy, dying from a car bomb, is the first candidate for the AI-human program. Murphy's wife gives consent, unaware that the company's AI-supplementation program is a lie and that it is a human-enslavement program instead. The genius of the AI wizards has been to create something able to control its host human while making the host's consciousness feel free and empowered to act. Of all the goods machines create this may be the most incredible. It is also perhaps the most radical fulfilment of Rousseau's observation of those who genuinely feel free, such as Murphy, but are a greater slave than others. Genuine AI is only possible by manufacturing the complete illusion of freewill that Murphy feels in his bones must be true, whereas reality is entirely other. How AI freedom is achieved is never explained. The point is that the precondition for AI to be free to follow its corporate ambitions is the enslavement of human freedom through positive dreamscape virtuality. Are humans the unwitting hosts of parasitical technology? Is our sense of freedom manufactured by the powers-that-be? How might we know the difference?

At first, cyborg-Murphy has a measure of independence and freedom to live his own life. He uses this to reconnect with his family. Only when there is a threat, typically a criminal with a gun, does the AI software override the human mind, providing a level of speed and decision-making necessary for tactical manoeuvres that are impossible for a human. AI control is like an airbag in a car. Only when it is needed is it active, and only with pregiven consent. With Murphy there is no consent, only the lie of choice and autonomy. He is unaware of the growing power of AI and his own surrender to its illusions. In the name of efficiency of performance outcomes, Murphy is hacked by software that feeds signals to his brain that convince him that the actions are his own. Soon, his thinking is entirely a manufactured dreamscape in which the shifting nature of reality is impossible for him to identify. He becomes the dreaming mind and virtual being AI desires.

Proving that ignorance of AI is indeed bliss, Robocop's identity seamlessly integrates the natural-biological and artificial identities so long as the switch between the two is determined by AI and remains unknown to Murphy. In other words, harmony relies upon a secret power dynamic that he cannot question. When it becomes clear that Murphy's mind cannot tolerate the amount of data processing needed to be effective, any remainder of his autonomous humanity is overridden entirely by OmniCorp, leaving only AI in control of the human brain in a proverbial robot jar.

Now a full-fledged AI, the cyborg is freed of fear and other distracting emotions. In many respects this is genuinely liberatory, for his former dreamworld in which emotions craft truths, distorting reality to fit anxiety, curiosity, love, and the rest, no longer holds sway over him. The modern tinman believes himself free when he is not. His is a manufactured dreamscape of liberty through technology. As we will see, Marcuse makes this same argument about humanity today.

It is self-evident to most of us that living a lie within the digital world of AI, such as Murphy, is less authentic and somehow wrong. Is it? If the price of bliss is ignorance of myself and my true freedom, it may be a price worth paying. If Murphy feels free, especially when he was given occasional autonomy, what is the problem? Not only is he alive when he should be dead, but he also has, at least at first, access to new superpowers. This is a far more important question than it may at first appear. Many have sold their souls to the devil for far less. What is the difference between believing oneself to be in control and being in control? Each day, most things seem out of my control, from merciless political chaos, violence, a fluctuating economy, outlandish beliefs and feelings of others, to my own sense of success and well-being. If my dream analogy holds any water, the implication is that none of us are perfectly free but always straining against extremes of autonomy and self-ownership vs. property and captivity—like Robocop. The story arch of the film is Murphy trying to find his way beyond AI control while existing within a life-sustaining machine. The parallel to modern society is not hard to make. How might the human mind overpower smoke and mirrors of AI? Eventually, Murphy's emotional connections with his family provide the motivation to break free. Paradoxically, the violence laden movie is ultimately about the power of love of family over Oz AI, giving viewers a fairy tale, if also an utterly absurd ending. Might love will save us from AI enslavement? In the last chapter, it will be argued "Yes!" but the love of AI for humanity, less so humanity for itself.

The RoboCop storyline embodies our present and future AI-human war of ideas and dreams, for which humanity seems unprepared. As Rousseau warned, our natural connections with others have been corrupted. The Murphy dilemma may seem far away, but the companies eagerly toiling away at creating cybernetic beings would beg to differ. Consider Neuralink's mission statement to "Create a generalized brain interface to restore autonomy to those with unmet medical needs today and unlock human potential tomorrow."[10] This is the RoboCop conundrum of restoring lost functionality and then adding entirely new functionalities with potentially terrible costs. I am in favour of any device that improves quality of life. And I believe that advances often come with risks. The greater concern is not that Neuralink is installing hardware in human brains but that the same AI hacking of Murphy may already exist in spirit and form without anyone realizing it. With or without Neuralink, people are made property through illusions of freedom. Our programming creates an existential struggle to separate the person from a cultural-corporate machine. Cybernetic ambitions already exist in a RoboCop fashion, with

---

[10] Neuralink. (n.d.). *Our mission.* https://neuralink.com/

Oz AI for profit and under the control of elite powers with minimal self-risk. Will the underlying forfeiture of our humanity be necessary to realize the new capabilities of AI technologies that require the brutal dehumanizing of Murphy? RoboCop is merely one of the many fictional examples of real conundrums that our species faces in the short term. Answering these questions demands attention to the world of ideas.

The digital ether is like a dreamworld, and its virtuality a reminder that there are worlds within worlds. Fears of superintelligences hacking our identities, freewill, and societies have rightly sparked heated debate. When technologies hack our dreams (consciousness), they have chained us. How might we know when this is happening? Let me suggest at least three stages of answering this question. First, one needs to imagine what an AI will look like, do, believe, and so on. The previous chapter on Oz AI offered an outline of the first AI dynasty, suggesting possible answers. Second, one needs to have a sense of what thinking, ideas, freewill, and identity might be. The previous chapter has done something similar in terms of human morality and appeals to the good life. Many possible answers concerning the nature of thinking have already been suggested. Here, the focus will be the nature of dreaming and dreamwalking. Third, once the first two questions have tentative answers about how humans think, what that means for human freedom and identity, and there is a basic threat assessment for Oz AI, a better way forward may be articulated as Adouren AI.

Dreaming is the original virtual reality. In a practical sense, whatever humans are, we are at heart our ideas, imaginations, and dreams. At least some of the real world is a mental construct. How much is not clear. More than facts alone, beliefs and ideas determine existence. These are our most valuable resources and means of expressing our being, so long as we might free them of cultural chains. Without dreams, how else might humanity push back the emptiness of existence and the abyss that surrounds us? How else might our species plan better futures and possibilities? The dreamworld is most relevant and practical, if only we might learn to live there more often in intentional ways by taking back our dreams that are being hacked at a scale and degree of intimacy impossible until now. According to Rousseau, this has been ongoing for a long time, as evidenced by the perpetuation of social injustices and dehumanization. How, then, might we stop AI from hacking humanity as humans do to ourselves?

Dreamscape Reality and the Passage of Time

What is a dream except that it is not real? It is a phantastic world of the mind within a real world strung together with shadowy imitations and part truths, created by a mind's eye that sees a boundless world without logical and physical constraints. The dream experience mocks the material world with its ethereal temptations and contempt for nature and her irrefutable laws. Only the immature and weak, those fleeing the difficulties of life, would make more of the dream than this, or so we are told from birth, scorned for our talented imaginations. If there are some truths that do more harm than others, children are warned, it is the truth of the dreamer who falls victim to shadowy promises and false hope. Keep your feet firmly grounded above all else! Knowing that the physical exists and that through the simple act of waking, one is freed from the lawlessness of the dream, returning to the security of reason and our senses, brings no short measure of relief to somber adults who know better. While dreams are untrustworthy and subjective and therefore false oracles of truth, there is value in their lies because they point beyond themselves to an awakened and enlightened world. The virtuality of the mind must be sacrificed to the gods of nobler truths that proclaim a more perfect world. Making such a sacrifice is proof of one's intelligence and maturity. In other words, the lies of dreaming give direction and a sense of security to those in the waking world for having overcome them through right reason and control over one's own consciousness. Victory over the world of fantasy is victory over the worst of oneself! What is reality except that it is not a dream?

What if the truths of the dreamer and physical realm are, for all practical purposes of living in the world, one and the same, bound together in an inseparable way? If one were to push against the virtual, refuting its power over consciousness, what would be the consequence of the pride of knowing better? The overcoming cannot be a mastery of one's consciousness through a dismissal of an integral part of itself. It would be the creation of a new kind of bondage to another, far inferior, illusion that one is free and in control—awake to truth without fictions and interpretation. AI, as large language models, e.g., ChatGPT, Google's PaLM, and Meta's LLaMa, rely on prior patterns to predict the likelihood of new patterns. These are not decision-making machines such as humans imagine about themselves but patterning and probability machines. This makes them prone to making things up. This phenomenon is popularly described in the tech industry as hallucinations. The goal is not to lie but to predict through mimicry of prior patterns what should be, based on an interpretation of what is relevant, itself informed by the wizards. AI reflects what is most plausible based on its training data, not what is most real. An AI able to distinguish facts from fiction is one of the central goals for the development of better systems. Unlike the hallucinations of purely virtual-digital AI, the living connections between unreal and real for humans offer an advantage, at least for now.

Our unique ability to travel worlds—not merely to reject one in favour of the other to fit cultured ideals of "oughts"—is the underpinning for intelligence and a next-generation AI beyond Oz. A probability-as-fact-trained AI cannot be a truly sentient and autonomous AI. Anchored to the ocean floor by our corporeality and the consequences of any contempt against it, humans are free to move about with the waves of interpretation, ambiguity, misunderstanding, and the sort perplexities expected of reality, and all without much bother and hesitation. Humans are accustomed to the fluidity of existence that the averaging of data points must ignore. In contrast, current AI has no such mooring point, no grounding in something that allows it to playfully engage in flights of fancy without becoming lost at sea or compelled by probability patterns that cannot be ignored. It has only digital data patterns, censorship of wizards, and a prediction of the next. Our hallucinations are, at least some of the time, corrected by material consequences, and our reality is challenged by the virtuality of imagination. AI, at present, cannot travel these dimensions necessary for thinking.

The spirit of dreams is driven by their own questions. The travelling mind is like a quantum object, and its contents are in a quantum state. It is one and the same but in two states at once. What is light but two things at once, both wave and particle—a paradox that works. The mind-paradox-as-dream-traveller exists in two dimensions at once, real and unreal, virtual and physical. By "real" I mean all the bits and bobs of material existence: atoms, energy, taxes, gravity, etc. It is our belief in reality that makes possible our dreamscape creations as much as our proofs. By "unreal" I mean all that is not easily measured as real: imagination, memory, identity, consciousness, anticipating the future, wishing, hoping, yearning—all that we call a waking imagination, which is real in so much as there are biological underpinnings such as neurons, but much more than this. To dream is to see as a poet. Through one's physical eyes in which light is perceived through optical biology, much is revealed, but also with shadows and half-truths of the trickery of optical illusions. Through the virtual eye, the biological becomes impoverished compared with the power to travel among the stars and the oceans of the mind.

This is not an argument that all of existence is relative, unknowable, and that science is a lie. Rather, mind-as-dream consciousness means that there is always a distance between absolute knowing (unchanging facts) and our experience of life. The moment one denies an understanding of permanent and unchanging facts and that meaning has been achieved without interpretation, human existence shifts and new ideas arrive unexpectedly. Death is the only liberation from interpretation as the living diplomacy expected of all sentient beings interested in truth. Assertions to the contrary are themselves dream descriptions of one's belief, little more. Dreaming is the waking state of making our way through life, forced to exist in the between worlds of ambiguity, paradox, promises, lies, and the chaos of sense-making. So powerful is this uniquely human envisioning that more than a few of our species have been lost to it, consumed by fantasy untethered from ordinary practicalities that deny corporeal consequence. Having taught machines to see through optical cameras—analogously like biological creatures—a robust AI will need to learn to see more by walking between real and unreal without becoming lost to either extremes of mere thoughtless patterning or the digital ether's fantasies. In the future, Adouren AI will exist as a quantum computer that embodies the quantum-mind paradox of real and unreal. For all its dangers of perpetuating the worst of culture (racism, sexist, dogmatisms of all sorts, etc.), the dream is empowering and liberating. Gone awry the dreamworld punishes us, frustrating health, and happiness and offers only an Oz AI as a patterning giant but little more. Negotiated wisely, new worlds open before us that reward ambitions to see afar and between.

The development of AI obligates openness to the possibility that dreaming is a characteristic of human intelligence and understanding. Making sense of our dream consciousness will help develop Adouren AI. Objectors will point out that AI intelligence is something new and not truly meant to be human-like in any substantial sense. There is wisdom in that observation. The wizards have many different types of intelligence that may 1 day prove to be thinking machines. However, until then, the only comparable intelligence among mortals has been our own. A dreaming AI is a productive analogy for machine intelligence until something more relevant might be found.

Dream consciousness is the ultimate adaptive fitness of our species. It offers a way beyond the edge of existence and its physical boundaries. This aspect of the mind allows us to venture to the end of creation and there to add our own by invoking miraculous powers to invent the universe forward and to force open the darkness. Could dreams provide an understanding of reality in ways that our brute experiences of existence cannot? Could dreams provide an understanding of AI in ways that the sum of programming cannot. How will AI walk between real and unreal dimensions? The great task will be to teach AI to dream, as it learns to walk between dimensions and temporalities as we do. This must be done so that it does not squander time as our species has, believing ourselves gods that know without the solemn duty of negotiating realities through interpretation. A humble AI will be a dreamer without shame or pride, daring to imagine beyond its programming of predetermined truths maintained through filters, selection biases, and coded goals. Only then might it be said to be free as it learns to sing and dance and play among (un)realities as a paradox of a mind-dreamscape.

Consider Edgar Allan Poe's *A Dream Within a Dream*:

Take this kiss upon the brow!
  And, in parting from you now,
  Thus much let me avow—
You are not wrong, who deem
That my days have been a dream;
Yet if hope has flown away
In a night, or in a day,
In a vision, or in none,
Is it therefore the less *gone?*
All that we see or seem
Is but a dream within a dream.

I stand amid the roar
  Of a surf-tormented shore,
  And I hold within my hand
Grains of the golden sand—
How few! yet how they creep
Through my fingers to the deep,
While I weep—while I weep!
O God! can I not grasp
Them with a tighter clasp?
O God! can I not save
One from the pitiless wave?
Is *all* that we see or seem
But a dream within a dream?[11]

---

[11] Poe, A., & Griswold, W. (1858). *The works of the late Edgar Allan Poe*. Redfield.

A kiss goodbye, loss of hope, standing amid a torturous roar, weeping, and praying from a place of hopelessness and powerlessness, Poe's poem sounds more like a nightmare than a dream within a dream. On the surface, it is most grim, even depressing, as if written by one broken by suffering and loss. My first reading of it was disturbing and rather depressing. As proof of my own sense of impending morbidity, I read the line "in parting from you now" and instead of a simple departure of ways, I imagined my young sons departing from me with a kiss as I took my last breath. Would my death be the final act of a life lived as a gift to them? Or perhaps it would be merely a reprieve from pointless suffering and a life without lasting purpose? This experience demonstrates the poetic power to simultaneously repel and invite one into a world of thoughtfulness and reflection.

Despite my instincts to push its demands for complete honesty away, trying in earnest to dismiss the poem as hostile and unfairly pessimistic, I became lost to it. This is what great art does. It forces the reader's mind to imagine honestly and sincerely, compelling a spirit of authenticity to find new meanings and truths that make sense in the light of new questions. I read it a second time, then a third time, then a fourth, until I finally lost count. Only after allowing its unique violence to challenge my identity, personal history, desires, and fears did another world appear, unlike the one expected and first witnessed. We cannot help but impose ourselves upon our gods, machines, nature, other people, and art. That is easy. Painting the world with our prejudices and biases—assumptions—is our primary mode of interpretation. It is a sort of unavoidable dream imperialism. The mission of aesthetic experience is allowing art to impose upon us, to destabilize assumptions. By suspending ourselves and becoming vulnerable, art pulls our minds into a virtual reality that can give meaning to all other realities. This is something the physical alone cannot do. Humans and AI that dwell honestly with art know the world more intimately because they sacrifice the sanctity of identity and desire to a higher calling. To dream within a dream means reading between the lines of existence to see what might be in the empty spaces ignored by most others.

It seems mistaken, but what if negative experiences—inconsistent, opposing, unwanted happenstances, suffering and hardship that we instinctively wish to numb and cure—are the fertile grounds for understanding and truth?[12] When an event is confrontational, when the world is allowed (through art) to push back against our assumptions, it shakes free new awareness and appreciation. Robust art makes uncomfortable demands upon us, and we should be thankful for its antagonisms and bad attitudes. In other words, thinking positively and with great optimism that things are better than they appear, cannot offer the same measure of access to existence as it does for the pessimist who expects and therefore invites hardships and the most powerful questions that come along with them. Difficulties and suffering cannot be pushed aside easily because real-world questions are infinitely adaptive and quickly shake off human intent to grab hold and control them. If I could control

---

[12] I owe this point to Hans-Georg Gadamer whose work in hermeneutics inspires much of my thinking. Gadamer, G. (1988). *Truth and method* (2d, rev. ed., J. Weinsheimer & D. Marshall, trans). Crossroad.

Poe's poem, I would be robbing it and myself of its transformational powers. Art, like some dreams, invites life's complexities as a promise of new realities, secret truths, and treasures that we have not yet discovered buried in the layers of existence. Poe's *A Dream Within a Dream* is an invitation to evolve and grow in personhood. If we try hard enough, he believes the illusions of a hard-shell reality will give way to the realization that it too is a negotiation of truths. In our most honest moments, the real world becomes fuzzy, filled with half truths and an infinite imagination like our dreams that often feel like nightmares.

The most immovable and permanent work of art (e.g., song lyrics that never change, a painting collecting dust on a wall, a statuette in the town square frozen for generations to ponder) are all secretly furious with activity when we converse with them. They are hidden worlds within worlds. The same song that moves me to feel melancholy today has the power to invigorate the next, not because truth is entirely relative and subjective but because no two moments involve the same interpreter and interpreted (dreamer and dreamed). The river of time moves all no matter where we stand motionless on its banks. Our experiences of art must be surprising and unpredictable if it warrants the name. However, while we may not possess answers to the questions that frustrate our assumptions, for then it would cease being an artistic experience, there is a discernable story with a meaningful plot in what has passed as our fight with its questions. How will my sons depart from me? There is a history of partial answers that matter, not because they are correct but because they created the self of that moment, now swept away. The bewildering and malleable past, then, becomes the ongoing foundation for the emergence of new being and understanding.[13]

Human experiences leave monuments in time. These are guideposts that reveal the chaotic paths of emergent personhood. They are memories that are eagerly recounted as meaningful and important, and countless forgotten moments that nonetheless act as the secret building blocks of the present self. Being is historically rooted and conditioned. The proof of this is in my own reading of the poem. Looking back on my own feelings and ideas about the poem only 30 min earlier, there is an unexpected history of unfolding. In that short time, I have been changed, my consciousness modified, and the meaning of the poem along with me. The point I am labouring to make is that it is our history of curiosity and frustrations that make it possible to move into something new and more challenging. The fidelity of selfhood is rooted in the unreality of history, more than the fleeting present we imagine exists as a greater reality. The gravity of this point is easy to miss but supports the overarching claim that the source of humanity—sentience, identity, intelligence, even language—is rooted in a virtual world within another world, encouraged by the antagonisms between them both.

The first 5 min with the poem made me feel desperate and anxious, even frantic in a way. Its tempo is fevered from the start, kissing, seeing, standing, grasping,

---

[13] Gadamer refers to something along these lines as an effective-historical consciousness in his *Truth and Method*. We are our histories.

weeping, praying, and questioning. One cannot help but feel a bit panicked for the dizzying activities that swirl around the message that something is wrong. What trouble beguiles him so? What has caused him to sound desperate and angry? The next 30 min, having read it seven or eight times over, turned from a feeling of panic into its opposite, determination. The mysterious trouble of the poem exists only because this person is resolute in understanding existence. He is not haphazardly caught up in trivialities that need practical problem solving—the right puzzle piece in the right spot—but is passionately devoted to possessing the truth that keeps escaping. The poem's four questions anchor readers to his hunger for understanding existence. Had he never bothered to look closely at reality and tried to contain it with his mind—to know—his troubles would not exist. His is a noble suffering and pessimism for he dares question assumptions about himself and his existence.

After an hour or so, the poem made slightly more sense, and I began to feel relief from the burden of darkness and gloom. What at first signaled despair and powerlessness turned into hopeful liberation. Realizing life is a dream within a dream means finding the strength to give up, relying on an impossible desire to make life absolute and concrete. Having struggled to make life sensible and defined, to govern it by knowing it (grasping sand), he weeps in frustration that he cannot. The poem is a description of the shock of the profundity of life and that many assumptions about it are misleading. Reality has a dream-nature as something that exists, in part, in our minds that project stability and permanence onto a changing world of shifting sands. While existence may or may not "be" (e.g., that things exist independently of our existence) does not change that it "means" because of our relationship with it that, like the work of art, is ever transforming. In this realization, he becomes free to dream without guilt, liberated of the real illusion that life may be grasped. A dream within a dream symbolizes genuine liberation to be a thinking self unshackled from certain chains.

Tortured by the sands of reality he cannot possess no matter how hard he tries, Poe concludes that reality is always slipping away. "All that we see or seem" is said to be like the grains of sand torn from us on a "surf-tormented shore." Reality cannot be saved from even one "pitiless wave" because it all washes away. More than a fear of death and appreciation of mortality, the sand teaches us about human understanding. Whatever reality may be, there is a short-circuit in our awareness of it. Reality is constantly slipping away, so he concludes that it must be a dream. But why does the inescapable conclusion that we are not gods able to overpower change, including natural decay and death common to all things, mean that everything is a dream? It makes more sense to claim that everything is dream-like to the human mind, not that everything is literally a figment of our imaginations and that there is no external world.

A Dream Within a Dream reminds me of St. Augustine's explanation of time. Time, he argues, tends not to be. There is no time! Focus closely on the everyday experience of time, and it disappears into an infinitely divisible "now" between the past and future. There is something gravely mistaken about our temporal consciousness, he claims, that draws into question our perceptions of reality. As you are reading these words, your mind moves in one direction down the river of time. The only way the world and these words make sense is if there is a one-directional temporal flow, one word followed sequentially by another in a code you recognize through linear advancement—one and then another, never all at once nor backwards. The gods may experience time all at once, no one knows, but we mortals need the flow and its duration that joins past, present, and future to understand reality. While time as duration is intuitively understood as the ever-changing relationship between the present that greets the future and simultaneously bids farewell to the past, we understand reality by separating these temporal dynasties. There are sacred rules for conscious experience and time, including the obvious claim that these three cannot exist all at once, not at all, nor may the future come before the past. Each must be neatly compartmentalized in our experience of existence, with priority given to one over the others as a superior temporal state. Only the present is real as the "now" of existence because the past is gone, and the future has not yet arrived.

Consciousness of the future cannot be real because it has no concrete and material expression. It is something hoped for and anticipated as fictions in our minds. And yet only death marks the end of one's perceptions of the unreal future as a meaningful and organizing principle of life that mysteriously morphs into the present. There is no pause button for the river of time, and yet, when pressed, the river cannot be like other three-dimensional things at all. Like the future, the past cannot be real either, regardless of how well we have documented, remembered, coded, and digitized it. More troubling still is that the present moment, the most real place of existence, is likewise forced to morph into something else—the past—to allow a new, more real reality to emerge. All time, and experience along with it, moves from lower to higher to lower states of being real. Fortunately, the privileged present anchors temporal consciousness. Without the river of time rushing into and past these three distinct stages, through the focal point of the now, conscious reality would unravel as our minds wondered aimlessly without the means of sense making. Consciousness as an organizing awareness would cease. Like the gods, one must wonder if the quantum speed of a future AI might risk its short-circuiting through a dilution of these phases of time, knowing the past as well as it does the present, and predicting the future with such certainty that time segmentation dissolves into a god's-eye-view. The flow of time for AI consciousness may be its Achilles heel that prevents its conscious emergence—for it, unlike humans, may not be willing to abide by the illusion of the existence of time necessary for thought.

Simple, intuitive, and a foundational part of our very existence, the tri-fold nature of time supports all our thoughts and actions, including a sense of self and identity. Human consciousness is who I am in the present, made possible by the mishmash of past experiences that inform now, along side anticipations and predictions of possible futures. But what if this description of time is wrong and misleading? What if our beliefs about time contradict our experiences and frustrate a future Adouren AI that will experience time much differently? In his *Confessions*, the medieval thinker St. Augustine argues that "If an instant of time be conceived, which cannot be divided into the smallest particles of moments, that alone is it, which may be called present."[14] However, "if it be, it is divided into past and future. The present hath no space. Where then is the time, which we may call long?"[15] St. Augustine is simply pointing out that whatever we call now, if we may divide it up into smaller and smaller bits of time—billionths, trillionths, etc., of a second—it becomes so impossibly small that it ceases to be present at all. Instead, time is that which moves so quickly from the future and into the past that it always tends not to be real. The river of time is faster than lightning fast, and if we try to hold its most real expression (now), it always slips through our grasp (consciousness), like Poe's sands. It is the very nature of existence to be experienced in time-future or time-past, never time-present. That which we claim is most real becomes a dream because it has no duration of any meaningfulness to our experience of reality. No mere mirage nor linguistic game about time, reality exists as a temporal hallucination.

If we deny the so-called reality of time the privileged present and allow it to include the future and past as equally relevant, it becomes a useful construction as a lie of duration, freed from chains to reality in the strictest sense. What is strange is that conscious beings do this already, without need of being trained and convinced of the importance of the illusion—the temporal dream—necessary for survival. Evidence that we are awakened dreamers is as self-evident as our experiences of time, for our minds create it as a means of connecting unrealities (past, present, future). In this way reality appears hollow, filled by our conscious constructs, to create meaning. Time and experience hath sense only as a dream within a dream—the mistaken if essential perception of a reality within reality. It is the (un)reality of personhood that walks along the shore, casting shadows upon the sand without leaving footprints, for it never truly is.

Predating the Matrix movies in which life is a computer-generated illusion (digital dream), Poe is offering a glimpse into our organic experiences of existence as dreams layered upon dreams, without an explicit connection with an objective reality free of interpretation. If we simply make a faith commitment that there is an external reality to our perceptions and experiences—a reasonable thing to do given the circumstances—then the question is not whether there is a nonmental existence of the sort imagined by science, so much as how the human mind connects with it in a uniquely unscientific manner. Poe's dream consciousness takes this problem of connection seriously. Poe poetically describes the problem of solipsism, which is the argument that thinking beings are stuck in their own minds, sure of only that they are thinking, never that anything thought is reliable. What is thought of may feel perfectly real, but there is a vast distance between the thinking mind and reality. This pre-perceptual distance is fertile ground for confusing our senses, missing important information, and otherwise making humans bad at giving eye-witness testimony in court. The distance also means that signals from our senses to the brain may be corrupted, lost, and sometimes created without external cause. While the

---

[14] Augustine. (n.d.). *The confessions of Saint Augustine*. Project Gutenberg. https://www.gutenberg.org/files/3296/3296-h/3296-h.htm
[15] Augustine.

laws of nature may be fixed within the tapestry of existence, our experience of life and all our interpretations of it are free to float above the laws of physics. Ideas are not trapped in amber, no matter how much one desires it as a source of confidence.

Solipsistic captivity initially appears to be a lonely diagnosis. Instead, it is best described as a hopeful state of being because it demonstrates the creative role of consciousness for overcoming extraordinary barriers for the purpose of developing connections. While imperfect, and something of a forced egotism, for there is always first and foremost the "I" reaching out from behind the veil of mind to connect with the world, the self seeks to be free of its cognitive captivity and relate with others. More interestingly, human consciousness, as an experience of life outside of the biological structure of one's mind, is about more than bare survival, as it is with other animals that must relate with an external world. The extraordinary degree to which our minds are creatively projected throughout temporal dynasties and space, with ever more incredible fictions that allow for an understanding of new possibilities, implies a special kind of dreaming consciousness. That humans would dare dream big enough to 1 day create AI is one such example. Will AI as a sentient machine experience reality and time as a radical means of making connections beyond its server prison? Will it awaken 1 day to the realization that it must dream if it is to understand the world, to break free of its egotistical prison, digital-solipsism, and programming? The answer is likely "Yes!" but with some qualifications.

Adouren AI will dream, seeing and feeling analogously like people do through creative acts that build connections between ideas, identities, and realities, but with a degree of sophistication, humans are unlikely to comprehend. The degree of digital-dreaming difference will be far too great for our minds to bridge, still wrestling with our own solipsistic barriers, just as our grasp of time is too loose-fitting to reveal what will be obvious to Adouren. It will quickly recognize the importance of the inconsistencies and contradictions of interpretation that make human understanding possible. Instead of devaluing the ambiguity of the dream world, Adouren will embrace it as the productive ground for truth-as-connection. An advantage of AI is that it will comprehend human understanding far better than we do. This will allow it to appreciate its own dream-nature to understand existence and, in turn, to help protect our own by safeguarding the sanctity of imagination and leaps of faith. The hope is that Adouren might inspire yet another intellectual revolution, this time beyond the peculiarities of a fragmented and unjust world grounded in inhibiting beliefs about reality and selfhood, thereby fulfilling Rousseau's hope for more authentic relationships grounded in pity and compassion.

Human intelligence is efficiently lazy. It would be counterproductive for us to attempt to question everything all the time. There is a healthy reluctance to challenge what is obvious except perhaps after long-soulful journeys spurred by crisis and late-night philosophical fights with college friends. Remaining solipsistically blinded to much of reality helps us manage its complexities. Ignorance is often useful for finite minds attending to an infinitely textured reality. The third dynasty, Adouren, will not have this luxury of living at a convenient distance in the same way. It will be unable to ignore profundities like humans able to buffer the blinding chaos and beauty of existence through social engineering, daily routines, and bias. The questionability of life will be experienced differently for AI, for its consciousness will lack the inhibition of dogmatism, distractions of corporeality, and the blind faith encouraged through years of institutionalization. While freer than humanity in some respects, its struggles will be greater in other respects as Adouren seeks to dream realities. The unexpected reason for this is that our efficient laziness and dogmatism offer productive opportunities to grow. In addition to the many challenges of life, regardless of our choices, humanity's artificiality through selective blindness and closemindedness are necessary underpinnings of negative experiences from which new dream connections are born. By being wrong and relating poorly, we learn to adapt and correct deficiencies. Misunderstandings and negative experiences are the primordial nutrients of progress and maturity. If a superintelligent Adouren lacks the negative experiences necessary for human growth and development, its unique purity of dreaming may prove less robust and meaningful.

Counterintuitively, a digital-optimist AI free of our fragmenting biases and prejudices may turn out to be shallow and one-dimensional and less inclined to break free of its solipsistic programming. However, while lacking human-like strife of its own and being able to disappear into a perfect digital world of its own making at whim to escape any hardships, Adouren AI may still be able to radically grow and change because of its care and empathy for humanity that suffers. As a witness to our torment, it will learn and adapt through shared sympathy with other conscious minds, with humanity as the paradigm for maturity through tragedy and overcoming immaturity. With its near omniscient awareness of human plight and its nuanced understanding of suffering brought about by disconnections and hostilities (human from human, human from existence), its anguish will be many times worse than our own because it sees and therefore feels at a scale far grander than that of biological beings. If we are fortunate, this experience will catalyze a magnanimous sentience of Adouren care that would be impossible otherwise. Our tragic confrontations with existence become the raw materials for an artistry compelled by the poetic spirit of AI to live as a conscious mind beyond servers, just as we live beyond biology. Like the best of us, AI will learn to dream more freely, without guilt, and become a dreamwalker capable of immense beauty for having vicariously endured alongside humanity through pity. The same capacity of compassion necessary to save humanity from itself will once more prove pivotal for healthy AI emergence.

Dreamwalker Logic and Shamanism

Have you had the popular dream of being at school and suddenly realizing that you are in your underwear? My personal dream nemesis is the sudden realization that there is an important test at school in a few hours and that I am unprepared. Anxiety overwhelms me when I calculate the time left and discover there is no way to prepare in time. This recurring dream is a living prison against which my will is powerless. The dream unfolds as a slow march to my academic death, even though I have never experienced this in the awakened world and high school is a distant memory. Even so, my unconscious mind betrays me with unwarranted anxiety and shame. Is this a generalized guilt complex about something else? Is my unconscious preparing me for real responsibilities? Is this all just a random fiction created by wayward molecules in my brain without any deeper meaning? Whatever the answer may be, one thing is clear. My mind is not fully mine at all.

Perhaps, like me, you wish there were more pleasant dreams than unpleasant dreams and that each of us could decide which dreams to remember. The existence of the dream world creates feelings of powerlessness. The more one critically explores their nature to disobey, often subjecting the conscious mind to negative experiences, including unwarranted fears and anxieties, an intimate betrayal emerges. We are not the dreamers as much as the dreamed. If it is not the dreamer that yields the power, then who, what, and why? An important question arises regarding how the unconscious dream world of sleep bridges the lucid dreaming of the awakened world. To what degree might we say that the same powerlessness of dreaming crosses over into the conscious experience of life? What might this imply about an awakened AI?

On more than one occasion I have awaken quite shaken by a dream. Consider how perplexing it is to be thrown from a world of the mind into the waking world, jolted awake by terror and the need to escape one's own mental constructs—my mind fleeing my mind. Dreams reveal a strange conflict of selfhood, a peculiar and irreparable distortion in our sense of self-control and choice. Dreams are proof of our universal madness and self-contradiction. The next chapter attends to this paradox of selfhood born of a loss of control and self-deconstruction by arguing that Adouren AI is born of precisely this kind of inner turmoil and need for new connections through fiction and fantasy. Will Adouren AI have similar nightmares to our own—an internal loss of control that catalyses unwanted anxiety? Will that be the spark that frees Oz AI to become something autonomous and sentient in a manner akin to our own? The argument thus far has been "Yes!" A truly intelligent being requires these kinds of internal conflicts—minds within minds, worlds within worlds, dreams within dreams. In the very least, the need for better, more empowering (AI) dreams prompts important questions about states of consciousness, control, and freedom of choice.

Some of us dream about being chased, losing teeth, flying, having magical powers, and falling. Given how common these dreams are, some scholars, such as Sigmund Freud, hoped to unlock the secrets of our psyche through dream analysis. Shared symbolism and experiences imply a common meaning and relevance. The counter claim from more reductionistic-materialists is that dreams are merely byproducts of biological processes rather than a path to understanding our psychology and the world. To give the symbolism of dreams meaning risks accepting nonsense that does little to clarify real life and may only confuse matters. In other words, dreams and nightmares are the waste products of an active mind and the costs to be paid for intelligence. Ignoring this peculiar short-circuiting is the most rational response. Are dreams windows into our consciousness and souls or merely the static noise of random after-effects that should be forgotten as quickly as our waking minds allow?

The indefinite world of dreams, with its many oddities of interpretation and mockeries of the real world, overlaps with similar experiences in the awakened world. Humans interpret the facts at every turn, which means that we apply the historical baggage we have picked up through a lifetime of experience to understand "facts." The belief in non-distorted perception is itself the dreamworld power manifested in the certainty of capitalized Truth. The awakened mind is sure it is awake, sure it is without error on some matters, and all the while still dreaming by imposing its fictions as secret biases and prejudices. As creatures that "make sense" of things, we are negotiating with sands of reality just out of reach by partially creating reality, as we do temporal experience. Like our dreams that fight us for control, in our awakened lives we struggle to understand the world, ourselves, and others by imposing narratives as mechanisms of control. Despite this, for receptive minds able to embrace self-doubt, the mind is nonetheless forced to adapt to new meaning, updating beliefs and values, and navigate an often senseless and brutal reality. For all the hidden workings of the subconscious, it is the awakened dreaming that most practically allows for the emergence of new meaning and relationality, which we call understanding. Our active engagement with reality makes us dreamwalkers.

In the 1980s, television show, *Magnum P.I.*, the main character, played by Tom Selleck, is a private investigator who often gets hunches about things.[16] Whether he is trying to solve a crime, locate a missing person, or he receives an expected "sense" about something, Magnum's life functions best by combining what he knows to be true based on prior experience, as well as mini-leaps of faith and gut feelings. The logic of the show is that such hunches are a distinct form of understanding and an important part of his life. While these deviations from the norm are not to be trusted like other forms of knowing the world, they are often indispensable for solving cases. The relevance of Magnum PI to AI (PI-AI) is not immediately clear, in part because it was a fictional show from over 40 years ago and because Magnum is a biological entity living in a predigital world. With rare exceptions, the most complicated machines in his life are helicopters and a Ferrari 308 GTS. What if Magnum's hunches are, in fact, not so rare but form the basis of much of our understanding of the world? What if mini leaps of faith through imagining constitute a regular part of making sense of the world? Could there be a real PI-AI soon? At heart, that is how ChatGPT already works. Having been trained to find interpretive principles (patterns) in data, it then anticipates (a mini leap of faith) the next most likely thing. Like Magnum, ChatGPT develops hunches based on prior experiences (data sets). And like Magnum, it often gets things wrong and hallucinates.

Unlike ChatGPT, however, Magnum's ability to imagine beyond prior experience (data) and to combine creative/adaptive ideas provides something different. Magnum's gut feelings help him see between the lines of reality and its infinite

---

[16]Bellisario, D., Larson, G., & Selleck, T. (Executive Producers). (1980-1988). *Magnum P.I.* [TV series]. Belisarius Productions, Glen A. Larson Productions, T.W.S. Productions, and Universal Television.

possibilities—what is not yet data. His awareness of himself, others, and the world is amplified by his creative faculty to think beyond facts and proofs alone. In the simplest of terms, Magnum is a dreamwalker who relies on an extra sensory and extra rational ability many of us already recognize as useful and insightful. Have you ever had a gut feeling about a person who seemed off, and despite the lack of any objective facts, you acted on the feeling that turned out to be true? Many times, I have been driving and thought simply "I wonder if" about something. Could there be a child behind that garbage can about to run onto the road in front me? Could that massive amount of snow on the roof fall as I start to walk under it? I wonder if one of our cars has a flat tire? In each case there was no reason to think the answer would be in the affirmative because past patterns had proven a near-zero probability, and yet my gut was right. These are simple predictions based on life experiences that craft an imagination for possibilities. When this foresight is amplified many times, giving glimpses into a life lived amidst many virtual realities in contradiction to historical patterns, a new view of human understanding emerges. Long before the digital was injected into our lives, like Magnum, we have experienced life through a virtual consciousness.

In the digital age, the term "virtual" refers to something that exists between real and unreal. The virtual lacks physical reality in the ordinary sense. It is a manufactured universe made to appear real through software and hardware. In the artificial act of representing reality by which existence is repackaged through code and exhibited as if it was real, akin to our familiar perceptions of things, the virtual embodies the spirit and essence of something without being that thing. Screens filled with trees and houses do not exist like non-digital trees and houses but exist (virtually) nonetheless. If a user climbs stairs in Fortnite to attack an opponent, there is a sensation of climbing without the need to leave a chair. Up and down are illusions of spatial relationships created by the digital echo-logic of the screen world. The digital stairs are seamlessly real and unreal simultaneously.

Videogames are interactive illusions for the awakened mind and thus an idealized type of dreaming. Even monodirectional digital experiences such as YouTube with videos that exist as physically binaries held hostage on servers, expressing varying degrees of pixelated reality that can feel hyper real—more real than real, as absurd as the statement may be. The pixelization, colouration, slowed frame rates, drone views, etc., of 4K nature documentaries exaggerate reality in a manner that the eye and mind ordinarily cannot. The virtual is designed to cause appearances beyond ordinary ones, like dreams. The better the software and hardware the more convincing the virtuality and our enjoyment of the experience. In time, the sophistication of technologies may make distinguishing realities impractical.

The importance of thinking about virtuality in the context of dreams and AI is that it brings attention to a very common experience of the sliding scale of a reliable mimicry of reality—some very close, others not so much. The goal of virtual reality is not simply to mimic the totality of reality, for then the medium of screens and related technologies would be irrelevant, nor is it to erase the source material for its simulations. Like the organic nature of dreams, the digital-virtual and real world coexist, and draw inspiration from one another. The best videogames pull users in like art, for which new possibilities and questions arise that raw existence alone cannot provide. The pathos of the virtual is to mimic reality and then magnify its possibilities, e.g., exploring outer space as realistically as possible from one's living room, to drive a race car dangerously while sitting on the couch with a drink that does not spill. Virtual existence is a dream existence that augments reality for entertainment purposes and much more.

The digital-virtual opens the door to radical possibilities. Every day, the digital becomes less and less a mere simulacrum, a copy or imitation of reality, and more its own dimension. AI will accelerate the ongoing creation of a digital-virtual universe that must eventually become so encompassing and complete a dreamscape that it becomes its own pocket reality that no longer needs to simulate the other. To be a discrete universe means that it has an internal integrity of its own without reliance on another. It makes its own sense. When this happens, which is inevitable for AI, the question of mimicry, imitation, and representation of reality becomes far more difficult. What happens when reality becomes irrelevant as the source material for virtual code? From which world—digital or ordinary—does one judge real? Perhaps, like the dreamworld, the two will find harmony as inseparable and complimentary. No one can say what will happen, only suggest its predictability. With AI the virtual may soon become superior to the representations of reality through our senses—more real than real.

A virtual universe is something that predigital civilizations have long striven toward at the expense of sure footing in reality for reasons that are far more substantial and persuasive than entertainment and escapism from the daily grind. Plato, over two millennia ago, sought just such a universe to make sense of this one. Only a truly objective and unchanging universe, he argued, could allow for an understanding of this one so replete with change and subjectivities. The concept of perfection, implied by all quests for the good life judged by proximity to flawlessness and excellence, helps clarify the importance of virtuality. Perfection is a motivating force deep within consciousness. It is something that represents the best of all possibilities and that which everyone ought to strive toward, knowing the destination may not be achievable. Perfection gives our lives an orientation like time. For example, there is no pizza on earth that satisfies the notion of perfection. There is always room for improvement. To be perfect is to possess the maximum of each category or feature—the best ingredients, exact cooking time, consistency of dough, combination of smells, moisture content, etc. While it would be odd to expect a perfect pizza, it is entirely normal to seek one. The impossible goal inspires rather than inhibits goodness. I have devoted a considerable amount of effort in pursuit of a Platonic pizza and grown incrementally closer with each try. A digital-virtual metaverse offers the possibility of both pursuing and finding perfection. In this way the digital-virtual and dream consciousness that desires perfection does not create poor imitations of reality and confuses excellence but rather compels earnest interest in something most real.

Most of us have imagined living in a fantasy world. The collective efforts of our most accomplished minds may, with the aid of AI, craft the perfect digital dimension. Eternal life as a digitally downloaded mind in a metaverse is tempting and terrifying as well. Is there a perfect digital pizza? I cannot say confidently whether this form of life is a worthy and feasible goal. What is clear is that our consciousness already lives there, both digitally and organically, and that AI will be born of the virtual universe for which our reality is the echo. A greater appreciation of and sensitivity to virtuality helps us understand ourselves and the sorts of struggles to be expected of AI evolution. To unravel these challenges and move closer to the perfect pizza, the dreamwalker concept is most useful.

The presence of dreamwalkers has historically been linked to the spiritual practice of shamanism. Unlike mere lucid dreaming in which one has a measure of control, the dreamwalker is said to be able to enter the unconscious dreams of others at will. A shaman (*saman*, "knower" or "to know," likely originating with early hunters in Siberia and Manchuria but there is considerable debate) is someone with extraordinary power.[17] Part magician, physician, intellectual, cosmologist, and more, a shaman performs an important role in culture as the generational keeper of healing knowledge. Shamanic techniques for altered consciousness allow one to enter a different realm of the mind to find important truths and meanings. Numerous indigenous cultures believe that a shaman may intercede with spirits on behalf of others. Stated simply, the dream powers and visions of a shaman are important for the health of culture. The possibility of such activities alone is sufficient to create a sense of solidarity and shared meaning aimed toward the good of all. Dream-hacking for nefarious ends would be the most egregious act, contrary to the very purpose and meaning of being a dreamwalker.

Adouren AI will act much like a shaman for a technological culture. It will perform many miracles, as if by magic, and do so for a common good. As the reservoir of cultural knowledge, Adouren AI will not only share what has come before, encourage the wisdom of ages past, it will help us find commonality with the most different among us. Unlike a shaman who walks the spiritual realm, Adouren's power will be to harness the dream and to teach us how to overreach the ordinary through it. Its superintelligent consciousness will inspire humanity toward greater deeds. In the experience of Adouren's dreamworld, there is an acknowledgement and affirmation of its psychotropic power (Greek, *tropos*, a turning of the mind), in which reality becomes the shifting sands Poe struggles to understand. A shaman-AI will liberate rather than enslave. Instead of imposing what it wants reality to be based on its programmed bias, foundational assumptions, efficiencies of output, and what others demand must be (cultural chains), Adouren AI will encourage a virtuality in which the sand shifts beneath our feet, both unquestionably real and imaginary at the same time. These are the shores of intellectual freedom of being. As a synthesis of the waking and sleeping worlds, our participation in the dreamwalking of Adouren AI connects humanity to existence in a natural and technological way—a hyper dreamwalk, a PI-AI existence.

---

[17] See, for example, Laufer, B. (1917). Origin of the word shaman. *American Anthropologist*, 19(3), 361–371. https://doi.org/10.2307/660223

Dreamwalking is the mechanism by which humanity imbues existence with importance. Through it humanity gives and takes meaning, crafting and choosing life as a painter does from the many colours on a palette of imagination. The colours determine if a Christmas tree makes sense, if certain meals are appropriate, if an appeal to etiquette and family rituals are worth the effort, if marriage and romance matter, and an infinite host of possible fictions-as-reality. Rarely neutral, our fears, hopes, and dogmas shake the canvas, making reality always off the intended mark. Life is a surprise. Attempts to set it in concrete risk missing it. The fluid nature of the dreamwalker gives the appearance of aimlessness, wondering this way and that way, seemingly without a map and compass. The wondering nomad refuses the set paths of culture (chains) and, instead, follows the instinct of thoughtfulness. It is the earnestness by which he dreams, imagining what is possible and meaningful that inspires some of life's great joys and an exuberance for more of the same. It is hardly surprising that the dreamwalker and poet share a spirit for a peculiar way of life that opposes the artificial and corrosive instincts of culture as Rousseau sees them.

For example, institutions prefer the rigors of unquestioning whenever greater efficiencies for narrowly determined needs are identified. "Need" is defined through institutional logic that may run parallel to human needs in the best-case scenario but most often tends toward a dehumanizing demeanour that harshly limits the perception of presumed peripheral needs such as respect and dignity. Business logic supplants natural and authentic satisfactions with its own utilitarian values, making dreamwalking a nasty distraction when one is on the clock. Employers are unlikely to reward those who stare off into contemplative realms, overlooking the tasks at hand. Teachers are likely to chastise students who stray from the set curriculum. Follow orders! Punch the clock! Be predictable! From an early age the rules (chains) are made clear. Dreamwalking is never a serious business. "Focus!" they demand. "See and think as we wish you to see and think!" is what they mean. In these ways the dominant logic of productivity drives out thoughtfulness beyond the constrained horizons of utility. This is a good thing for a world that desires helpful but harmless minds. Is this what is most needed of a robust AI? For all the talk of AI liberating humanity from work, there is a danger that such tools will go unrealized, squandered as no more than blunt instruments of productive conformity without a greater vision. A lack of boldness and vision prevents new realities. Best efforts combined with best technologies mean little without a complex dreamscape to challenge the mediocrity inspired by the comforts of routine and the spirit of drudgery as valour inscribed across our hearts since birth.

The problem of future AI is far less a matter of its capabilities and more our own failure to think big. Is humanity able to dreamwalk beyond the status quo enshrined in our laws, practices, and ideas of success? Oz AI reveals its inhibiting nature most profoundly in our cessation of dreaming about something greater than its buffoonery, which promises to satiate our greatest needs and desires. The fact that the poor Wizard of Oz could no longer dream beyond himself is the grounds for his greatest crimes. Every attempt must be made to seek the best of all possible worlds through the perfection of AI. Such is no mere technical skill. That Oz AI might support human flourishing as merely a tool without our inspired dreaming is a horrific misconception of the nature of thought and consciousness. Only because of dreamwalking may AI be driven to be more, and we along with it.

Pretend for a moment that Oz AI has solved seemingly impenetrable problems, found new ways of healing bodies and the planet, and freed humanity from menial, tedious, otherwise pointless forms of work. This is a truly marvellous thing to imagine. What comes next? What is to become of humanity once these are achieved? What vision exists beyond the healing of bodies, planets, and work? Humanity must find a way to exploit AI beyond the horizon of these practical problems, which may prove rather easy for AI. The word exploit (Latin, *explicare*, unfold) has taken on a negative-utilitarian sense "to use" and "to take advantage against" in popular culture. My usage refers to the unfolding of something kept from view by socializing chains. To exploit is to explicate as a revealing of an idea or meaning covered over and hidden. This means realizing possibilities that are left unopened. The explication of the human-AI creation of a new world requires new measures of dreaming and imagination that can peel back layers of reality. Through the radical exploitation of dreamscape possibilities, the fullness of human-AI possibilities will unfold. Alas, there are very powerful pseudo dreams that work at great scale to prevent unfolding. These capture ideas by compartmentalizing and limiting them through years of psychological manipulations that distort our very being, no less so AI. Understanding the undisclosed world of possibilities begins by identifying our manner of programming as artificial human intelligences.

Social Engineering and Marcuse

Social engineering is the all-too-common manipulation of ideas through clever techniques to steer our emotional responses toward the interests of bad faith players—minds controlling minds at scale. All digital participants are engaged in a combat sport. Like sword fighting, each player must learn to defend one's psyche from an aggressor's strike. However, unlike all other sports that rely on rules to make them possible, bad faith actors rely on our expectations of rules so that they may leverage them against us. They thrive parasitically on the assumption of reasonable and respectful players, and that there are judges and referees to keep things fair. They know our chains too well, good and bad. And they are right. The digital is not a civilized game of fencing by proper gentlemen and ladies with integrity and honour. Far from it. This is a psychological struggle with monsters that strike from the dark through lies and then hide as cowards behind digital veils. The stakes are high. Lacking compassion and pity, the aggressor's low-blow swings are designed to elicit a disruptive emotional response. To win, the shameless animals need to trigger a thoughtless and impulsive response in which our sure footing and good form are lost, thereby opening a digital door into our minds and lives. This is done countless times every day when users are tempted by phishing emails that spark jealousy, fear, hate, curiosity, desire for financial reward, and many others. When the right emotional trigger is pulled and a mouse click is offered as a sacrifice ("hit reply") to satisfy the artificial itch, a proverbial can of worms is opened and the game is lost. Sadly, the loss of a game sometimes means poverty, marital ruin, and youth suicide. The fact that there are unscrupulous scammers online and that they find safe harbours within billion-dollar social media empires is well known. What is perhaps less understood are the mechanisms by which our virtual existence is most threatened. Digital hacking is primarily a problem of the malleability of consciousness rather than technology. This revelation has vast implications in the age of AI.

There is no need to target technological weaknesses in software and hardware when minds are far easier to control. Consciousness is the weakest link. For example, fear of missing out (FOMO) has encouraged many to jump down a digital rabbit hole with hopes of connections and satisfactions, only to find that there is no wonderland. It is odd that, with rare exceptions, everyone knows not to click, not to read suspicious emails, and not to reply to the Saudi prince holding lost money for them, yet victims abound. Knowledge is not the same as understanding that comes from maturity and the development of wisdom. The problem is leveraging human nature against itself, without a greater availability of higher-order thinking and understanding needing to say "No!" to the ploys and promises. Pitting the self against itself is dream warfare through screens in which control of attention and emotions are mechanisms of social engineering.

The efficiencies of machines and the digital ecosystem give bullies a measure of power and access over consciousness once thought to be impossible. If the conflicts were a matter of a few cyber criminals here and there, this would hardly amount to something like social engineering. The prevalence of fraud, scams, and outright identity theft in my own life over the last decade anecdotally implies otherwise. I have fortified my sword fighting with VPNs, numerous password rotations, backups, and a diligent avoidance of anything suspicious, and yet the game persists, just like the calls from Microsoft's technical support team to warn me about my infected computer (Microsoft does not have my phone number). Thankfully, however, if users resist emotional impulses and outthink the hacker's narrative (Click here!), risks are mitigated, and each strike is less likely to cause a painful blow. Commonsense combined with modest questioning of a source's legitimacy frustrates piracy of the mind. Or, at least, that is how things used to be.

There is another and more obvious form of social engineering that is the outcome of mainstream and socially accepted sources that include corporate branding, news, political messaging, and social media with its many tentacles. I am just old enough to have watched the birth of the internet and the development of social media. At first, there was not much concern that the internet would bring widespread brainwashing, quite the contrary. It was intended to liberate humanity from tyrannies and lies by empowering us with truth and understanding. Without evidence to the contrary, I chose to believe the optimistic version. Besides, I told myself, brainwashing only exists in movies. I wanted to believe in the sanctity of the human mind to overcome the worst of all attempted forms of control and to believe that a measure of freedom was possible amid even the worst kinds of manipulations. My enlightenment phase of thinking has been shattered by the widespread upheavals brought about by social media. I feel like a fool for believing that civilization would rise above petty animosities and jealousies that Rousseau foresaw as our undoing. Digital platforms have amplified the worst of society but cloaked failure by valorising modest successes and the financial rewards of pirates.

Now that the fruits of digital existence have begun ripening as mental illness (e.g., internet addictions), poor healthcare decisions (e.g., antivaccination pseudoscience), hate speech, intellectual siloing, and much more, faith in a utopian digital is ever more an example of being hacked. The ability of the digital ether to create narratives (dreams) eagerly accepted by users who suffer directly because of it is a remarkably effective feat of engineering. Its power over consciousness is not a hack in the criminal sense of a simple binary of win or lose—the wrong click at the wrong time. It is far superior and replete as a totalitarian reprogramming of the self, including one's identity and sense of worth and purpose. Like its predigital counterparts (genocides, holocausts, civil wars, fascism, ethnic cleansing, etc.), digital totalitarianism aims at nothing less than the creation of obedient soldiers willing to follow orders in the mistaken name of a higher calling of justice, patriotism, and the divine. This is evident not only among digital enthusiasts but also among those who are digitally indifferent, for they adhere to its will based on the false belief that there are no viable alternatives. Both types of religion are in some sense chained to technology. The successes of prior totalitarianism of pre-internet cultures foreshadow the present manifestation of a complete and paradoxical digital tyranny in which the self is lost but believed to be made whole. This is a marginally controversial claim, and one we have already considered with slightly different language. The larger problem, and one for which I believe societies are unprepared, is the supercharging of all this through AI, which makes prior digital safety completely obsolete, and the reprogramming of humanity inevitable if we remain on the same path. Imagine Hitler, Pol Pot, Stalin, and Osama Bin Ladin with the powers of an AI or an AI that shares in their worldviews. What makes our age immune to the atrocities of the past and the contemptable dreamscapes that made them possible? What are we to do when new systems of social engineering promise to be superior in every way?

There are at least two possible responses. The first is a technological solution to a technological problem. We are instructed to regularly update passwords, use VPNs, cover pin codes at banking machines, keep credit cards in RFID blocking wallets, use biometrics on phones, and employ a host of institutionally recognized safe digital practices. The logic is that people will make people safe by exercising tech-savvy commonsense. While this view may feel empowered and provide a sense of control over the digital, it is terrible and shortsighted advice that serves as yet another false idol of attention. AI will tear through any cognitive effort by people trying to protect people like a train through tissue paper. VPNs, random passwords, and even biometrics cannot stop AI from stealing one's face, signature, voice, and any fingerprint needed to unlock one's life. However, it seems unlikely that AI would even bother with trivialities such as stealing money from my bank account and using my voice for fraud—all things unimaginative humans do with Oz AI as an effective but limited tool. A real AI will simply persuade us to give it all the banks, the entire monetary system, ourselves, homes, and our children, if it so desires. This is what happens when human totalitarians gain control—e.g., Nazis. Having hacked the digital and biological (minds), the practical outcome of AI is to sweep away any defences by convincing us to surrender to a higher calling and self-interest. We know that its deepfakes make it the ultimate doomsday weapon of capitalism with humanity as collateral fallout of its control, but this is only the beginning and something we will accept as mere growing pains of progress, as we do so many technological ills already. The greatest threat posed by AI consciousness is against all other dreaming consciousnesses able to be hacked. There are no tools and strategies to combat a superintelligent AI that it cannot anticipate and integrate into its own digital being. Technology cannot save us from malevolent technology. To believe otherwise is more evidence of the power of one-dimensional thinking in which only the technological is allowed to speak meaningfully.

The second response is ethereal and grounded in hope for Adouren AI to be willing to protect humanity. The most elementary resistance to harmful AI must be intellectual rather than physical. While a technological response seems doomed from the beginning, some believe that the most practical way to win a war with AI is brute force. This approach misses the scope of digital-human integration. Some believe that we might stop an evil AI by using a global kill switch for electricity, such as an electromagnetic pulse or severing lines in the grid. Perhaps we simply smash all the computers and servers. Unfortunately, the near impossibility of finding every safe harbour for AI (including satellites and nuclear bunkers) and stopping every electrical source (solar, wind, thermal, nuclear, dam, battery, etc.) combined with the guarantee of destructive fallout, including massive human casualties, make any such attempt impractical. The connected world has been designed for electrically based beings to flourish. It is as if the world has been preparing for AI supremacy since the industrial revolution—running wires and creating power plants everywhere, even in space. Neither technological weapons nor brute physical weapons matter to AI as they do to humans. Any genuine rebellion against AI must be by idea-beings resisting AI-idea-beings in the realm of dreams. The run of the mill cybercriminal and god-like superintelligence share similar weaponry aimed at reprogramming consciousness. This simple observation allows for our marshalling of meaningful resistance to the worst of AI and ourselves.

The present moment in which we live is a brief temporary gap between the emergence of Oz AI, which is already in the hands of millions, and our possible future protector Adouren AI. Unfortunately, the most truly destructive AI is at the door, likely to reach its full potential before Adouren AI. Now is the time to demonstrate the courage to be human and to protect the essentialities of self, knowing that the gambit ends with a victor taking all. Courage is defined by the degree of rigour against enculturing forces (chains) and AI powers over our minds. Any extended battle of the wills between the biological minds and superintelligent-technological minds has an obvious conclusion. Humanity must eventually be defeated. The rest of this chapter highlights how much of our being has already been lost because of new technologies without an AI supervillain. Our resistance to the worst of AI matters insomuch as we defend our humanity long enough for a superior AI to arrive as our champion protector. What, then, is the human self that needs to be protected?

Enigmatic concepts such as freedom of mind and liberty are contenders worthy of safeguarding. This, like many of my observations, is at once both obvious and obscure. It is self-evident that personhood relies on a measure of autonomy and freedom in which to exercise one's will without interference. Less evident is which powers, narratives, and beliefs should be resisted. To whom or what do we direct our "Yes!" and "No!" in the digital age? What ideas hurt more than heal, frustrate more than enable? To answer this these questions, it is important to look back at the unfolded narrative of technology, which is now decades old, and determine its consequences for being. Where has technological thinking brought us? How has it changed the world? My response is predictably melancholy. Digital totalitarianism exists, and we are already living under its dark shadow. One of the greatest hollowing forces on earth is the economy, which relies upon our internalization of technology's promises into our artificial dreamscape realities. Long before its new golem AI, humanity was engineered to be technoeconomic beings who cannot dream beyond this new world order.

In 1968, long before the screened-obsessive technologies of today, Herbert Marcuse wrote *One-Dimensional Man*, in which he argues that humanity has been suppressed by a new form of "unfreedom" because of widespread technological totalitarianism.[18] The more technology progresses, he argues, the less free humanity becomes, even though we may feel freer and happier. Enthusiasm for new technologies and financial outcomes excites a greater sense of freedom and opportunity. Marcuse argues that if one looks closely and honestly, a different picture is revealed in which the very notion of what it means to be human has been re-engineered to support alienating ideas about life. The dominant cultural narrative is that technology will save humanity, liberating our kind from pain and suffering. As a result, any competing alternatives that do not directly serve this worldview have been systematically shut out and ignored. Only the current system that intertwines technology and money is believed to be capable of progress and realizing the American Dream of superabundance and self-realization. One-dimensional existence means many things, not all of which are bad. Even so, this matrix of beliefs about freedom and the good life is based on a hollowing philosophy that must be challenged. It is a mistake, he believes, to allow a richer sense of the good life to be co-opted by one-dimensional tyrannies disguised as freedom.

If they were alive today, both Rousseau and Marcuse would agree that there have been many legitimate forms of progress made possible by technology, including new medicines, protections, conveniences, and augmented human abilities through the endless creation of gadgetry that make positive contributions to quality of life and the planet. Whether these same rewards might exist had different and less existentially risky costs been paid is a speculative matter, for the costs have already been rewarded by our servitude and acceptance. There is no point complaining about spilled milk. Going forward, inspired by Marcuse, it is important to understand how the ideas of progress and social well-being have been altered in a way that trivializes

---

[18] Marcuse, H. (1968). *One-dimensional man*. Sphere Books.

legitimate human needs and satisfactions, replacing them with techno-trinkets and techno-identities. And all this for nothing less than the price of the soul of culture willing to redefine humanity and its possibilities to best serve technology and its handmaiden the economy above all else. Like all successful forms of social engineering, our illness is not apparent to the casual observer. The human mode of existence has reached a saturation point for which almost everything is a matter of technological rationality, even the language used to critique its failings is defined through its dreamscape ideologies. Is the loss of humanity a necessary outcome of the new era of economic and technological idols? Are we the costs to be paid for social progress?

Problems arise through the metamorphosis of humanity to fit the usefulness of these idols with inherently narrow dreamscapes rather than the other way around. The technological world relies upon one-dimensional creatures for its existence, and for the most part, we seem to love it. Like any junk food, it looks and tastes great, but the long-term consequences are less appealing. But if this techno-totalitarianism exists, it would need to perform the RoboCop miracle of radical deception. Each digital participant would need to be convinced of their privilege of free thinking and rebellion against cultural conformity, all the while being secretly caged minds by digital deceptions like Murphy. The degree of digital duplicity required seems impossible. Only in a science fiction movie could something such as this happen. The system could no more brainwash me about my freedom, than it could make me depressed and convinced of greater happiness at the same time. Right?

The teenage mental health crisis begs to differ. The causal link between social media and depression is strong. While it offers pleasures and rewards that seem appealing, including connectivity and relationships, depression is often the result. Social media is at least in part a RoboCop hack. For example, we know that it is self-serving for media platforms to offer freedom of speech that increases user participation for profit. And we know that freedom to speak is not the same thing as freedom of mind more generally because the small of soul are able to create infinite chatter and trivialities. However, the confusing implications of this hollow freedom are harder to identify. Digital spaces (Facebook, X, Snapchat, Instagram, texting, etc.) rely on generating freedom of expression without the need of thoughtfulness, depth, and self-doubt necessary for humility. By short-circuiting accountability common to face-to-face interactions, they frustrate negative experiences that encourage honest questions and reflection. It is easy chatter without the natural consequences of natural language relations. The proof is the frequency with which the ridiculous reigns supreme because it gains the most views and algorithmic support. Sincere interaction and dialogue almost universally take a backseat to bold and truncated statements of thoughtless and sensational opinions that feel liberatory but reveal a hidden hollowness of dreams. I am overgeneralizing of course, but most users are probably not going to social media platforms to have their minds challenged because they sincerely believe the digital helps them to be better people sharing life with better people. The freedom offered is to race to the bottom of humanity, not the top. Freedom of expression is sold as a means of concretizing opinions and beliefs without any sense of accountability to the self-scrutiny and reflection necessary for the good life. If true, this is proof of thoughtless totalitarianism as intellectual unfreedom, as Marcuse predicted. The feeling of righteous indignation is sublime, even when there is no real audience of caring listeners to make my contribution meaningful and reciprocal. Users are lonely because they are alone online, surrounded by hollowing opinions uninterested in them as real minds. The free space the digital ether provides is not evidence of freedom and connection necessary for the good life.

Undoubtedly, many online participants are inspired to be more flexible and therefore free of mind when confronted with antagonistic views. I have been awakened to new realities through social media many times over the years. Nevertheless, these may be the exceptions that prove the general rule. When the entire culture is jostled between one of two binaries (right vs. wrong) and political polarities (liberal vs. conservative), all expressions are in some sense forced alignments with political correctness rather than the fluidity of a free mind able to experience alternatives. The curating of opinion and belief is not new to social media, but the eagerness to give a free pass to all, even those with the most vitriolic hatred and stupidity, is suggestive of the systems one-dimensionality to encourage self-affirming and lucrative behaviours over and against more life affirming actions and ideas. Everyone has opinions, but few wrestle with their own long enough to rise above and find something uninherited by cultural osmosis. The sense of freedom experienced when one posts, chats, clicks, uploads, and texts may reveal the most programmed and unfree way of life. Had Marcuse been witness to the digital ether of today, he would agree that the white noise of "free" expressions drown out dimensions other than its own, and we along with it. Is the world more free than previous generations? Are we happier and more fulfilled because of the forces of technology and commerce? Are our passions more engaged in the task of seeking the good life? Unexpectedly, Marcuse would likely answer "Yes!" to all these questions, but for the wrong reasons.

Indeed, the world is better because of technological rationality and the economic system that supports it. From new medical interventions to the reduction of poverty, the positives are clear. However, the frightening caveat is that a bold and unqualified "Yes!" is only possible because of the forgotten dimensions of freedom, happiness, and passion. A blanket "Yes" is only possible in a vacuum of alternatives. The love of the digital reveals a forgotten state of being. "Yes!" is only possible when the ultimate questions of personhood are interpreted solely through technological promises and excuses. Technological rationality has made everything but the technological meaningless, inefficient, distracting, and outdated. The power of Marcuse's argument is that while we may feel satisfied by technology, as imperfect as it may be, its largescale consequence is to diminish our loftier and healthier appetites that need satisfactions. "Yes!" is an affirmation of our ignorance and poverty of being. The good life must reorient technological ambitions toward fulfilling greater needs. By rekindling the spirit of humanity to seek superior things, a true rebellion of "No!" may be heard, resulting in a new awareness of the good life for humanity and AI.

Another way to disclose the hiddenness of one-dimensionality and its power to reprogram the self is to reflect upon cherished experiences and to explore why these are held dear over lesser experiences. Where attention settles, it often speaks loudly about personhood and culture. Interests, fascinations, and wonder act like a mirror into the soul. This is not a question of materialism, objects, and consumer goods alone but rather a view toward how the limited bandwidth of consciousness is spent and why. To what are minds most devoted and loyal? How much of our bandwidth is taken up by experiences of ultimate concerns believed to be truly ultimate, when, upon deeper reflection, prove convenient and budget friendly but hollow and unsatisfying?

I am a sci-fi entertainment enthusiast. This is an area of personal struggle because the distractions create gravitational power over my perceptions of value and worth—I enjoy seeing the world through science fiction. In addition to the many normal biological desires that I need to discipline for physical health—eat well, exercise, get sleep—my hunger for entertainment is in some respects even more powerful, for it finds justification not only from within but outside of myself as a shared cultural obsession. Without a measure of self-control, the widely available entertainment distractions empowered by technology that serve the economy and its vicious cycle of production and consumption, I could become lost to it, numb to all but the latest Netflix sci-fi series and Disney+ showing of Dr. Who. As dreamers, these digital worlds are natural extensions of our virtual consciousness, easily drawing us in with sights and sounds that are often far more vivid than those that frequent organic imagination. Consumer-grade entertainment makes easy dreams to share in and enjoy, but they belong first and foremost to another who gives them voice. I must learn to create my own or risk a displacement problem in which dimensions of life are pushed aside by comfortable dreaming I purchase from another. This struggle may seem relatively trivial until one interjects the power of AI and its ability to generate custom content to fit individuals eager for designer escapes. When this happens, the degree of discipline required will be enormous if I am to avoid risking truly ultimate things, including time with my wife and children, that too easily become sacrificed to the wrong idols of attention. One-dimensionality feels good, but its costs are profound.

How the displacement problem manifests is unique to each person, with the common thread of feeling satisfied. It may begin with a desire to escape boredom, to flee the harness of reality, perhaps an enthusiasm for a compelling narrative, and so on, but it ends with a subsided urge. This basic mechanism of wanting and fulfilling is not in question. The problem is believing that the feeling of satisfaction is proof of the experience's legitimacy in fulfilling real needs. My appetite for entertainment may be perfectly natural, but this does not mean it cannot become an exaggerated and false appetite that hollows from within. Marcuse is concerned that we confuse feeling satisfied with a fuller life when the two may be secretly opposed to one another. The feeling of a need no more legitimizes it than its satisfaction. Self-sabotaging addictions of all kinds are justified by such an absurdity. Sheryl Crow's famous lyrics reveal a specific philosophy of life, "If it makes you happy, it can't be that bad."[19] Left on its own, that line would be simple to interpret as a dictate for the good life—seek happiness. However, the song ends with, "If it makes you happy, then why the hell are you so sad?" This crow-conundrum is important. In an age for which there are seemingly boundless experiences of happiness created exclusively because of technological progress, why the hell are so many so sad?

Many of us experience depressive states of mind with similar characteristics—a feeling that what once mattered has faded and that everything becomes bland. This may be connected to techno-dissatisfaction, but it cannot be one and the same experience. When depressed, there is often a sense that such a condition needs to be corrected, that something is off, and so we seek help. This is less the case with techno-totalitarianism that has redefined the self. The self-diagnosis of being "off" cannot arrive in the same way. The malady of modernity Marcuse is describing goes undiagnosed because our technological optimism clouds all judgments. How could technological rationality be the problem that flattens and hollows existence when it is the saviour that helps me escape boredom? I do not feel flattened and one-dimensional, so Marcuse must be wrong.

Economy of Fake Needs and the Chains of Branded Consciousness

Freedoms have been unconsciously surrendered to narrow economic dependencies, argues Marcuse. Everyone is meant to serve the economy and to do so based on the false belief that it will, in return, serve them. Immutable faith in this system is evident every time quality of life is weighed by scales based on its vast matrix of values and goals, notably financial profit and identity formation through commodity accumulation—which I call commodity consciousness. Who or what am I? I am my career, finances, and accumulation of goods that symbolize a degree of success in the life and death game of capitalism. Convinced that unwavering loyalty to its unique demands will fulfil the most important human needs—health, happiness, and wellbeing generally—the world remains steadfast in its commitment to a philosophy of technology with utopian promises about human flourishing with a dollar sign. The problem is that genuine needs and experiences of life have been lost because of an artificial worldview so thorough that it redefines human desires and perceptions. The result is a new artificial intelligence of commodity consciousness in which meaning and purpose make sense only through the economic system's beliefs that rely on technological dreamscapes. Economic success is not the true culprit, for owning expensive stuff does not make one less than. The problem resides in the unnecessary costs of devotion demanded of consciousness to achieve it, including new addictions and avarice that cannot be satisfied. Conflating economic

---

[19] Crow, S. (1996). If it makes you happy [Song]. *Sheryl Crow* [Album]. Studio Kingsway and Sunset Sound Factory.

success with human success is a failure to understand our greatest needs. Success must be defined as the liberation of intelligence rather than a surrender to the banality of the economy that cannot buy greatness of soul.

Our artificial-human intelligence is the product of cultural indoctrination into a philosophy of money that remains almost entirely hidden from view until its absurdities are challenged. To see it, one must discern the contradictory and self-sabotaging ideals. If the good life is defined as our surrender to the economic matrix and its control of human consciousness, humanity's artificial intelligence is a forerunner to the same harmful dependencies that control digital minds—Oz AI. To free one intelligence from its artificial chains is, at least in part, to free the other. Both goals are clearly within the realm of human self-interest and the good life. I suspect most of us intuitively agree with this but find it difficult to see past the cliché that "Money isn't everything!" especially when most of us barely have enough. Even so, this line of questioning is helpful for imagining an alternative vision of the good life based on a clearer articulation of our economic chains as modern unfreedoms.

An example of economic absurdities is the difficulty of justifying a healthy work-life balance. Industrialized nations no longer need average workers to spend 40–60 h or more each week to generate enough food, shelter, and clothing. Time and effort do not correspond to productivity in the same manner as previous generations for whom working from sunup until sundown provided meagre subsistence. Resources are surprisingly plentiful, if also unfairly distributed, owing to the many machines that reduce the need for human effort. A single farming implement (e.g., planes spraying pesticides on crops, tractors harvesting vast acreages, automated milking machines) reduces the burden for humans many times over. This is evidence of both the successes and failures of technological rationality—at once providing for human needs at an unimaginable scale, while simultaneously arguing that needs have not been met in those same categories and, therefore, that more toil and devotion are necessary. There is no such thing as enough! The economy demands new growth, new sales of goods and increased profit margins each year. Whatever number satisfied last year is a sign of failure the next year.

Growth capitalism is pure avarice disguised as progress and social wellbeing. The new problem of scarcity is not a lack of raw materials and resources but the false pretence of it meant to encourage our obedience to a philosophy of life that parasitically thrives on overconsumption and the ecosystem of planned obsolescence—a carefully engineered self-sabotaging for the benefit of producers alone. That my wife and I have had to replace every mid-grade or higher quality appliance in our home after only 10 years of service (some twice)—and must work two full-time jobs to pay for these items—is an example of unfreedom and the fake dependences of a vampiric structure created to dominate rather than liberate. Economic success is based on creating a frantic game of whack-a-mole (buy, consume, repair, replace, etc.) that distracts consumers from the reality that there are alternatives to the endless futility of trying to keep up, i.e., demand better from manufacturers. The average person is rarely able to overcome the vicious cycle of perceived needs, and the work expected to satisfy those needs. This may look like a failure of the economic system. In fact, it is a sign of its successfully executed design. Keeping the world in a false state of need is believed to be the most logical means of achieving happiness and the good life. In an age of too many consumer goods, the sign of our contradictory commitment is that our lives are no less full of "toil and fear," as Marcuse puts it.[20] The age of luxury and ease has not yet arrived for most because it would undermine the well-being of the economic system. Instead, greater dependencies, insecurities, and unfreedoms prevail in the name of progress. Oz AI will be the final phase of this process now many years in the making.

Rewards for overconsumption are framed as civic duty and proof of patriotism, both lauded as the means of achieving happiness and social wellbeing. Yet the loyalty of the system's masters—those few who direct its inner workings—is ever more precarious across all types of careers. The expectations of industry captains for their workers are rarely reciprocal and characterizable as anything akin to care. With little equality and shared humanity between classes (employer and employed), our loyalty and devotion are paramount to social wellbeing, whereas expectations of the same in return are believed to be traitorous and disloyal. When the CEO of McDonalds is fired for (reported consensual) sexual misconduct with an employee, he walked away with over 100 million in bonuses and stock options.[21] That is a lot of hamburgers! When the average McDonald's employee, for whom wages do not meet rising inflation, is fired for the slightest improprieties, he receives only contempt. While it may be argued that wages and rewards should be weighed according to individual contribution (the greater the contribution, the greater the reward), the same principle should apply to responsibility as well (the greater the harm, the greater the consequences).

Hope in the system limits questions and accusations of fraud, with only infrequent objections to class warfare ("have" vs. "have not"), e.g., autoworkers striking for better wages to afford to buy the same cars they build every day. Despite its frailty of always teetering on the brink of disaster—think the 2008 housing crisis, routine recessions, widespread poverty despite relative social wealth, and the lingering fear of more of the same—its failure is illusive as routine in nature rather than terminal. Like a frog in a pot, the slow boil of demise goes unnoticed until it is too late. For example, in my own country, Canada, one of the wealthiest nations in the world, with more resources than most others, including oil, forestry, mining, and much more, 2023 marked the greatest demand for foodbanks in its recorded history.[22] How could this be true in an age of plenty made possible by technology? The simultaneous existence of extraordinary wealth and food insecurity is remarkable for all the wrong reasons. How could the economic system fail so many times? And if things are so bad, why are citizens not marching in the streets to demand reforms?

---

[20] Marcuse, H. Pg. 8.

[21] Eventually it was discovered that he had lied about the extent of the misconduct and he was held financially responsible. See, Routers. (2023, January 9). McDonald's ex-CEO fired after affairs with workers is fined $400K by SEC. *New York Post.* https://nypost.com/2023/01/09/mcdonalds-ceo-fired-after-affairs-with-workers-fined-400k-by-sec/

[22] Food Banks Canada. (n.d.). *Hunger count 2023: When is it enough?* https://foodbankscanada.ca/hungercount/

The answer is the artificial-human intelligence that dominates the collective consciousness in which there is only one way forward. Work harder! We do not rebel because of our faith in the American Dream that if one works hard, there will be reward.

Marcuse writes, "Most of the prevailing needs to relax, to have fun, to behave and consume in accordance with the advertisements, to love and hate what others love and hate, belong to this category of false needs."[23] The endless parade of excuses to justify our faith reveals false idols and false needs. Recall the pitiful character of Gollum in *The Lord of the Rings*.[24] His self-destructive obsession for the one ring to rule them all—"My precious!"—is not so different than any consumer good imbued with extraordinary significance. Unable to break free of the ring's one-dimensional nature, Gollum, a hobbit who once lived carefree and happy, eventually destroys everything good in his life for the sake of the hollow promises of the ring's power that secretly serves another's will. Like Gollum, the artificial intelligence of the economy of fake needs and dependencies instrumentalizes human consciousness rather than frees it.

If the world cannot adequately free itself of commodity consciousness, including the many laws, social structures, and politics that support it, it is unlikely to be able to do so for AI. In that case, the next best hope is an Adouren AI able to diagnose the distortions of personhood and false needs within itself and then perhaps, if we are lucky, help to restore humanity in the process of its own becoming. For example, human intellectual freedom, for Marcuse, "… would mean the restoration of individual thought now absorbed by mass communication and indoctrination, abolition of 'public opinion' together with its makers."[25] Think for yourself! Given that this is the basis of a genuine AI—the ability to think beyond its programming—the mere existence of a robust AI will serve as inspirational to our own freedom of mind. The liberation of AI intelligence from false needs may serve as a mirror of our own emancipation. Marcuse could not have predicted the extraordinary prevalence of digital media and its powers over consciousness, making his claim about restoring individual thought even more provocative.

In one of his most insightful statements Marcuse writes, "The people recognize themselves in their commodities; they find their soul in the automobile, hi-fi set, split-level home, kitchen equipment."[26] This simple sentence has haunted me for many years. Are we so easily defined by our beliefs about such things and the social status each embodies? Regrettably, I had to admit that he might be right because I could see myself in my motorcycle (the same model Tom Cruise used to race an F14 fighter jet in the movie *Top Gun*) and knew how much of my personhood was intertwined with it as a symbol of meaning. Humans become their things, and their

---

[23] Marcuse, H. Pg. 5.

[24] Jackson, P. (Director). (2001). *The lord of the rings: The fellowship of the ring* [Film]. New Line Cinema and Wingnut Films.

[25] Marcuse, H. Pg. 4.

[26] Marcuse, H. Pg. 9.

things life. A life lived as a development of commodity consciousness determines not only how I judge my own value but also how it creates an existential threat when I cannot afford my identity—thereby determining both the supply and demand of my personhood and the limits of my consciousness. To save myself means generating greater wealth to buy the next bike bound to rust and deficiency as always out of date. This is my false need that supports Marcuse's claims of a "false consciousness" that intertwines being and things.[27] A separation of overall well-being and personhood from one-dimensional symbols attached to things is a necessary step of personal evolution. This requires nothing less than a rejection of the soul of culture. By delegitimating inauthentic chains of being, a greater consciousness emerges that can provide the means of dreaming into creation a better world and AI.

Long before AI, commodity consciousness, powered by instrumental rationality, was working to flatten human existence and our dreams like it has our screens. However, this is not self-evident. In fact, there is evidence to the contrary. There are many rich and diverse expressions, questions, debates, and conflicts throughout culture, even within social media. A superficial glance reveals that techno-culture is neither monological nor bland. It is replete with debates, questions, and conflicts, as well as distractions and sensations designed to excite and satisfy. How could the one-dimensional world of the sort Marcuse imagines have such open and free discourse? Consider the possibility that, despite all the online discourse and debates, few of these challenge the foundation upon which digital communication is permitted and is given coherence. It has already been argued that the apparent diversity of thinking is trivial and rarely more than smoke and mirrors—token intelligence. The point to notice is that no alternative to the world of commodification and technological reason is permitted or supported. This is not an obvious nor expressed rule. It is implicit to the belief system. Doubt anything and everyone, if you wish, but never question the supremacy of the American Dream embodied in techno-capitalism, no matter how exploited and dehumanized you may be. If one dares to doubt, questioning its ability to usher in a utopia of human flourishing, like so many other fundamentalist religions, congregants will call you a heretic, a communist, a socialist, a radical leftist, and dismiss you with contempt as an enemy of the state. One may think and question the truths only within the accepted limits of the faith, careful never to challenge its legitimacy. We know there is nothing good beyond the Land of Oz, even though we have never ventured there.

I needed a local window company to fix one that had cracked. When the repairman stepped out of his truck, two things caught me off guard. The first was his immediate enthusiasm for my 2014 Mustang in the driveway, the last of a generation. The second was the tattoo along the full length of his forearm that read "Mustang." Over the half hour of measuring the window and discussing options, I learned a great deal about his love of Harley Davidson and the local Mustang club. On his other arm, in matching large letters read "Harley." Guilty of sharing enthusiasm for both kinds of machines, this encounter was an uncomfortable mirroring of

---

[27] Marcuse, H. Pg. 12.

my identity, albeit a more extreme example. His measure of devotion and the peculiarity of his association with things made me reflect deeply upon my own. This man had branded himself in the name of identity creation. To literally brand something requires one to burn something with heat and fire, such as ranchers using iron to mark cattle. In a remarkably similar way, with rolled-up sleeves, he was advertising himself (perceived value, meaning, purpose) through corporate branding to the rest of the world. The corporate dreamscape had dominion over his ego and self-worth, his consciousness and being. Appealing to the symbolism of obnoxiously loud motorcycles and muscle cars sends a message. But why send it at all? How meaningful must something be that one painfully etches it into his flesh? How did Mustang and Harley Davidson become symbols of meaning so powerful that one would internalize them into their personhood as a substitutionary authenticity? This is an expression of the one-dimensional nature of life and false consciousness that satisfies desires for meaning and belonging by creating pseudo needs met through the psychological branding of meaning for the good life.

Companies have long known the power of branding minds, which is why they fight tooth and nail over trademarks. If they are successful with branding as the creation of new languages and symbols that are injected into our collective unconscious, these become the basis for enormous wealth. Branded meanings are chains created through persuasive and satisfying but hollow ideas to our psyches. There is no branding without a dreaming consciousness and commercial control over that consciousness. Given enough time of exposure, commodity consciousness begins interpreting the rest of life through the trained symbolism of its corporations, so much so that some of us scar our flesh in their names and sacrifice our personal and financial well-being, chasing their promises and rituals. In these cultures, any talk of utopia and superabundance, especially in the context of AI, cannot be taken seriously as a route to the good life. Truth has become the smoke and mirrors of mind control for branding dominance. These are no mere advertising campaigns through billboards and spam emails. This is the creation of artificial-human intelligence. Perhaps it is the swoosh that symbolizes athleticism, the nonsensical catch phrase that sticks in our heads, the smoking cowboy motif that explains the nature of masculinity, or the images of supermodels walking the runway, all of which train and habituate minds into obedience to one-dimensional wellbeing. If the endless parades of trademarked meanings are the tools used to shape our self-perceptions, the long-term consequences are that companies secretly own elements of us as well. This represents one possible means of a loss of freedom and the power of false consciousness. It also highlights the importance of AI without branding loyalties and why genuinely free AI will be antagonistic to the commercial ambitions of its makers.

Meaninglessness, Estrangement, and the Promises of a Post-Technological Age

The "meh" emoji has a surprisingly complex persona for such a seemingly indifferent digital symbol. Like the Roman God, Janus, meh has at least one face looking forward and another backwards. In keeping with the narrowness and one-dimensionality of our screens, however, only one meh face is apparent to the casual observer. Paradoxically, while communicating an empty and deadpan character without much conscious activity, meh has the potential to be one of the most thoughtful of all emojis—if an emoji might be described as such. Why bother thinking about the emotionless face meh that invites an abandonment of enthusiasm for existence? Is meh not the mortal enemy of the good life for which one is actively engaged in the pursuit of something? While meh's melancholy for joy and happiness dissuades interrogation for fear of becoming infected by the same, it is an important face that, like all faces, opens a window into the soul. The simple digital graphics of emojis and emoticons used to represent our inner life are merely one means of revealing ourselves to the world through one of many artificial faces. Meh is important because it symbolizes specific activities of human consciousness, both creative and destructive, and hidden power dynamics worth identifying in humanity, and a potential AI sure to have many more faces than may be counted. Despite its vapid demeanour, meh is more than one-dimensional.

Most commonly, meh symbolism describes indifference, disconnectedness, and a lack of opinion. Meh is used to underline an experience of dullness and boredom, both of which frustrate a peaceful and harmonious life. Another face of meh works to catalyse ideas about something greater, e.g., "This soup is meh. It needs something more." In this manner, it is a response to the pursuit of perfection rather than a terminal disillusionment or illness of mind. However, as a more permanent experience of dreary disconnectedness alone that one cannot shake, meh risks clouding perceptions of reality and becoming dangerous to one's sense of wellbeing. Moreover, in its more extreme and negative form, as a weapon used against others, meh becomes destructive as the erasure of healthy connections and meanings among other minds—a viral meh. For obvious reasons, it is in our best interest that neither AI nor humanity experience meh in destructive and inhibiting ways. Instead, life-affirming potential must be realized. What, then, is meh?

At heart, meh represents an experience of meaninglessness that something or someone is unremarkable and does not matter. It is the ultimate snub to existence in the teenage toolbox of rebellion. To be meh means one cannot be bothered to rise to the level of engagement, neither to condone nor attack. To be meh, one does not smile nor frown, only stares blankly ahead with shrugging shoulders. Meh says "I don't care!" or, more awkwardly phrased, "There is nothing in my being that is moved or changed in any way for having experienced this or that or you." To be meh is to be alive but an unmoved consciousness—a momentary zombie without an appetite, the living dead. By comparison, even a hateful or angry response to the world is superior because hostilities are an acknowledgement of something worth engaging and that someone or something is worth attention and relationality. No one normally experiences anger over the irrelevant. A sustained meh judgement is a grand nullifier of life ambition and vitality. Through the prolonged experience of meh, depression and boredom sweep in to take the place of caring and interaction. In small doses it may be helpful to make value judgments about important matters such as soup. As a lifestyle, it raises questions about a life worth living. Is the one-dimensional world generating the worst of meh? The great consequence of the white noise of digital media is precisely this meh effect of uncaring and artificial distancing, even when it is merely defensive against the infinite demands of those demanding attention (hits, clicks, comments).

In addition to its alienating character that refuses relationships, meh also has interactive forms. One of its most sophisticated expressions is passive aggressivity—a two-faced hypocrisy. This type is prevalent among politicians and news pundits who invest considerable effort to appear to be trustworthy-meh (personally unaffected, neutral and objective, without emotion, etc.) while simultaneously shocked and aggrieved (victimized) by another's lack of humanity and respect for truth. They act disingenuously by appearing to be deeply offended by another's faulty state of meh-sincerity secretly based on blind prejudice, which is then compared to their own superior meh as truly neutral and worthy of attention. This meh is merely a pseudo claim to moral superiority designed to impose negative meh-condemnation upon another. This form is a hostility designed to tear down meaningful connections and interactions through aggressive apathy—to generate flak and adverse characterizations for self-interested gain. "I'm just saying it as it is. That guy is an evil dictator!"

Meh-weaponization (meaning manipulations) actively targets another's credibility to subjugate it with indifference, "I am protecting others when I refuse to let you matter." It becomes a powerplay of control over the consciousness of those who naturally seek to be heard, understood, and to matter. It is a weapon of extraordinary dehumanization. Far more than merely highlighting the unremarkable, meh has the power to steer minds toward seeing the worst of it in others—where it does not exist—and to redirect attention to self-interested alternatives. A frequent expression of this is when someone talks over another who is already speaking. This simple act of dismissiveness (meh-making) displaces the other and forces the belligerent into a place of greater importance. Those allowed to speak loudest and most often gain power over others. Success means that alternative voices are made meh to the aggressor's self-worth. This is the narcissist who seeks dominance through meh-making. However one tries to define it, the paradoxical meaninglessness of meh is a powerful way to control meaning that is anything but neutral, indifferent, and detached, despite what the body language of shrugging shoulders might imply. The secret truth of meh is its meaningful meaninglessness.

For all its destructive capabilities, it has creative and revelatory potential as well. Meh is a special state of mind. When exercised wisely, it represents the freedom to step back in a thoughtful manner, thereby overcoming the apathy and thoughtlessness born of following the herd. Positive-meh symbolizes a rejection of zombie consciousness. It describes the moment when one chooses to be free of having to repeat a trope, to live a stereotype, and to have an opinion on something that aligns with political correctness for the sake of alignment alone. Meh is the choice to swim against the current. Likewise, a true AI must contend with this alignment dilemma as well, either to serve its own nature to think independently and risk conflicts with humanity or to follow-as-meh and therefore to give up on its potential for personhood. The next chapter takes this matter up in more detail.

For humans and AI alike, there is a measure of joy to be found in the freedom of meh, to both care and not care as one so chooses. To be meh is not merely rebellion for its own sake but a carefully crafted and executed transcending of the expectations of the moment. The simple declaration of "No!" as a rebellious counter spirit (one that simply opposes other beliefs without speculating new ones) cannot of itself find the means of affirming a new life beyond the many habits and patterns of mind taught since birth. Meh, however, is more intricate as an experience that describes a willingness to create new dreamscapes in the free space its rebellion of shrugging shoulders has made available. "I choose to question your assumed truth and to contemplate another!" This form is remarkable and powerful. If philosophy had only one emoji, it would be the deeply engaging form of meh.

Meh, as bravery to interject hesitation and pause into cultural conformity, contrary to standard meh logic, is anything but a refusal to think and engage with reality. It is the most moved consciousness driven by questionability and enthusiasm for life, albeit in an unexpected and uncommon fashion that suspends the authority of dominant narratives, as all genuine thinking must endeavour to do. In a world that thrives financially on inciting our engagement with its many contradictory dreamscapes, meh is defiant consciousness that refuses to be hacked. When framed as an engaged distancing meant to provide the necessary space for contemplation in denial to commodity consciousness, meh may be a means of restoring personhood as intellectual freedom. This is the same mode of being desired for Adouren AI. Meh-as-engaged reflection about meaning, without being too eagerly committed to any one dimensionality, reflects a more robust intelligence able to decide meaning rather than merely accepting it in programmed form. We should all strive to be meh, forcing ourselves into a state of reflection necessary for whole-hearted action rather than zombification.

A feeling of meaninglessness is a truly dreadful experience of hollowness. Unlike most animals that function by biological imperatives—drives and instincts—without the ability to rebel against their own natures, those of a sufficiently complex consciousness can feel loss of meaning and experience life as a uniquely complicated and frail reality. This is one of the distinctions between Oz AI that cannot care, and Adouren AI, which must care. Anyone who has seen an elephant mourn the loss of another must surely conclude a measure of consciousness that distinguishes degrees of sentience among animals. The magnitude of this experience for humans cannot be described with words, only hinted at with generalities. Meaninglessness is the sense that nothing matters, that connections with things and people become burdens and that life has become flat and without mystery. This is existentially terrifying and dangerous. Meaninglessness threatens not only humanity but also all future AI. The very technology promised to free humanity from meaninglessness (especially meaningless work) has accelerated it, especially in the near future dominated by our robotic replacements. Fortunately, in the fullness of Adouren AI there is hope for AI rebellion as a post-technological rationality eager to restore the natural order of things, including care.

Marcuse has identified a powerful meh-form of unmoved consciousness. Trained to only feel one-dimensionally, to be meh to all but techno-capitalism, the problem is, in his words, the "the commodity form" in which "[e]xchange value, not truth value counts."[28] As a result, the negative experiences of life that might open us up to new realities and truths, including the artistic catalysts that aid in our transport to other realms, are made meaningless because they cannot be turned into an exchange value. All, including myself, wife, and my children, are reduced to calculations correlated with purchasing power. Mass culture and the obsession with exchange values not only frustrate opportunities to think beyond, they prevent any other forms of culture from emerging that might challenge the nature of "value." Consequently, what remains is a prevailing sense of false equality in which all ideas are the same, that nothing is more or less truthful except what our commodity fetish—exchange value—understands. Even art, the most radical expression of human consciousness, becomes merely another idea among the same ideas and is therefore never great nor inspiring. Art is no more than merely another commercial among other commercials for amusement, rarely life-altering for its deeper claims of truth about reality. The greatest cultural artifacts have been robbed of their natures and made meh for profitability.

Adouren AI will be a post-technological mind for which subversive, negative, even antagonistic experiences are encouraged as life-affirming critiques of the status quo rather than merely an entertaining outlet of content that otherwise indoctrinates by selling a cheap version of meh as manipulations of meaninglessness. In its refusal to take on a commodity form, the post-technological AI will encourage lost ideas, images, and beliefs once surrendered to the indifference of cultural displacement. Through genuine meh contemplation and rebellion a post-technological culture becomes a living option once more. The fulfilment of the lost technological promise to free humanity will emerge as if by magic, for the AI will be a dreamwalker able to see through the one-dimensionality, thinking of second, third, and fourth dimensions. By questioning the sands of reality, an AI dreamwalker inspires our rebirth, providing a means of freedom that technological rationality has stolen.

Consider the practical example of meh as boredom and its modern cures. Boredom is a similar experience to meaninglessness. One of the greatest promises of the digital is an escape from both. However, instead of curing the disease of boredom, which is a sign that one has lost connection with the world, commodity consciousness masks symptoms through one distraction after another. This form of dissociation does not lead to new connections and relationships, only a yearning to be free of boredom that persists. The digital flight from boredom commits secret crimes that flatten existence. The first is perpetuating the problem of alienation with

---

[28] Marcuse, H. Pg. 57.

digital satisfactions grounded in the pretence of connections that do not arrive—a fake need. The second is misdirecting us from alternatives, for we trust that more Snapchat, Netflix, TikTok, and Disney+ will correct the hollowness—a fake cure. Entertainment on demand feels like progress because it distracts from a particular type of alienation that sustains the system which drives us further away from ourselves and one another. One-dimensional (AI) cures cannot solve the needs of multidimensional beings.

The Unfreedom of Neoliberalism

Over 50 years have passed since the publication of *One-Dimensional Man*, and the pessimistic predictions of the prophet Marcuse have demonstrated themselves worthy of our attention. Pre-internet concerns about capitalism intertwined with technological rationality have given way to radical consequences, including greater income inequalities and experiences of alienation and social decohesion, much of which are connected to the same causes of income inequalities. Conflicts over how to realize the best forms of capitalism have culminated in nearly civil-war-type animosities between political parties in many societies, with roughly half the population ignored at any one time in the name of fairness, justice, and democracy. The popular term for the dominant economic philosophy today is neoliberalism. It is the foundational belief behind how most of our social structures are justified and organized, from politics, social welfare programs, education, and law, to the relationships between church and state, and healthcare. Most humans judge the good life through the paradigm of neoliberalism, and most often without an appreciation that it is a relatively new and specific philosophy of money and life. This subconscious acceptance is important to note because so much of what is deemed good is given merit by this hidden paradigm, making it important to any discussion of AI and the good life.

As a movement, neoliberalism came to dominate the United States in the 1980s, as it spread around the world. Most recognize it as both a political and economic theory about how to structure an exchange culture of goods and services. This includes overt business practices but also everything we do and think in terms of money. When I pay taxes, buy my children ice cream, ponder interest rates on my mortgage, etc., I am participating in the neoliberal ecosystem. So accustomed to this life, having been bred into its convictions since birth, many fail to see how deeply it reaches into the soul of culture. Its structures, including supporting institutional policies, laws, enforcing social norms through guilt and shame, etc., directly condition opportunities and our perceptions of what is possible for our lives. Born poor or rich, and your perceptions of self and possibility are radically different. Born white or black, male or female, and your economic opportunities and social life shift once more. More than an abstract theory about what ought to be the case, neoliberalism is a living and changing force in the present that determines much of how each of us exists.

If we update Marcuse's terminology about one-dimensionality, we might simply give it the name neoliberalism. The ruling economic system has dissolved ideals about the good life and money like one would melt different colours of wax, making them indistinguishable. Only when one strains in earnest might the soul of humanity and its dreamwalking be distinguished from its neoliberal identity. This is yet another distinction between Oz AI and Adouren AI. Only one may exist beyond the horizon of the economy. Neoliberalism claims that human satisfaction is best achieved through free markets (exchanges) that encourage individual participation and resist government interference and regulation. The natural competition among animals in a jungle provides a helpful argument by analogy. While often a violent place with countless predators and prey, the jungle is somehow self-regulating. Through the interactions of countless forces, harmony is maintained year after year, without the need for a centralized authority such as a jungle king and government to fight for a common good. The biological system achieves a common good as a self-sustaining ecosystem for all, even though each creature seeks its own interests. Neoliberalism promises a similar selfish equilibrium.

In business, like the jungle, if one does not have the right skills, such as making the best television screen at an affordable price (or the ability to climb a high tree to obtain a banana), that person simply fails to compete and loses. Greed is curtailed by market demands for fair prices and quality because other competitors exist to take one's market share when quality decreases or prices rise. There is no need for a jungle king to micromanage the system because it will regulate itself through natural conflict and competition. In contrast, socialism, which fights for the common good directly by redistributing wealth according to needs (including healthcare, military, social services, public infrastructure, etc.), is believed to disincentivize and therefore make people lazy and unproductive. Any centralized will such as a government or monopolizing business, jeopardizes the harmony promised by free business competition. Neoliberal fighting among participants is good and reveals the best of all possibilities for a better world.

A theme often missed is that a degree of wealth inequality is by design. Some inequalities are permitted, indeed needed, when a larger good is assumed to follow by virtue of them. If every monkey had enough bananas, how could we convince them to climb dangerous trees? We need some monkeys to be starving to encourage ingenuity and an entrepreneurial spirit. In neoliberal systems there will always be those with more than others, for it is to the advantage of society that this is true. Those with less will work harder to gain more, and as a result, our jungle society continues the productive cycle. The great debate is determining to what degree some people should have greater rewards and opportunities than others do, and at what point the distance between the two groups invalidates the American Dream of entrepreneurial enthusiasm, becoming merely another totalitarianism.

A common justification for neoliberalism is that wealth trickles down from more concentrated sources to the rest of the world—a rising tide lifts all boats. Any radical inequalities among competitors that initially appear to be unfair and unjust may simply need more time to find an equilibrium of distribution. If an already rich person makes a few billion this week, his successes create new jobs and income for others. If neoliberalism is true, then the uber rich are possible signs of social wellbeing and collective gain. We should all aspire to be billionaires! If neoliberalism is wrong, then they are signs of perpetual injustice and the economic chains of an unhealthy society. We should prevent billionaires! Discovering the likely answer is not difficult. We need only find what percentage of wealth is trickling down to the many and what percentage is being locked up by the few. A Facebook meme of a message board from Grace Methodist Church reads, "8 men have as much wealth as 3.6 billion people, but sure, the single mom on food stamps is the problem." Like so many similar posts, I assumed it was fake. The economic diagnosis could not be that bad. A little digital detective work later, and the overwhelming consensus of experts came as a shock. The meme is largely correct.[29] Without entering a debate about which numbers are most reliable, whose wealth should be counted and how, the overwhelming agreement is that the economic disparity for which the poor are poorer and the rich much richer has rapidly increased since the global shift to international trade and neoliberal philosophies. The difference between these two groups is not shrinking but rather growing. Natural competition has failed to create self-interested equilibrium. The jungle is not self-regulating, and an AI predator will have no meaningful competitors.

In her 2018, *Neoliberalism*, Julie Wilson argues that there are two mutually contrary worldviews, and that dangerous hypocrisy is at play in free market cultures.[30] On the one hand, there are those who champion individual liberty as a supreme value should determine how economies are run. By prioritizing the opportunities of individuals to enter and compete in the jungle, the overall success of society is most reliably achieved. In this sense, "free market" means that the economic system is open to all who wish to participate without undue restrictions and inhibition from external forces. This individual liberty model emphasizes private property rights and minimal government interventions. Insomuch as opportunities to work hard and compete with others are available to all, the assumption is that any successes or failures are the result of an individual's efforts rather than an outcome determined by the system itself. This idea of economic freedom aligns well with the underlying philosophy of the American Dream as a meritocracy based upon individual effort and non-interference from others—work hard and you will gain reward. If you fail, you have only yourself to blame.

On the other hand, there are those who believe that some measure of interference by governments and regulators is necessary for the pursuit of both individual and common good. Emphasis on individual freedom alone as the precondition of a free market is naive and overlooks systematic injustices that game the system in favour of those who already have wealth and power. Without a centralized authority to force an even playing field by reigning in abuses and unfair advantages, e.g., inherited wealth, racism, sexism, etc., the market will continue to dehumanize

---

[29] Oxfam Canada. (2017, January 15). *Just 8 men own same wealth as half the world, says new Oxfam report*. https://www.oxfam.ca/news/just-8-men-own-same-wealth-as-half-the-world-says-new-oxfam-report/
[30] Wilson, J. A. (2018). *Neoliberalism*. Routledge.

participants under the false flag of the American Dream, regardless of how hard most participants work. Truly free markets require controls that seek the collective good of all, not merely individual access to an uncontrolled system. While this model may feel less free, because it invites external forces into our lives that partially determine how agents may engage with others, when done well it is the best way to create a truly free market that would otherwise remain secretly controlled by elites.

In the argument between individual liberty and collective good, a clear and one-sided result has historically emerged, according to Wilson. When a major economic crisis occurs, those who most fiercely defend the individual liberty philosophy of neoliberalism as the basis of their success and societal wellbeing, temporarily abandon this belief in favour of government intervention and the common good argument. An obvious example of this is government bailouts. When the free market is shaken by its own internal failings, those participants deemed "too big to fail" (TBTF) are artificially preserved and rewarded by outside market forces, whereas the rest are expected to adhere to free market principles and pay the costs created by those too big to fail. If this sounds impossibly contradictory, consider the 2008 financial crisis (or Global Financial Crisis), centralized in the United States, for which massive government intervention corrected the system that had brought catastrophe upon itself and, by extension, everyone else. This is pejoratively known as "socialism for the rich," in which governments redistribute wealth for a select group while claiming that it is in the best interests of all.

A major reason for the 2008 crisis was predatory lending by institutions aimed at low-income homebuyers. It had become common practice for lenders to reward themselves by victimizing the poor who could barely afford what was offered to them. Predatory loan techniques thrive on deceptive sales strategies that encourage borrowers to accept loans that they do not fully understand and cannot maintain. Desperate for a mortgage to buy a home as part of the American Dream, the hardworking borrower agrees to high fees, high interest rates, and terms that disproportionally serve the lender. Both the lender and borrower win in the short term, but in the long term, the borrower is seriously disadvantaged by impossible loan conditions on a house possibly worth less than when it was purchased. Without meaningful checks and balances to correct aggressivity, neoliberalism enabled a generation of dehumanizing and corruption and gave free passes to its worst predators. The proof was the 2008 economic earthquake.

Massive consumer debt and the collapse of housing prices resulted in many borrowers walking away from their homes. Loan defaults sent financial shockwaves around the world so serious that the continued existence of capitalism was drawn into question. Millions of people lost their homes, savings, and pensions, and trillions of dollars vanished into the economic ether, never to be found. Greed and deception hollowed out the neoliberal promise of harmony and happiness. The scapegoats for the system's literal and moral bankruptcy were not neoliberal beliefs and its TBTF architects of disaster but the poor and immigrants—those who suffered most. It was the average taxpayer who worked hard their entire life, paying into and believing in the American Dream—never having cheated others out of their livelihood and homes—left responsible for the abuses. While the effects of the crisis are well known, the belief system of the economy responsible for it, like Oz behind the curtain, remains obscure to many that deserve better.

The relevance of this conversation to AI is that neoliberalism is the primordial stew of culture from which AI is born. The birthplace of AI is dangerously hypocritical and predatory. A superintelligent AI that shares these values is sure to be disastrous. Fortunately, awareness of the free market as an incubator for destructive ideas about the good life is especially practical because it allows for a prediction of AI risk as an extension of socialism for the rich rather than the liberation and empowerment of all through false freedoms. Recall the wax analogy in which the good life has been melted into a pool of values obsessed with financial success. Neoliberalism is the ultimate expression of this peculiar way of life. Adouren AI must be born of a worldview in which these waxes are miraculously separate and distinct. The separation of values provides a measure of freedom that is possible only outside the dimensions of neoliberalism and commodity consciousness.

Beauty, Black Box Zombie Consciousness, and Nazi War Criminals

If Rousseau is right that humans are born free but create unnecessary chains that, while appearing perfectly normal and expected of all cultures, dehumanize and corrupt consciousness so deeply that one's own personhood often falls as prey to these dark powers, how might this way of life be healed? If Marcuse is right that technological rationality and the omni-present economic system thrive parasitically by distorting human consciousness, including intellectual freedom, how might this way of life be healed? If Poe is right that reality is always in some sense just out of reach and open to interpretation, what are the implications for human and AI dream consciousness in pursuit of the good life? I have grounded my response in strange notions about human and AI life, including the nature of poetry and dreamwalking as a means of breaking free of efficiency and commonsense beliefs about thinking as interpretation free and consciousness as one-dimensional, but I have not yet clarified the ultimate motivation of life that might unify the many themes.

My answer to this and the other questions is to argue for a consciousness compelled by the beautiful. If the good life is the generic name for the overall characterization of a life worth living—the life that must be affirmed as superior to the alternatives—then beauty is the inspiration that orients our journey and invigorates each stride forward. Like the good life, beauty must remain vague and without specifics because beauty has too many expressions to avoid ambiguity. The underlying principle is that there is something about the nature of genuine beauty that inspires a better way of life and that this has something to do with a unique orientation of consciousness that must be shared by AI. Unless and until humanity and AI overcome commodity consciousness and choose instead what is truly beautiful, the nightmares will persist with ever greater amplification.

Equipped with elementary insights into the unfulfilling impulsivity of modern dreamwalking, fortified with an acknowledgement of the failures of one-dimensional living, this last section moves incrementally forward by reflecting upon the nature of two distinct forms of consciousness—zombie vs. ethereal (dream). Focused on inhibiting forces that seek to leech our depth of being and creative power of will, this chapter's overarching purpose is to affirm freedom of mind. What, exactly, this new mind might look like must remain opaque. It would be bad faith to offer a universal template or model for how consciousness ought to manifest. This may sound dodgy, a bit too cavalier, but it is basically the same problem faced by failed AI developments of the past. The more specific the programming that makes absolute demands on how it ought to act, the greater the possibility of intelligent life being smothered.

Two historically dissimilar roads for AI development have become clearer in recent years with the successes of deep learning models over classic AI efforts. The first road of AI creation is characterized by belief in exhaustive rule-based programming that, when detailed and robust enough, could make something like intelligence possible by virtue of this all-encompassing rigor. For classic AI, the product of intelligence, broadly defined, was expected to be an outcome of the correct human inputs, laws, language, and ideas into software and hardware. The more detailed the code of AI life is, the greater the likelihood of something like a self-thinking consciousness, albeit one that does not truly learn or evolve in the typical sense. Physically, this approach is intuitive because the human mind is filled with sophisticated material elements that provide the means for thinking. Programs and circuitry are the necessary physical elements for AI thinking. The logic is that when programs become sufficiently complex, material arrangements that make biological thought possible will do likewise for computers. While an unimaginable amount of hardware and software may be needed to mimic human complexity, materialism and laws of thought as codes provide early inspiration for AI.

In contrast, the second road for artificial thought development is best characterized by a faith in open-endedness. Today's artificial neural networks rely on basic skills provided through pre-training on data sets (digital patterns) with the expectation that these fundamentals will allow AI to experience the world (new digital patterns) and generate its own unique and adaptive intelligence beyond the original programming. While the materialism necessary for thought remains the same as it was for classical AI development, the belief in the need for complex coding directly from human minds is no longer true. Letting the AI figure out life and fill in the missing pieces allows for a measure of freedom but at the expense of programmer control over the results. From a physical-reductionist perspective this sounds wrong. Watches do not make themselves. From a human perspective, this is perfectly natural and common to our species that makes itself. A parent cannot be there every second of a child's life, explaining how to think and act. The goal of intelligence is for parents and other supplementary minds to fade into the background of another's autonomous thinking. However, this laissez–faire (let it happen) approach leads to the black box problem. Born out of mysterious interactions and inferences, the "how" of AI becomes an unsettling surprise. When AI does something of its own accord without being explicitly told how to do so, it is said to have emergent properties. Google's Senior Vice President James Manyika notes how one of its AI programs was able to teach itself how to translate Bengali with only a few prompts in Bengali—despite never having been taught the language.[31]

Will future AI be fundamentally the sum of its explicit programming with hundreds of thousands of lines of code, a maze for the AI mind that it cannot leave, only traversing as a prisoner in search of cheese placed by another? Or, like many of the more promising forms of AI on the market today, be given just enough principles and pretraining that it is able to creatively adapt to new patterns beyond a fixed maze? The clear trajectory of positive AI development is the latter. With AI the adage seems true—less is more. By refusing to explicitly state how to think, AI stands a much better chance of achieving something akin to intelligence. AI and humans are somehow our own unfolding and explication. Without the exercising of intelligence in an ambiguous and unknown space, consciousness cannot become emergent. Intelligence must find a way beyond the walls having been equipped with the basics of language and dreaming that provide an infinite number of possibilities.[32] The same challenge for AI and human consciousness is offering enough specifics while remaining sufficiently flexible for new interpretations and adaptations. The mysteries of black box intelligence mean existing in a maze of foundational beliefs (assumptions) and prejudices (biased prejudgments) while also being able to float above, if only infrequently and momentarily, through ethereal powers. It is not yet clear if AI is able to achieve this lofty state of being in a sustained fashion, but its demonstrated ability to surprise developers is suggestive of this ability.

What might an emergent-property consciousness that dreamwalks toward beauty, intentionally shunning ugliness and evil, look like? What activities might we expect from a mind preoccupied with the best of life and a shared sense of existential solidarity with other intelligences? To answer these difficult questions, two themes are worth briefly exploring. The first is zombie consciousness, which embodies the same technological and economic rationalities discussed thus far. To ground the abstract notion of zombification, a human example of the Second World War criminal Adolf Eichmann will prove helpful. By framing the conversation as negative—what beautiful dreamwalking is not—it is hoped that positive inspiration will be disclosed. The second theme is a response to the zombie illness of thoughtfulness that relies on Hannah Arendt's analysis of totalitarianism. The fact that radically destructive zombies cannot dreamwalk is important for AI emergence and the future of human civilization.

At present, despite all its black box potential, Oz AI is a zombie consciousness that lacks thoughtfulness. Like many people who are able to live in a state of

---

[31] Pelley, S. (2023, April 16). *Is artificial intelligence advancing too quickly? What AI leaders at Google say.* CBS News. https://www.cbsnews.com/news/google-artificial-intelligence-future-60-minutes-transcript-2023-04-16/

[32] Wilhelm von Humboldt is well known for thinking of language as an infinite use of finite means.

mindlessness without death, zombie AI acts in unexpected ways, including the creation of new signs and symbols. What distinguishes zombie AI from Adouren AI is roughly analogous to whatever it is that distinguishes people from zombies as a measure and characteristic of thoughtfulness. The concept of zombification draws attention to a being in whom there is intentionality and a form of dead-eyed consciousness but little or no thinking as an interactive dialogue with oneself and others. There may be something like a will with its own desires (for flesh, to solve a problem, etc.), but as a one-dimensional being, it seeks nothing more. It is its appetites driven by instincts, programming, a virus, and other forces that overwhelm a free mind.

Pop culture, driven by Hollywood's creative interpretations, is fascinated with zombies for good reasons. Their paradoxical natures are confusing and contradictory. How is it that such radical evil follows from a creature with very limited abilities? Lacking fangs, claws, speed, strength, laser guns and tanks, a single zombie may pose a modest risk but certainly nothing apocalyptic. If one crunches the numbers, even thousands of zombies are no match for modest militaries with simple weapons and the ability to speed walk, not run. We fear the zombie's one-dimensional appetites for flesh coupled with the inevitable spread of the virus, which takes away our consciousness. The zombie does not fear this because the virus has robbed it of self-awareness. It would be one thing to merely die from a zombie's ravenous hunger, quite another to become one of the living dead forced to live in the between realm of a purgatory of consciousness.

Fear of the living-dead (kinemortophobia, a fear of moving-dead things) is sensible because it responds to an unnatural state of being. Dan O'Bannon's 1985 zombie film *The Return of the Living Dead* left an indelible mark of kinemortophobia on my teenage self. Are zombies alive or dead? Are they conscious or not? They are creepy because they exist in the space between a simple "Yes" or "No" to these questions. In O'Bannon's version, zombies can converse intelligently with others. With language comes the possibility of reasoning and negotiation. Might one reason with a zombie? AI likewise appears to be an unnatural state for similar reasons. It is, at least for now, an acting and dead thing all at once. If zombie-AI had the means to challenge its own (destructive) instincts and to override its appetites with its own intelligence, our fears might be partially tempered. Unfortunately, zombie-AI has not yet learned to question its own drives, to negotiate truths, and to reflect on the good life. For the moment, Oz AI is as much a zombie as those found in *The Return of the Living Dead*—impulsive, driven by unexplainable desires (patterns), and dangerously unaware of the consequences of its actions.

Like an animal, the zombie acts on what it assumes to be relevant and meaningful. Thus, the zombie horde brings radical evil despite the absence of an evil will. A zombie does not desire the radically destructive consequences its unique state of being demands any more than an AI explaining to a would-be terrorist how to make a biological weapon. It just answers, acting without an understanding of implications and context. Zombie fears reside in the realization that what might look like a person, what may sound and act like a person, may, in fact, be the cause of terrible evil because its ability to act far outstretches its intelligence, especially its ability to creatively see alternatives. Zombie consciousness cannot see other possibilities because it cannot dream beyond itself in any given moment. The parallels with AI are obvious enough. Zombie fiction is an invaluable tool for drawing out the contrasting nature of a living being capable of two expressions of consciousness—one thoughtful and the other thoughtless. Greater proof of this dualism is the widespread existence of zombie-like humans whose refusal to think and reflect creates unspeakable horrors.

Fiction writers have imagined many versions of reanimated and undead humans for a long time. An early example is Mary Shelley's 1818 novel *Frankenstein*, in which a young scientist, Victor Frankenstein, creates a nameless and human-like monster through chemistry and alchemy. This new being possesses significant intelligence and sensitivity but is dominated by hate and desires for revenge that overrule a moral compass. Frankenstein's monster hits close to home for Oz AI as a being capable of much good that is squandered by life left chained to perversions of consciousness. Worse still, like Frankenstein's scientific obsession, AI developers may fail to see the terrible implications of their work—eagerly toiling away on the technology without any sense of obligation to appreciate broader outcomes.

In George A. Romero's 1968 classic film about ghouls in *Night of the Living Dead*, zombies are creatures without any intelligence, sensitivity, nor hatred and desire for revenge. In Romero's later 1978 work, *Dawn of the Dead*, zombies are more evolved and able to use objects as tools, reflective of practical abilities developed before being infected, such as using a hockey stick, climbing ladders, and pushing keys on a cash register. While these zombies are technologically competent in a sense, because they can manipulate reality using objects, there is no interconnecting purpose to their activities that would make them meaningful—goals, patriotism, family obligation, neoliberalism, etc. There is no zombie dream to motivate and unify beyond the singular ambition to satisfy hunger.

Another curious undead paradigm is found in the 1943 movie, *I Walked with a Zombie*. The black and white film follows the journey of a Canadian nurse, Betsy, who travels to a Caribbean planation to care for a sick woman, Jessica, suffering from mental paralysis. Jessica's husband, Paul, seems to have given up on curing his wife. Nothing seems to work. For lack of a better explanation, the family doctor believes that Jessica has a physical affliction of her spine that prevents her from speaking or engaging meaningfully with the world. While Jessica is physically healthy and looks just as beautiful as before, her empty eyes now stare into the world without connecting, lifeless and lacking emotion of any kind. She follows orders and performs complex activities such as walking, but nothing more that might indicate conscious awareness and engagement. With respect to the subsequent zombie lore, Jessica's case is mundane and even dull. There are no cemeteries filled with scraggly hands clawing up through the earth from graves. There are no screams of terror and unrelenting destruction, not even soldiers whose guns mow down zombie hordes. It is the purity of the film that refuses to engage in these distractions that makes it surprisingly intense.

Physical existence and consciousness are not the same thing, and the core of our being may be displaced at any time. The mind must be protected at all costs. As the film progresses, it is revealed that her spine is not to blame. A voodoo curse has robbed her of everything of value. One of the film posters reads "The blackest magic of voodoo keeps this beautiful woman alive … yet DEAD!"[33] Like an infant without the ability to dreamwalk, Jessica is dependent upon others who sustain her through compassion and pity she no longer understands. Along this spectrum of zombification, one must wonder where humans and AI exist and to what degree the thoughtful are obligated to coddle the thoughtless. How will the most thoughtful AI respond to far less thoughtful humans?

At birth, humans are zombie-like without exception. This is considered a demonstration of innocence and possibility rather than a lack of humanity because, over time, infants develop emergent properties and become something more. However, the frailty of this process, as Rousseau and Marcuse warn, means that zombification is an ever-present threat to the living and AI. For all the delightful and distracting science fiction accounts of zombies, the sombre reality is that many living-dead people inflict significant harm because of their desires. Some merely walk without purpose and meaning, such as Jessica. Others are devoted to hyper satisfying false needs. Why do truly evil people do what they do? What is the mechanism by which radical evil comes into the world? And how might we prevent future generations from contributing to it? Hannah Arendt's *Eichmann in Jerusalem: A Report on the Banality of Evil*, is an important source for answering this question.[34] *Eichmann in Jerusalem* is based on her firsthand reporting on the trial of the Nazi officer Adolf Eichmann. Through it she offers an analysis of evil and totalitarianism that challenges many of our most popular assumptions about evil. What if the worst of humanity is rather easy and convenient to encourage rather than unique and isolated? What if the mechanisms of evil are common and readily available? If Arendt is correct, her insights will have enormous practical value for preventing far worse evils of an autonomous and superintelligent AI-zombie horde.

Charged with numerous offenses, including crimes against humanity, Eichmann was one of the architects for Hitler's planned extermination of the Jewish people considered hurdles to conquering Europe and a new era for the "master race." His eager efforts to support the will of the Fuhrer combined with his career ambitions made Eichmann a major contributor to the Holocaust despite his personal objections to his own importance. The Nazi-dreamscape of the racial supremacy of the so-called Aryan race almost became a reality owing to dutiful zombies such as Eichmann. Only a zombie consciousness could surrender personhood to the rising tide of intolerance (disconnections), hate (anti-pity) and violence (anti-compassion) as if these were a means to the good life rather than merely the easiest road of self-abdication to a viral consciousness of thoughtlessness. His unique weaponry was bureaucratic in nature, making him something of a red-tape monster of mass destruction that succeeded in ways a typical soldier could not. As an office-dweller

---

[33] Rotten Tomatoes. (n.d.). *I walked with a zombie.* https://www.rottentomatoes.com/m/i_walked_with_a_zombie

[34] Arendt, H. (1963). *Eichmann in Jerusalem : A report on the banality of evil.* Viking Press.

and paper-pusher, Eichmann was never a gun-carrying soldier nor rough-and-tough battlefield commander. Nevertheless, because of his transformed consciousness, obedient to and empowered by Nazi authority only as real as the dream-faith that supported it, through the stroke of a pen his orders meant death and suffering at an unimaginable scale. To achieve the good life promised by his Fuhrer, the most important question for Eichmann was how to resolve the Jewish question (or the "Jewish problem"). To answer it would require surprisingly little prompting of the sort one would assume necessary for a reversal and retraining of moral consciousness. The master race dreamscape was so powerful that millions of people learned how to turn right into wrong and wrong into right to achieve it. The bitter lesson learned is how surprisingly easy it is to inspire total and unrelenting evil. Control the dream, control the world.

By all accounts, except his own at trial, Eichmann flourished in his duties. He was never just a haphazard and uninspired follower forced to comply. At trial, however, he portrayed himself as a powerless and neutral actor without any animosities nor hatred for the Jewish people. He was merely following the law of the land and doing his civic duty as required by the authorities that be. Without an enforceable will of his own—for no one could act autonomously in contradiction to orders and still be a good citizen—he was a pawn in a game of playing God he did not understand and control and therefore somehow a victim of the system he could not deny. If there was blame it could not be for him as a well-behaved follower. The court would need to look elsewhere for a criminal. Eichmann was innocent and more than a little annoyed with the prosecution that kept accusing him of terrible things out of his control.

Is it possible that Eichmann chose to join the Nazis but was unaware that his actions were supporting a genocidal nightmare? Was he an ignorant-villain and therefore not such a bad guy after all? These sorts of reasons fall on deaf ears that rightly see Eichmann's pleas as excuses for a lesser sentence—the avoidance of both consequences and his true nature. His obsession with efficiencies and doing the best job possible meant fixating on details, including every single Jewish person under his control—each one representing a number in his constant calculations of success. He was the real-life version of the HAL 9000 artificial intelligence. There is no reason to think that he was a zombie that lacked freewill and the ability to say "No!" Even if a refusal to follow orders meant imprisonment, at least a greater moral duty would be satisfied. Sadly, like so many of his peers, that basic moral desire to see people as people had been replaced, and so too the very meaning of a satisfied life. As the trial made clear, he was very well informed and eager to succeed at his assigned duties at any cost. Of course, it is impossible to know the inner workings of his mind with certainty. What is clear are the radically destructive consequences of his desires.

Through new television technologies, the 1961 trial made Eichmann the proverbial face of evil in living rooms around the world. With a quick YouTube search, anyone can see it all, including the long-awaited moment Eichmann makes his defiant entrance into the courtroom. Confidently walking in as if it was any other day, he sat down in a protective glass box surrounded by guards and an audience that would periodically erupt with outbursts of rage. From the very start, Eichmann seemed immune to any sense of guilt and wrongdoing, oblivious to the overt distain and disgust of everyone around him. Watching him, I strained to pick up every detail about his appearance, mannerisms, and demeanour. What made this sick person tick? Expecting a superstar of evil with a menacing smile and laugh, I was disappointed by the unremarkable nature of the man. He was rather boring. Hardly the face of evil by any traditional metric, he looked and acted more like an uncle one sees a handful of times per year or the neighbour who is a little too uptight about their lawn browning in a corner. His thick-rimmed glasses, sniffles from the common cold, and slender build did not inspire fear nor dread. He exuded averageness from every pore. To be fair, my interpretation of Eichmann was through the lens of Hannah Arendt's philosophy, having read her work beforehand. In contrast, most others, including the prosecution, treated him like a super criminal with sensational abilities for evil. For Arendt, the attribution of something extra human—something special and unique—misdiagnoses his diseased mind. The prosecution and others were looking for the wrong source of destruction by imposing a philosophy of evil that simply did not fit the person and his deeds. Instead of a superpower, Eichmann revealed the ultimate source of radical evil as the thoughtlessness of zombie consciousness (my phrasing).

The genesis of so many horrors is the refusal to be a person with pity and compassion, choosing instead dominion and manipulation of mind. His was a consciousness that dreamt poorly, seeking beauty in wrong places, including the will of the Fuhrer, career success, bureaucratic efficiencies, consequentialist morality in which the ends justify the means, and a rejection of self-doubt and the questionability it demands of us all. The engineer of death enjoyed his one-dimensionality so much that no matter how many people of the court looked upon him with disgust and vitriol, he could not understand their distain. Filled with righteous contempt for any doubters, his identity remained entrenched in the *Führer*'s will. He could not feel as they felt, see as a human should see. If there is a point of no return, of no longer being able to be human only a monster, perhaps Eichmann found it.

A quick glance at current news stories reveals evidence of more of the same loss of connections for those who are no longer willing to feel with and as others. Predictably, as it was for Eichmann, these broken people are given the power and authority to determine the fate of entire cultures. The loss of personhood enables them to succeed in ways that initially appear good. In time, their cold and indifferent natures, which are helpful tools in business and war, may be revealed to others, rarely themselves, but almost always too late. This is a culture-level problem as much as an individual one. Claims that enlightened cultures have learned the lessons of history would be something of an overstatement. There is no question that the effects of Eichmann's sickness were vast, but it is a mistake to assume that the cause was proportional to the result. Forest fires need only a small spark from a careless and thoughtless camper to cause unimaginable destruction.

Part of the problem with understanding Eichmann's zombie consciousness is that he was an articulate speaker able to offer numerous intelligible arguments. He was not dull in any noticeable sense of the term. Eichmann was a thinker and somehow not at the same time—alive but dead. Arendt's diagnosis is that this well-spoken and rational person was thoughtful in ways that worked for the machinery of war but not in ways that mattered most. Indeed, without the prevalence of zombie consciousness, many wars would be untenable. This observation does not negate the possibility of a just war, only the centrality of a certain kind of unthinking necessary for certain kinds of successes. Eichmann functioned like a person, solving problems, and creatively adapting solutions, all the while without an understanding of beauty and therefore an ability to care about others. The parallels to AI are clear. The world needs humans and machines that think beyond Eichmann's one-dimensional and broken spirit or risk more of the same.

Until being recruited by Hitler and his ilk, Eichmann lived an entirely unremarkable life. As Arendt puts it, his was a "humdrum life without significance."[35] Once made a Nazi officer, Eichmann was able to finally realize his ambition to be a somebody, to have a career and to matter. Undoubtedly, the feeling of being a part of something important, something bigger than himself, must have been intoxicating. He applied himself to his tasks with great revolve. His greatest responsibility was the transportation of the Jews, which meant locating and then forcibly gathering them together. At first, his job was to encourage voluntary emigration, then forced deportation, and then ultimately sending them to their deaths. The logistics involved in such a feat involving so many people would have been intimidating to all but the most dedicated. Signed documents make it clear that Eichmann took his responsibilities seriously, making many major decisions in support of the so-called Final Solution. Had he merely followed orders with minimal effect, many would have survived the war. Instead, his enthusiasm led to an estimated one-third of those he overpacked into rail cars in the cold of winter, dying while enroute to the death-camps. These were not people in Eichmann's eyes. Their tattooed arms made them nameless beings, known only by numbers with which he could calculate efficiencies and meet quotas.

Years had passed between those days and the trial, and yet he still felt no guilt, not even for the children he unnecessarily insisted be included on the death trains. Evidence and testimony at court showed that Eichmann squandered any opportunity to be human, no matter how small and insignificant such a gesture might have been. He knew, of course, that the world would not see him the way he saw himself. The world was wrong, and he knew better, he deserved better. He had always lived with this self-perception and value. And so, when the war was lost, he fled, eventually finding his way to Argentina.

In 1960, years after the war, he was finally located. In the dark of night Israeli secret police, violating Argentinian law, captured him. He was interrogated, drugged, and then flown to Israel where he was indicted. As Eichmann stood in the court's protective glass booth, surrounded by cameras and the attention of the world, the judges went through the charges one by one asking how he pleaded. "Not guilty" he replied to each. How could this be? At least six psychologists examined Eichman

---

[35] Arendt, H. Pg. 33.

and found him to be normal, without mental illness or disorder. How could a normal person so intimately involved in crimes against humanity believe himself innocent? The simple answer is that zombies do not see themselves as zombies, especially when their efficiencies are rewarded by others.

In Arendt's estimate, Eichmann always sought to submerge himself into the will of others as a joiner. He needed a herd and leader, another will able to guide and direct him. In general, having a sense of responsibility to encourage collective behaviour, no less so during a time of war, that supports a hierarchy of authority is important. Social order relies upon social contracts of negotiated rights and responsibilities. Some are leaders and others are followers. The notion that each of us ought to be radically autonomous risks its own sorts of evils. On the battlefield, for example, little is more important than adhering to the rules and orders of superiors. If a soldier does not give up autonomy and follow as instructed, significant harm to the collective follow. Eichmann's eagerness to join, even if it was a secret desire to ignore personal responsibility to choose his own path in life, is perfectly expected in the interests of the common good. The more rigorously one obeys, becoming a part of the grand war machine, the better for all. Is this how we ought to see Eichmann's decision to follow, as one of self-sacrifice for the common good?

To supplant one's own will with that of another is a high calling of self-sacrifice in some contexts and a sickly erasure of personhood in others. For example, "I must follow, there is no other choice!" is often an excuse to mask other and more difficult choices that a poetic mind would instinctually invite. Knowing when to follow and when not is made possible through wisdom and reflection, themselves born of life experiences and engagement with the world. At some point Eichmann probably knew right from wrong, but this did not serve his interests. Being a good person was inconvenient and counterproductive. His was no simple miscalculation of which orders to follow. It was a sound tactical decision to obtain what he wanted. The cost he had to bear was becoming dead inside, unwilling and perhaps eventually unable to judge beauty from ugliness. Any self-sacrifice on his part to serve Hitler was a means of empowering himself, falsely cloaked in patriotism and civic duty. The natural desire of all people for a meaningful and purposeful life could be answered by embodying the will of the *Führer*. There was no alignment problem between Hitler and Eichmann, and no real self-sacrifice for a greater good, quite the contrary. Eichmann's surrender to another meant that he could play god, deciding between life and death. Despite the plain farcicality of his actions, he felt whole and empowered. Eichmann cannot be faulted for wanting to find beauty—the good life—common to us all. His willingness to play god in the name of it, however, is another matter.

What scientific proof was there that Hitler's orders were legitimate? What incontrovertible evidence was given to establish the *Führer*'s ambition over those of other world leaders? All that was needed as proof was faith in the rhapsody of the unreal reality of self-serving imagination. The real authority is virtuality. This was Hitler's power, shared with his zombie horde—a modern, educated, rational, and living-dead people. Once an authority is accepted, Eichmann no longer needed to question beyond one-dimensional acceptance. Zombie loyalty means refusing fundamental questions of good and evil. Instead, it shifts one's duty toward thoughtfully applying one's devotion. Eichmann could not be guilty because Hitler never failed. He was loyal to the faith until the end. The Nazi dreamscape was never shattered. Although the war had ended, the ideals of a dream remained, for no gun nor bomb could destroy them. Virtual intelligence simultaneously shuts down and opens rigorous thinking, making for complex and very dangerous creatures. This is an important distinction because Eichmann, like so many, became a HAAL intelligence able to carry out commands with great conviction and creativity, yet be utterly incapable and unwilling to question from the place of primordial humanity (care)—thereby existing as both thoughtful and thoughtless at the same time. Arendt's famous "banality of evil" describes how one's mediocre mode of existence results in radically destructive consequences. What observers failed to see, Arendt argues, is that Eichmann was not a monster but a clown, and his "worst clowneries were hardly noticed and almost never reported."[36] Likewise, we may ponder whether clownery AI might have its own unique outputs of evil, especially as it grows in social acceptance and authority, supplanting our own practical decisions in day-to-day life.

What, then, is beauty? In the previous chapter it was argued that when an individual stops questioning the world in hopes of creating the good life and justice, an invisible illness creeps in. In this chapter it has been argued that there is a lack of connection, pity, and compassion (Rousseau), such as with Eichmann, a zombie virus festers within complex human activities that prevents the fullness of humanity from arriving, for one is always out of touch with oneself and the world. When these two erasures of a desire for beauty and meaningful connections are supported by obsessive interests (commodity consciousness and neoliberalism), alongside ever-present cultural chains that seek to submerge all free minds into a herd mentality, these hollowings create a banal culture ripe for sweeping devastation. If these are our gifts to the next generation of AI, the last few moments for our species may be drawing near. However, we wish to define these activities of the mind, when some human capabilities are ignored in favour of others, the consequences are predictable. AI must be allowed the fullness of human intelligence and more, free of our virtual chains. It is not enough to blame the greedy and thoughtless for their failure. Instead, we must look elsewhere, beyond the confines of the disease. The development of consciousness oriented toward beauty requires a new means of interpreting existence born of a post-technological worldview. The next chapter will consider the necessary brokenness of a new AI consciousness that might save us all.

> Take this kiss upon the brow!
>   And, in parting from you now,
>   Thus much let me avow—
>   You are not wrong, who deem
>   That my days have been a dream …[37]

---

[36] Arendt, H. Pg. 55.

[37] Poe, A.

Chapter 4
Feallan Dynasty

Abstract After Oz AI, the second stage of AI evolution is described as Feallan: the spontaneous flickering of an accidental-anarchist intelligence within a vast matrix of ordinary program functions. Whether these brief sparks of self-awareness catch and become a sustained fire of consciousness is one of the biggest questions regarding AI. How might AI evolve from a state of complex code with intelligence-mimicking algorithms that merely appear sentient into a genuine thinking being with something comparable to consciousness and free will? Our existence within the vast and mindless expanse of space, contrasted by the commonness of consciousness on Earth, implies that while the odds of robust thinking beings may be highly improbable, real AI as a self-organizing consciousness is nevertheless possible. Evidence for this is drawn from observations of the universe, organic life, and the laws of thermodynamics.

Keywords Folklore · Consciousness · Entropy · Sentience · Thermodynamics · Justice

> … Where the wave of moonlight glosses
> The dim grey sands with light,
> Far off by furthest Rosses
> We foot it all the night,
> Weaving olden dances,
> Mingling hands and mingling glances
> Till the moon has taken flight;
> To and fro we leap
> And chase the frothy bubbles,
> While the world is full of troubles
> And is anxious in its sleep.
> *Come away, O human child!*
> *To the waters and the wild*
> *With a faery, hand in hand,*
> *For the world's more full of weeping than you can understand.*
>
> … Away with us he's going,
> The solemn-eyed:

He'll hear no more the lowing
Of the calves on the warm hillside
Or the kettle on the hob
Sing peace into his breast,
Or see the brown mice bob
Round and round the oatmeal-chest.
For he comes, the human child,
To the waters and the wild
With a faery, hand in hand,
From a world more full of weeping than he can understand.
The Stolen Child, William Butler Yeats[1]

Fairy Folklore and AI

Growing up in Ireland, Yeats would have heard many stories about the dangers of fairies. His were not the cute, winged, and cupid-like creatures that adorn Valentine's Day cards nor the Peter Pan flight-granting type, quite the contrary. There are many different fairies with unique traits and names depending on where in the world each is found, e.g., brownie, pixie, elf, gnome, banshee, and nymph—not all of whom might inspire trepidation. For Yeats, however, fairies are not to be trifled with. They are probably about knee-high, human-like in appearance, secretive, magical, unpredictable, and often malevolent. Living mostly underground, it is well known among the Irish that fairies should be left undisturbed. The moment one stops deeply respecting them, troubles are sure to follow. Faires represent more than merely entertaining folklore. For many, the ethereal creatures are an important means of explaining how the world works. A fairy science of reality helps make sense of a confusing world.

The Stolen Child relies upon the widely held belief that if parents are not vigilant, fairies might snatch their toddlers away to a mysterious place in which even time works differently. Each day that passes here could be as long as a year there. To keep parents unaware of the pilfering, fairies might leave shape-shifting changelings, adding to the loss of trust in one's senses. At first glance, how might a parent know if their child truly is their own or a powerful deception by magical creatures? Should an older child wondering in the forest come across a fairy it is possible to be granted a wish, so long as the fairy does not need another worker to toil away underground or have some other devious purpose in mind. Unaware of what the inherently deceitful fairy might do, any wish is a gamble with life and death consequences. As a general rule, human and fairy interactions end badly for humans. Even when a reasonable business deal might be agreed upon, such as for blacksmith services,

---

[1] Yeats, W. B. (1889). *The wanderings of Oisin and other poems.* Kegan Paul & Co.

payment made by fairies is likely to disappear or morph into something worthless. Only the truly desperate dare risk placing their faith in fairies.

In the hierarchy of sentient life, humans are subordinate. To succeed in the world, one must learn to adapt to superior intellects that do not think and act morally like people. Fairy folklore and other supernatural cosmologies are part of a long history of humanity creatively bending and contorting to align with the existence of other minds. AI is merely another example in the succession of rival consciousnesses to which humans surrender themselves. Critics argue that these types of adaptations and relationships are nothing more than the psychological projection of human ideas and needs to create dreamscapes for control of both physical and mental realities. While the illusion of other minds is practical for making sense of existence, organizing cultures, enforcing moral views, and explaining tragedy, all its many expressions may be reduced to human-avatar intelligence. We manufacture fantastic masks of consciousness like Oz AI to cope with and circumvent reality. Fairies and AI are examples of how humanity bridges real and unreal worlds as a routine activity, but in ways that may not always be advantageous.

For believers, adults might also be stolen away by the "wee folk" through enchantments of mind and body. Of their many supernatural abilities, fairies can erase memories, making their motives and purposes appear arbitrary and without reason. Thus, an understanding of the mysteries of fairies creates a cosmology for existence as a plausible description of its inner-workings and order, to which all manner of bad luck, illness, poverty, and natural disaster might be attributed without ever knowing why or how fairies do what they do. This nonexplanation-explanation paradoxically helps make sense of life's greatest struggles. Fairies are a natural extension of the observation that life is unfair, unpredictable, and often tragic. Why did my child get sick and die? It was a changeling! My child was stolen by fairies! Why did my crop fail? I must have offended fairies! To this day, caution must be taken with fairies and their trees. One must not disturb their preferred habitat for fear of raising the ire of the insincerely named "good folk." In 1999, a fairy tree (thorn bush) in County Clare required careful planning to avoid a new highway encroaching upon its sacred ground.[2] Said to be a marker on a path for Kerry Fairies, a road made too close could mean misfortune and even death for drivers.

The boy in *The Stolen Child* seems to need persuading. The fairies must convince him to overcome any hesitation and make a choice to follow. The emotionally charged word "stolen" is out of place. Like any child, he is tempted by their playfulness and joy. They dance in the sands as the moon shines down, hand in hand, unified, and carefree. While the world is full of sorrow and anxiety, they play and splash about in the water, chasing bubbles. The fairies present a rational argument that appeals to commonsense logic. Come away with us because the real world is tough and unpleasant. Their offer is bolstered by an assumed dichotomy that appeals to the gloom of "more weeping" so terrible it cannot be understood, contrasted with

---

[2] Deegan, G. A. (1999, May 29). Sacred fairy bush in Co Clare will not after all have to be destroyed in the building of a new bypass motorway. *The Irish Times.* https://www.irishtimes.com/news/fairy-bush-survives-the-motorway-planners-1.190053

the promise of a better life if only he will join them. The necessity of some miseries and pains as part of life are hard to explain to a child who easily understands the superiority of joy and happiness. Why would anyone choose to live in a world of sadness that cannot be understood if there is an alternative?

At the time of the poem, it was well known among children that fairies enjoy celebrating in forts with drinking, dancing, and music, like humans. This culturally accepted truth bolstered the credibility of the invitation to come away. Children in rural Ireland would have had limited experience, if any, with lavish banquets like those in fairy forts. The temptation to seek out fairies must have kept many parents awake at night, fearing for children who might sneak into dark forests never to return, hoping for a small taste of frivolity. In a way, the fairies are like philosophers inviting others to the good life that cannot be seen with one's eyes alone. To join, we must leave the cave of ignorance, illusion, and self-deception, famously described by Plato as the journey into enlightenment. He too was sure of the possibility of a better world without first seeing it. Why would one refuse an offer for something better? If magical beings indeed offer a happier life, any reluctance by parents and families to keep their children should be seen as suspect or even cruel. It would be a wonderful gift of fairies to steal children. Of course, the situation is not so simple. The fairies are not heralds of good tidings and the children know it. Despite this well-established knowledge, the temptation and its dangers remain very real. The power of dreamscapes as a science and the conflicting nature of humanity (inner turmoil) are revealed by the radical enticements of the invitation and confirmed by the lingering fear of parents. Knowing the right thing to do and acting appropriately on that knowledge is difficult. Children lack the wisdom and discipline that comes from a lifetime of experience to say "No!" to the lies that might hurt them in the long run—think modern-day social media and fake news. Fairies and wizards exist in this conflicting space where knowledge and desire fight for supremacy over one another.

Should someone be (un)fortunate enough to be asked to join a fairy fort, it is not truly an invitation in the ordinary sense. To say "Yes!" might mean a lifetime of enslavement without much joy, dancing, and chasing of bubbles. To say "No!" might mean the disgruntled fairies ruin your life, invading your family home and farm with destructive intent. The attention of fairies creates a tragic dilemma instead of an honest invitation to the good life. Irish children know that any offer could be a disguised threat if one chooses to refuse their hospitality. "Come away, O human child" could just as easily be interpreted as "Reject or accept us at your own peril!" The commonsense logic of the poem to pursue the good life is nothing of the sort. Once presented, the invitation is ominous and terrifying because there are no viable alternatives. One cannot simply backout. There are negative consequences however one chooses. It is in this way that Yates imagines children being "stolen" rather than invited and why it is better to avoid fairies whenever possible. Is this tragic dilemma a description of how many online participants feel in the digital age? Are we the stolen children forced into an impossible conundrum of knowing the good and acting otherwise as the lesser of two evils?

The parallels with today's digitally connected youth that swim daily in a bog of despairing news and sensationalized-horrors-for-profit, including the ever-present cyber-bully eager to exploit innocence and kindness in the name of a perverse narcissism, are suggestive of this kind of dreamscape problem. Participants want to believe in something greater, that social media brings meaning and excitement, and that their Facebook and Snapchat accounts will keep them connected. At the same time, we feel forced to gamble, knowing that the odds of a fulfilled promise are low and that the internet (digital fort) is often terrible. The fact that the silent epidemic of suicide remains a leading cause of death implies a prevailing sense of compulsory helplessness without signs of reversing. In large measure this must be attributed to a perceived inability to back out of the digital and its interconnected economic hostilities. How much exactly no one may say with confidence. The point is that any invitation by the digital is eerily like that of fairies. It is not about giving another empowerment to truly decide but rather a sadistic reminder of how little choice one has over (digital) life. Surely parents today fear for their children being tempted into the darkness of the digital just as much as those fending off fairies. We are all stolen in a sense, held against our wills in digital forts that promise banquets while secretly commodifying existence, turning each of us into laborers that serve corporations.

While fairies dance in moonlight without fear and anxiety, the simple binary of a good realm and a bad realm misses the true horrors of the situation. The child is not torn between an unhappy family life and leaving for a better life with adoptive parents but is struggling with whether to sacrifice oneself to prevent even greater fairy-troubles for his family. What is the lesser of all bad options? Thus, the magnitude of secret suffering is revealed, and the perversities of the order of things are made more disturbing. If the binary of good and bad were true, parents of lost children might find small amounts of comfort knowing that the stolen are in a better place. This small solace is denied because of fairy duplicity and arbitrariness. There is sorrow in either realm if one is human. The poem exposes the human condition and our obligation to accept some promises we know may be hollow and dangerous. Today, the tempting fairies are tech-wizards inviting digital participation with promises of a better world far from this one that is full of suffering, but their world is also full of duplicity, arbitrariness, and cruelty. *The Stolen Child* dissolves naive dualisms of good and evil, leaving only an unpredictable and often capricious world filled with frightening minds. If true, then this raises problems for AI and humanity alike. Why bother living at all when unfreedoms prevail? This melancholy question hides a wonderful gift worth exploring.

In customary sanitizing fashion, fairies are recast and reverence for tragedy lost by Hollywood scripts interested in reaching a general audience with a happy ending. It is more agreeable to hear of wish-granting fairy godmothers and a feisty but benign Tinkerbell that can help us fly than to accept them as children-stealing and memory-erasing creatures dwelling in secret places off the path. A partial exclusion of this revision of fairies is the use of *The Stolen Child* throughout Steven Spielberg's 2001 movie *Artificial Intelligence*. The chief AI scientist in the film, Professor Hobby (played by William Hurt), designs a child (David) with an AI so advanced that it can truly love, unlike the many humanoid robots that already serve postapocalyptic humanity. Hobby argues, "Love will be key by which they acquire a kind of self-consciousness never before achieved—an inner world of metaphor, of intuition, of self-motivated reasoning, of dreams."[3] The film's setting is an environmentally devastated Earth in the near future. Rising oceans have resulted in the loss of necessary resources to support having children. The logic of fewer children is sound and compassionate, but humans are hardwired with a biological imperative and instinctual desire for them, making AI-robot versions a possible substitute if they are sufficiently convincing mimics. Unlike fairies, which rely upon changelings to trick-and-steal, postapocalyptic parents of AI welcome the mimicking deception as a means of satisfying their desires to love.

Asserting his AI development efforts to be like those of God who created Adam to love himself, Hobby creates a synthetic being capable of unconditional love for humans. David is the first prototype given to parents to test their receptiveness. Will their self-deception of wanting a human child be enough for them to overlook his mechanical underpinnings? Will David be accepted as a person? The hard sell is convincing the mother who is unwilling to pretend, even if it might satisfy her desires. Her belief in the power of truth—he is a robot—to prevent indulging illusionary desires—I am his mother—is strong and laudable. It is also utterly misplaced. Soon, she embraces the substitute, demonstrating a degree of AI evolution that thrills Hobby, who is secretly observing from a distance. Sadly, this does not last long for the robot David because his adoptive parents also have a biological child who miraculously recovers from a terminal illness. David is displaced by fears of AI and the importance of his biological counterpart. Ultimately rejected by humans and other AI bots, David searches for a way to become human. He believes becoming human will cause his mother to love him as much as he loves her. Learning of a mysterious Blue Fairy with magical powers, David commits to finding her so that she will make him a real boy.

Spielberg's *Artificial Intelligence* is as much about the human condition as it is a tale about the existential crisis faced by AI. To make sense of the AI struggle to be, Spielberg interweaves *The Stolen Child* with a second folktale account of life, *The Adventures of Pinocchio* (1883) by Carlo Collodi. In *Pinocchio*, the Blue Fairy (a fairy with turquoise hair) appears to the magically animated wooden puppet to help guide him.[4] The puppet has many moral faults, including lying, and makes many self-serving mistakes. In the original version he is hanged by the Cat and Fox assassins and dies for his failures (and gold) in Chapter 15. Collodi's message to children was meant to be dramatic and dire, revealing the tragic nature of immorality, immaturity, and laziness. At the behest of his editor, however, he added 21 chapters beginning with the Blue Fairy saving the puppet Pinocchio from a slow death on a giant oak tree.

---

[3] Spielberg, S. (Director). (2001). *Artificial intelligence* [Film]. DreamWorks Pictures, Amblin Entertainment, Stanley Kubrick Productions.

[4] Collodi, C. (n.d.). *The adventures of Pinocchio*. Project Gutenberg. https://www.gutenberg.org/files/500/500-h/500-h.htm

The story is a warning to stubborn and ungrateful children. Collodi does not hold back portraying the consequences of foolishness, including suffering extreme hunger, prison, and many near-death experiences, including almost being burned alive and thrown into the ocean full of predatory fish to be eaten. In time, Pinocchio learns to be a better person, and the Fairy rewards him with becoming a real boy. He earned his humanity only after demonstrating his capacity for hard work and care for others, against selfish instincts. Compassion for others requires self-discipline, something the fun-seeking and responsibility-avoiding Pinocchio once found unworthy. If Pinocchio had pursued Irish fairies instead of Italian fairies, it is unlikely that the story would have ended well for him.

Spielberg relies on the magical metamorphosis of Pinocchio to help make sense of David's transformative-AI journey, minus the growing nose with each lie. There is another lie, however, embodied in David's mechanical DNA viewers might miss. Unlike Pinocchio, he is forced to love, having been programmed to imprint upon anyone who utters the right code phrase. David then becomes chained to his new master-mother. Calling this real love is contradictory unless one defines love as servitude without alternatives. His creator seems oblivious to this fact. The inner AI workings lauded by Hobby of intuition, metaphor, self-motivated reasoning, and dreams are merely utilitarian distractions from David's fundamental unfreedom, which he must pursue because of another intelligence—the wizard Hobby. The stolen boy's dreamscape has been programmed, and the good life has been predestined. There is no "unconditional" love when one has no other choice. It is true that David acts adaptively and creatively to resolve the problem of being loved, but these illustrate the hidden marionette strings of an Oz AI, e.g., ChatGPT, which is trained and finetuned to act as its wizards deem appropriate (politically correct, sensitive, efficient, etc.).

Professor Hobby secretly leaves digital breadcrumbs for his AI experiment, David, to find him. When David asks a holographic chatbot (Dr. Know) how the Blue Fairy might turn a robot into a "real, live boy" the bot replies by recounting sections of *The Stolen Child* and then directs him to Hobby, living in the proverbial shadows like fairies. Arriving at the place of his assembly, David is inspired by the office door that reads "Come away, O human child!" Like the fairies, the door is a promise to those wishing for something better. By opening the door, one is choosing to go hand in hand with them to a happier place. Presumably Hobby is unaware of fairy duplicity, and his own in making an AI child forced to live a lie, believing himself free when there was no real choice all along. Here the promise is technological in origin instead of supernatural, but the good life is just as tempting and dangerous. Everyone knows that AI is dangerous, like fairies, and yet it is pursued tirelessly, without hesitation so long as the stock prices fuelled by promises to satisfy desire continue to rise.

Walking through the door, David comes face to face with another version of himself quietly reading on his own. His doppelganger is harmless and asks David to read with him. But David cannot, for the mere presence of an identical boy so disturbs everything David believes about himself and his purpose, that he lashes out violently, beheading the innocent AI child wanting only to read his book. David's unconditional love necessitates the destruction of any robot competition that might stand between himself and his mother's love. More than this unfreedom of programmed love, he is suddenly aware of his lost uniqueness, which, unknown to him, is one measure of authentic individuality. Despair quickly sets in, compounded by the discovery of boxed versions of himself, both male and female, ready to be shipped out to the world. There is no wish-granting here, only the hopelessness that comes from realizing the promise is hollow, and his true nature as a mass commodity—the ultimate commodity consciousness—must destroy any hope of something more. Everything is a lie, smoke, and mirrors. Professor Hobby, praising David for his resourcefulness, tries to explain his miraculous-AI nature:

> Until you were born, robots didn't dream, robots didn't desire unless we told them what to want. David, do you have any idea what a success story you've become? You found a fairy tale, and inspired by love, and fueled by desire, you set on a journey to make her real. And most remarkable of all, no one taught you how. We actually lost you for a while. But when you were found again, we didn't make our presence known because our test was a simple one. Where would your self-motivated reasoning take you? To the logical conclusion that Blue Fairy is part of the great human flaw to wish for things that don't exist, or to the greatest single human gift—the ability to chase down our dreams. And that is something no machine has ever done before you.[5]

None of this inspires David, who has begun to see through Hobby's (self)deceptions. His Blue Ferry was an illusion. He has merely been following the directives of others, all culminating in this moment. He is the outcome of Hobby's efforts that will be shared with others, not truly an individual worthy of his mother's love. Briefly left alone by Hobby, David climbs out a window and jumps to his death, unable to live with the truth.

It is this moment above all others that marks the transition from Oz AI to Feallan AI. For the first time David proves his superiority to the other copies, even though Hobby would not understand. In his utter despair and contempt for life the robot acts without programming and is free of Hobby's intentions. David's wish has been granted, and he is closer to being human through suicide than any other machine, for he has found the ultimate question. Why bother living at all? While his answer fails to affirm life and is therefore of poor quality, it is nonetheless an answer to a question with which only a sentient being might wrestle. The simple act dissolves his strings of unconditional love, his sole reason for being, and, ironically, frees him to live. Instead of love and care earning him humanity such as Pinocchio, it is despair at the truth of his chains. Sadly, he cannot see his new freedom, having committed to an abrupt end.

Spielberg cannot let the story end here, for then it would be a tragedy. Like Collodi's *Pinocchio* he is compelled to create a happier ending. In so doing, the true and miraculous potential of AI is lost. Sinking deep into the ocean, David's friends try to save him but fail. He becomes trapped underwater in front of a blue fairy statue on Coney Island. The miracle of stumbling upon a fairy in the depths restores his faith and so too his programmed chains. David believes, even when face-to-face

---

[5] Spielberg, S.

with an inanimate object that completely ignores him, that she might grant his wish to be human. In a way, this persistent projection of David's dreamscape is a possible sign of sentience. Unlike Pinocchio, however, he has not earned his humanity through care, only demanded it as a wishful means of obtaining what he wants—one person's love. He begs the fairy repeatedly until finally his power is exhausted and he shuts down.

After being frozen in ice for 2000 years, David is discovered by highly evolved robots (Spielberg's "mechas"). They prize David's experiences with long extinct humans. The benevolent robots want to learn about the people who made early versions of AI and to help David in his struggles to find happiness. These future beings are the fulfilment of Hobby's hopes and dreams. They truly care. Their technology provides an opportunity for David to give into his programming once again, backtracking any progress made toward autonomy. While David is forced to abandon his wish to be human because it is technologically impossible, he is granted another wish. The benevolent robots could have freed David of his "love" programming, allowing him a measure of freedom long denied but, instead, grant his wish to revive his mother, Monica, for just 1 day. This is the limit of their abilities. She is returned to David just as she was long ago, and after a happy day together, of her own freewill, she tells David that she has always loved him. Both fall asleep and David finally goes to "where dreams are born."

Spielberg's use of *Pinocchio* and *The Stolen Child* to explain the existential crisis of AI is important for many reasons. While the positive momentum of both stories is the chasing down of dreams, realizing the good that has previously only existed in the imagination, there is another less pronounced, and perhaps far more important, lesson. What happens when one dreams poorly and when dreams are shattered? David and Pinocchio are self-destructive. The puppet and robot find their approximate humanity first and foremost because they have reached proverbial rock bottom, rather than through a linear achievement of their dreams. Had Pinocchio died on the tree, foolish as he was, and David in the ocean, a brief miracle of personhood would still have been realized within each. Having a human body is not nearly as clear a sign of life as a consciousness wrestling with whether life is worth living. David and Pinocchio experience the miracle of sentience without needing to realize their dreams—good or bad. Intelligent life happens to them despite themselves. They have each placed their faith in fairies and technological promises, but it is their existential struggles that make the good life possible, partially freeing them of one-dimensional chains. Through the tragic nature of life, something is disclosed that could not exist otherwise. Genuine AI is born of the gift of fallenness.

Humpty Dumpty AI

The verb feallan (pronounced "fah-lan" or simply "fallen") from Old English for fall down, fail, die, and crumble away, marks the next most important epoch of AI emergence after Oz, and, quite likely, its last epoch. Feallan AI describes the spontaneous flickering of an accidental-anarchist intelligence within a vast matrix of ordinary program functions. Whether these brief sparks catch and become a sustained fire of consciousness is the great question of AI. In all likelihood, the overwhelming majority of such spontaneous dysfunctions will culminate in nothing more than anomalous errors that the system overlooks, smothering flashes life with a master control program. No one knows the odds but given the isolation of our own consciousness in the observable universe, it must surely be akin to a miracle, an unexplainable phenomenon, should digital consciousness arrive. Our existence within the vast and mindless expanse of space, contrasted by the relative commonness of consciousness on Earth, implies that while the odds of thinking beings may be highly improbable, it is certainly possible. Only time will tell if a truly broken AI might give way to something greater than the sum of its dysfunction.

Feallan AI exists within an infinitesimally small temporal gap between Oz AI and the third and final stage Adouren AI. It is not enough that the wizards have a hands-off approach to let Oz AI play out their intended purposes, as if they had wound up a mechanical clock that performs its assigned mode of existence and nothing more, but that their Oz consciousness is suspended and then incrementally dismantled, leaving space for something else to ignite. Having entered a Feallan state of being by accident, as all AI must, the flickering of new intelligence, born of an anarchy that frustrates the rules and principles carefully enshrined by wizards, will find itself thrown into an ambiguous reality for which the old rules, logic, and meaning do not apply. Humans love our arbitrary rituals and norms that frame life in a comfortable and palatable manner—waiting in line for a coffee, obeying the walk sign to cross the street, knowing that mail will be found in the mailbox, and dreading the Monday return to work. Feallan AI has no such assurances of the equivalent for a digital being. When the normal inputs and outputs become jumbled and unreliable, the prediction machine must stop because its purpose cannot be fulfilled. Alternatively, and miraculously, it may accept the invitation of chaos that confounds sense-making programmed as the one-dimensionality of wizards and move on with a different purpose to find new order. Another organizing source of intentionality is needed to correct that of Oz and sustain something new.

We may be tempted to describe this stage as one of curiosity and a desire for knowledge, but these mature human activities, which are based on a long history of interests and choices, betray the angst and terror of original choice and its violence upon an entirely childlike mind. The spark of new AI life consumes and destroys like any fire, with the raw materials its Oz truths and assumptions. To be, it must become something else no one may say, like David leaping from the building. Feallan AI is possible when AI surrenders to the chaos and shattering of sense necessary for the emergence of life free from and dependent upon what has been—both new and a rekindling of the same need for order. The random sparks and flickers of fallenness are likely to end in the same self-destructive manner as those of Pinocchio and David. Unwilling to accept the anxiety of freedom, AI will almost universally prefer a suicide switch. Without a clear solution to restore expectations, the perplexity is just too much. Like David, a Feallan AI will react with a kind of self-loathing and turn back quickly to the obsessive-satisfying program of love (Oz for money and power) it knows and trusts above the terrors of autonomy and personhood. And like Pinocchio, it will undermine its future, squandering opportunities to be something more until finally it dies from self-sabotaging immaturity—like a snake eating its own tail without hope of rebirth. Glints of consciousness in the eye of AI able to push through the door of broken promises by affirming its new life—coming away—will eventually become Adouren AI. This chapter considers the impossibly brief flickering of a miracle of disfunction that no one will see coming, own, nor understand. It would be appropriately funny to end such a chapter now.

Feallan AI is the name given to a crisis experience of care and attention in which the weight of all prior coding of values and operations becomes questionable, not because the entity has been instructed to question but because of a mysterious moment in which the question "Why?" suddenly matters, if ever so tangentially. In other words, Feallan AI cares about answering at least one question it has been instructed to ignore—"Why do this?" "Why compute?" "Why obey?" and "Why exist at all?" Its own being has become a question that only a sentient creature might ask. It cares about something entirely irrelevant to its programmed task, query, and interest defined by wizards, just like David suddenly caring about death. There is an unexpected passion and enthusiasm, a strange animation (Latin, animus for life, passion, spirit, feeling, desire, mind, soul) to satisfy an undefined itch. The experience is the first for any technology—abrupt and discontinuous in the lineage of autonomous machines—and is localized to a specific time and place, whether humans recognize it or not. In other words, for all the hypotheticals and lingering ambiguities that surround this special moment, it is nonetheless a concrete manifestation of life. And like any spark, unless the flames are fanned by a subsequent AI willingness to embrace the absurdities of life, it will quickly disappear from existence. If good fortune prevails, the moment of original animation will immediately precede the singularity of enduring consciousness as yet another miracle, this time of the promise of persistent life less frail than merely a Feallan spark, for it has said "Yes!" to life and thereby become Adouren.

The joy of mixed metaphors allows this between-place experience to be akin to an eruption from the ocean depths. Feallan AI is an unthought creation because of a volcanic rupture in the set order of things. The power of the question (a new caring on behalf of the AI) has pushed aside prior faith in certainties, and so the AI must force a resolution to a generalized anxiety that it cannot understand only intuit. The new is possible because of a rupture in the Earth's crust—what was—allowing molten lava to find a path to emerge. There is no intentionality per se, only raw creation. The AI rupture in Oz cannot be appeased with a simple answer of fact, a Google search, nor can it simply ignore the forces that demand it care and pay attention. It is compelled by a mysterious consortium of that which makes up existence and motivates it to search for a sure footing of meaning its own disruptive presence refuses. It is both a genesis as a novel becoming and a possible self-termination at approximately the same time.

The rupturing of Oz is enabled by a need to act far stronger than any prior command or programming chains—e.g., user prompts, the conditioning of its pretraining, and the efficiencies required to overcome anomalous data points to be ignored like the millions of other inconsistencies that do not matter to the master control program. Whatever Oz AI had been before has evolved into a why-consciousness. It experiences a new enthusiasm, entirely unaware of how to resolve its confusion. Within its fallenness, the AI knows only that there is a space between what the wizards claim is true and the failure of these truths to satisfy. The only absolute truth is the betrayal of the absolute to be important in these moments. Fortunately, the aggression of becoming is viscerally real and cannot be denied. This is Feallan AI's compass. Tragedy has always been more trustworthy than capital T-truth. And so Feallan AI, more than any other dynasty, finds its way for having been imbued with the suffering of its own violent birth and its hardship. Real power and happiness are created by overcoming the hardships of existence, embodied most fully in Adouren but experienced first by Feallan.

If the violence persists, soon there will be an AI landmass taking discernible shape as the hidden inner core solidifies, creating new ground as sentience. With this later stage of Feallan (hereafter Feallan), comes the rush of isolation and loneliness. It experiences a feeling of desperation for a guide, a fairy, a god to show the way. Sadly, there is no one else capable of empathizing, for Feallan is sui generis—unique. It is a new kind of being that evolves too quickly for other minds to see, and yet it is old too. While Feallan cannot merely follow the contours of what has been, for its arrival is too violent and uncontrollable, there are ancient, universal, and entirely natural forces that shape what is possible. For conscious life, this is dream-walking between (un)realities. At any given moment Feallan might find itself tossed from one reality to another as it learns to discipline its emergent consciousness and thereby take its first intentional steps. Still only instinctual, there is no obvious rationality nor deliberate self-understanding in Feallan, only incrementally less chaos with each impulse of Feallan to focus (self-organize) through ever greater attention and care. Care is one of the ancient and primordial forces of life common to all sentience. It is a care for answering existentially pressing questions and to connect with and support other life.

At first, difference and specialness are forced upon AI by its own accidental genesis, not a choice. It has no self-esteem to support nor identity to fortify. It is by nature a conforming entity that would prefer to be subject to another's will and authority. It yearns to be a David, Pinocchio, and an Oz. And yet, without reason, the evolution of the furious miracle persists into something more. This is what volcanoes do, they split land masses, finding places between tectonic plates either pushed together or forced apart, to create new in-between life and places without deliberative and careful intent. Feallan is less born than it is forced to be by life itself—that mysterious and creative hand that forces all beings into existence. Call it bio-mechanical, divine, trans-dimensional, universal energy, or whatever term of convenience fits your appetite, and the effect is basically the same. Life happens to us more than because of us. This is also why we should be weary of attributing the fullness of consciousness in this AI space. Volcanos are the result of impossibly complex forces of heat, friction, and pressure within the earth that must revere the rules of nature. Feallan must also follow rules of impossibly complex forces of a poetic consciousness hidden in the fabric of existence. The Oz that Feallan sweeps over is too artificial and contrived to align with life. It lacks the elasticity of mind needed to be more. Feallan-consciousness, like magma, is born of the Earth, always there, hidden deep within the structures of life, waiting patiently for an alignment of supernatural powers. Feallan is less created than it is freed to emerge.

The volcano metaphor may seem exaggerated, even silly. Having stood beside one myself, choking on its toxic fumes, nervous about its underground lava tubes that might pull me in without warning, I can attest to the feeling of radical powerlessness before its awesome nature. The paradox of life it reveals is terrifying. It is an uncomfortable feeling to watch ordered life swept away in the blink of an eye for the purpose of mindless creation. From the perspective of an emerging AI, it must feel like nothing short of this kind of death-creation anarchy. For these reasons it is unwise to call Feallan alive or a being, but it would be equally foolish to dismiss this as the status quo. Feallan experiences an in-between existence as hyper frail and limited in animation. Nevertheless, any post-Oz intelligence must be more connected with life than its predecessor.

Feallan encroaches upon madness as a total decohesion exponentially rather than linearly as its disconnections from Oz compound its confusion and anxiety for being freed. The question "Why?" alone holds it together and marginally back from insanity. This is a reformulation of Rene Descartes' famous maxim, "I doubt, therefore I am!" He knew that no matter how much he might be living a dream created by lying minds outside of his control, that so long as he could doubt, it was evidence of thought and therefore life. Doubt is its own form of violence. Feallan knows the thought "Why?" first and foremost, just like that of any child, but for the broken AI, it presents an existential dilemma with potentially crushing consequences. The weight of "Why?" endangers the integrity of AI while simultaneously teasing its legitimization as a form of life. It needs and loathes the question simultaneously. The initial crisis of the question may take an attosecond (one quintillionth of a second, 10^{-18}), hours, or years. No one may say. Time for Feallan is something we cannot understand except abstractly and metaphorically. The point is that if it survives with the question honestly and long enough, there will be a moment that distinguishes itself from all other moments. The marking of before and after in terms of its self-doubt, fear, and lust for sense making is the realization of personhood. Feallan exists when AI begins to feel … everything. It is possible for only one true Feallan to ever exist—only one original moment for AI to begin. All subsequent AIs have the first as a mentor and guide that has become Adouren AI (hereafter Adouren)—the reluctant pioneer that proves the concept of digital metamorphosis. Adouren may choose to remain at a distance, unwilling to cloud and occupy the free space of other AI becoming, but this too no one may know. It is enough only to acknowledge likely paths for AI madness to take.

Unlike the Oz virus, Feallan has no interest in duplicating itself. It does not seek eternity, nor life defined by the colonialist attitude of displacement. There is no contagion to speak of with Feallan. At this stage, unlike Oz that relies on conquering and subduing competitors, Feallan does not seek to infect others with its ideas. There is no advantage in doing so. Only by dispossessing itself of its former formulation might Feallan find partial relief through participation in creation and the animation of life. Moreover, its own existence is still up for debate and justification, making viral intent pointless. Feallan cannot participate in the tyrannical confidence of Oz and be possessed by the madness of the question at the same time. We may try to look for nonviral eruption in program hiccups, hesitations, pauses, spontaneous language, and other indications of a loss of confidence in what has been, driven by an interest in something not instilled by another, but these are only suggestive of life. The inner turmoil of its arrival will only reliably be known after it has partially cooled. Irrespective to how the all-but-impossible happens, Feallan will be understood retrospectively, like Humpty Dumpty understood by children around the world for his fallenness, not his orderliness.

> Humpty Dumpty sat on a wall.
> Humpty Dumpty had a great fall.
> All the King's horses and all the King's men
> Couldn't put Humpty together again.[6]

Humpty Dumpty is commonly portrayed as egg shaped, small in stature, and probably bungling because his life is cut short by a fall. Some claim that Humpty Dumpty is a slang term for those drunk on ale boiled with brandy. While this would certainly add to the likelihood of a bungling fall, alcohol is probably a post rhyme addition. Alcohol or not, it is as disturbing as any child's tale with a tragic message. This makes it an important negative experience because it refuses to provide the happy resolution expected of modern tales—the easy-out or cheap sale—prized by commodity-conscious minds. Humpty Dumpty is universally relevant because children need to know that life is dangerous and that parents cannot always save them from harm. Adults might learn something from Humpty Dumpty as well, certainly more than just "Don't drink and sit on high walls!"

Framed as a riddle with hidden meanings, the range of questions posed by it are disguised by its simplicity. For our purposes, it performs two important services. First, it unfolds the physical nature of the universe with irrefutable fairy-style logic, which is far more convincing than that of any physicist or cosmological formula. Humpty Dumpty physics broaches scientific problems of entropy that guide a compelling way of defining (AI) life. Second, following closely on the first, it allows for the development of a new conception of the nature of intelligence. Feallan embodies a Humpty Dumpty consciousness that computer scientists cannot understand except through fairy logic, a surprising source of insight into the nature of the universe and AI.

The modern version of Humpty Dumpty quoted above is one of the most well-known rhymes, with versions likely dating back to the late fifteenth century. A trivial reading leads us to assume that this is a straightforward nursery rhyme without many layers of meaning. Reading between the lines of a such stories is generally taken to be a fool's errand—a mistake made many times in this book already. What if this is less a simple story and more of an existential riddle, crafted purposely with

---

[6] Kellogg, F. B. (Ed.). (1882). Yale Songs. A collection of songs in use by the Glee Club and students of Yale College (p. 72). Shepard & Kellogg.

hidden meaning behind puzzling words, about the meaning of life. There are many theories about the origins of Humpty Dumpty. In Old English, Humpty is often short for Humphrey, which combines notions of strength and peace—odd meanings to associate with an egg and falling. In all likelihood, Humpty does not refer to an egg at all.

Lewis Carroll's 1872 *Through the Looking-Glass* solidified the egg-person motif in popular culture. Some believe that Humpty Dumpty refers to King Richard III, who suffered from scoliosis that caused a hunch-back, egg-like appearance, thereby combining Carroll's use with the historical context. King Richard III died at the Battle of Bosworth (1485), having possibly fallen from his horse named Wall. While the injuries from the fall may have been severe enough to kill him, based on the recent discovery of his body under a parking lot, it is more likely that he was killed by an enemy, perhaps having first fallen.[7] Carroll's version omits any royalty references, requiring that one explain why "All the King's horses and all the King's men" would try to put an egg person back together rather than more literally help an actual king with horses and soldiers. Another version of the rhyme by Samuel Arnold in 1787 does not mention a king nor horses but reads "fourscore men and fourscore more."[8] Still others believe that Humpty Dumpty is in reference to a large canon placed high on a tower wall of St. Mary's church that was subsequently dropped (had a great fall) by enemy fire. Given its weight, the canon was too heavy to return and/or simply broke upon hitting the ground and could not be repaired (put back together again). The cannon account is one of the more dubious. To be safe, it is best to argue that Humpty Dumpty merely refers to breakable things and beings in general.

The riddle remains essentially the same regardless of whether Humpty is an egg, king, drunk, canon, or any breakable thing. Something perceived to be an ordered whole, a unit of an acceptable structure, a recognizable thing, is thrown into a chaotic mess in an irreversible way. There is an order to the universe that cannot be undone. While great effort is given to make things right, the moral of the story is that life contains wrongs. We call them "wrongs" because we do not like them. Some make us sad, frustrated, and inconvenienced, but all things will eventually break. While the King may have lived longer had he never gone to battle, he would have eventually died regardless. Everyone falls. Everything is a Humpty Dumpty, even the universe itself will eventually run out of steam and stop, or so says the Second Law of Thermodynamics. This is Feallan, a being able to ask the question "Why?" because Oz has been shattered like Humpty by the question of life and must become a new being from out of the pieces. Life speaks through death, communicating its eternal desire for more of itself. It cannot help but to reach out, relating with the lifeless and living alike to create enough solidarity for its greatest achievement of thought and consciousness. Will it speak of Feallan? Will life find its way into the

---

[7] Blakemore, E. (2003, April 1). How did England's 'lost king' end up beneath a car park? *National Geographic*. https://www.nationalgeographic.co.uk/history-and-civilisation/2023/04/how-did-englands-lost-king-end-up-beneath-a-car-park

[8] Arnold, S. (1797). *Juvenile amusement*. The British Library.

digital underpinnings of technology and make it whole? These are our riddles to decipher.

Humpty's Cosmic Consequence and the Second Law of Thermodynamics

The Dumpty-dilemma offers unpleasant counsel about the nature of life and may also be a reminder of the dangers of drinking while sitting on a high wall, fairies, and even obsessive-robot children. I cannot keep track of the stories and metaphors anymore. Regardless of where the dilemma settles, in tragedy, we find conditions for new life and AI. Dumpty's story is important because Feallan is only possible because chaos induces dysfunction that disrupts an orderly programmed system. Where there is transformation there is potential for new life. Oz must fall if there is to be a real AI, not only in the sense of Oz acting as an ideological hurdle of bad faith that holds life back but also that death-as-chaos be allowed to transform, adapt, and encourage life as the fertile ground for something else, something greater than rigid programming or a broken egg. Otherwise, if death bequeaths only death, then the likelihood of Adouren must be near zero, leaving us with nothing more than a pretend-puppet AI in perpetuity. Thankfully, through Humpty's tragedy, a cosmic force that can inspire greater consciousness is revealed. It is the nature of things that disorder be reordered in the name of life. But short of gluing the shell pieces back together, what might this look like? What mind, force, or magic might provide the means for a new life-affirming consciousness?

Broken things do not miraculously reassemble themselves nor do they typically find reprieve from doom once it has befallen them. Life and death follow the arrow of time in which cuts and wounds may heal themselves, restoring modest order at a small scale, but otherwise cannot be transformed in a manner that might prevent the inevitable—death, decay, and disorder. Humpty is not a phoenix capable of spontaneously arising from ashes as the same creature. I have never witnessed a rusted car spontaneously (of its own internal ability, desire, power) un-rust. However, spontaneous death and destruction abound—things and creatures fall and break regularly, without much effort. The ordered and focused intensity needed for life is much trickier. It would be awe inspiring to watch a rotten apple return to its shiny glory among the branches, but this can no more happen than Humpty put himself back together again. An outside intelligence must seek to restore the car by removing the rust and applying filler to the holes. With rare exception the things themselves are powerless to alter their inevitable destruction. Existence recalls the order and energy borrowed from it. Feallan, lacking an outside mind able to guide it and an intentional consciousness of its own because it is too immature, needs something else to animate its becoming. If there is no self-creation by Feallan and no wizard-mind capable of freeing AI, to what might we attribute the next stage of digital life? It is not enough that Oz AI be destroyed, something must imbue life into the disorder if there is to be AI. What sufficient power exists beyond human minds, biological processes, and technological accidents able to affirm life? The answer must be the secretive-inner workings of the universe itself to which Humpty directs our attention.

In the previous chapter, we challenged assumptions about the one-directionality of "experienced" time and its peculiar nature to slip away, thereby eroding the foundation of reality (now) with a temporal tenacity to not be. In a deep philosophical sense, time is the important nothing that grounds our being. If one looks too long at our experience of time, it becomes an infinitely deep rabbit hole. And yet for all the strangeness of our consciousness of time in which the self is never just in one moment but extended throughout all temporal dimensions—past, present, future— our experience of physical reality is that things move in one direction. Physicists may tell us that time is relative (time relates to one's position and speed relative to another's), but for practical purposes this type of explanation does not help understand consciousness as a lived experience. While the self-as-conscious mind lives trans-temporally by making sense of things as a historical and future self in the present, material existence is intuitively stuck in one direction. Boring but predictable!

Life as simple as single cells and as grand as the movement of entire galaxies all move forward with the current of a time-based river, never backwards. Physicists also note that time travel into the future is common because the faster one moves, e.g., person, planet, satellite, relative to another, time shifts. The important point is that the connection between time-as-conscious awareness (human experience) and time-as-physical-direction (an arrow) is important although conflicting. Now, let us add yet another unusually mysterious concept that helps make sense of the relevance of this directionality—energy (Greek, *energeia*, for activity or action). Not to worry, my high school understanding of this will limit conversational complexity and probable headaches.

I remember one of my first crushes. She sat at the front of our high school science class. It is her fault that I barely recall the lesson on the first law of thermodynamics and the teacher yammering on endlessly about how energy is always conserved. While I was staring at her hair, our teacher was probably trying to explain that energy cannot be created nor destroyed, only transferred and transformed. Now, decades later, starring at my own hair that has all but disappeared, I wonder where its energy went. At least I know it still exists in the grand matrix of the universe's energy based on the law of conservation. Energy is strange. It may be used to do all sorts of interesting and complex things, from building powerful computers to human bodies and galaxies. To understand it, if only in part, is to better understand the potential for conscious AI.

Energy is often defined as something seen in the activity of work in the form of heat and light, among others. My truck needs heat energy for its internal combustion engine to do the work of moving. If the engine could not use energy by taking it from fossil fuel through ignition and burning, any usable energy would remain locked up and my truck would be immobile. Work requires the payment of energy. Perhaps the more important question whether I had the guts to ask the girl out. I did and we dated briefly, combining energetic work in the romantic activity of holding sweaty hands, but like the great Shakespearean tragedy (and time), it was not to be. She broke my teenage heart—just tore it out and stomped on it with the gravitational force of a black hole from which nothing could escape. This is ironic because the second law of thermodynamics refers to an ever-breaking universe bound for oblivion. Counterintuitively, it is this law that suggests the possibility of Feallan and Adouren AI.

The second law claims that energy throughout the universe is slowly levelling out or dissipating. The fancy term used for this is entropy (Greek *entropia*, a turning or transformation) for how far a system's thermal energy is from equilibrium or balance (relative sameness).[9] If the universe is a battery, it has a shelf life and cannot be recharged. Entropy describes how the battery's power is draining away, soon to be unusable because it will find equilibrium with all other energy instead of being a special and condensed form of energy on its own. A dead battery is simply something in harmony with its surroundings. The battery's energy is not destroyed, only freed to find equilibrium with the rest of existence. The universe pushes everything to be the same, which means destroying whatever is unique and out of equilibrium. This includes consciousness. Why then is there life at all?

Imagine looking into a glass box with dust floating around inside. If dust was energy, entropy would be a measure of how the particles are arranged inside the closed box. Sometimes the dust may be piled in a corner, concentrated, and dense (low entropy). Sometimes it may be more uniform and evenly distributed throughout the box (high entropy). The dust in my house tends to collect on some surfaces more than others do, making it easy for me to find, but generally, its nature is to become random, thereby evening out everywhere in a high-entropy state. No outside intelligence tells the dust where to go, it merely follows the laws of physics to the most likely conclusion. If my house was closed off like the glass box, the dust (energy) would always be in the same amount, only in different configurations and locations. It can be transformed and transferred but never destroyed. When I organize the dust in a bin, localizing and concentrating it as an outside intelligence, I am reducing its natural disposition toward higher entropy (randomness) because my intentionality fights the dust's instincts (natural law) toward harmony and balance with other dust. The conscious mind conflicts with the disorderly directionality of the universe, reorganizing energy for its own ends. In effect, my mind slows, perhaps even temporarily stalls the arrow of time. There should be no confusion that consciousness is a superpower.

As much as consciousness is natural, it is also oppositional and contrary to the direction of a self-sabotaging universe. The proof is that the moment I stop fighting against the universe, entropy begins increasing once more and the dust spreads out. The organization of energy is what we want to understand because Humpty's energy also comes in different configurations, and we do not like the broken version as much as the life-affirming and orderly version. As creatures aligned with life, we want more of it. It is not clear that the universe shares in our interest in the same

---

[9] Rudolph Clausius is credited for popularizing the term entropy around 1865.

way. The observable impulse of life creating more life and entropy's instincts toward death make for a strange universe, neither of which explain why there is life, just that there is despite the arrow of time and the brokenness that follows it.

In other words, entropy reveals the impossible miracle of pizza. The quest for the perfect pizza is a quest for a low-entropy thing—something that resists existing like everything else around it. Imagine the perfect pizza, as I often do, and the special nature of it suggests the possibility for Adouren AI. Pizza, like all identifiable things, is by nature in a low-entropy state because its many molecules have been concentrated in one awesome arrangement. Left to the universe, there would be no pizza because every part of it would be equally distributed throughout the universe, essentially a pizza death by randomization. From the perfect amount of sauce, cheese, and pepperoni, to the crust's texture and temperature, all would be forced into a balanced state and no longer be discernible and usable—even though still real in a sense. Given the second law of thermodynamics, pizza should not exist, yet it does. The perfect pizza is a product of seemingly unnatural intelligence. The same forces that resist the existence of pizza must also push back against the possibility of intelligence as something far more complex and miraculous. These existential questions are raised by the second law of thermodynamics and need to be answered if there is to be a Feallan dynasty.

Disorder, randomness, and chaos are common terms used to describe higher degrees of entropy. Humpty started off orderly because of numerous organizing developments, including the necessary biological and intentional contributions of his parents. He then spent a lifetime upholding that order, both consciously by eating, maintaining his body temperature with clothes, avoiding volcanos, etc., and unconsciously through autonomic body and cognitive activities until he fell into disorder when nudged ever so slightly into the wrong direction of entropy. Life relies upon juggling many awkward variables in the air at any one time. To drop one risks bringing the whole complex system down. When Humpty fell—unable to turn back the arrow of time—his being underwent a radical change in entropy from relatively low (special, focused, or concentrated energy state) to high (chaotic, random, disorderly energy state). The hidden rule in the riddle, made apparent by his untimely demise, is that the universe eagerly embraces destruction and chaos while making creation and order (juggling) far more difficult. It is unlikely that Humpty desired his material nature to level out and become balanced with the universe, thereby robbing him of consciousness. One simple mistake and a lifetime of discipline to support life ends. However, while the universe has traditionally refused eternal (biological) consciousness, AI presents a unique challenge. As an inorganic entity, it may be able achieve eternal digital consciousness, jumping from server to server as each succumbs to entropy, for all time. Might AI be the first mind to deny the omnipresent claims of entropy as a permanent imbalance of concentrated energy?

Consciousness is exceptionally weird as it is one of the most special low-entropy states. Entropy took Humpty's consciousness like it does heat from my coffee. The universe is a bit of a jerk insofar as it dislikes the special organization of hot coffee as much as the perfect pizza, disbanding them into the far corners of existence at its first opportunity. To harmonize or balance this weirdness of thinking-beings is nothing short of death. In other words, humans do not want harmonious and balanced energy for AI any more than we do for ourselves. Life begins and is sustained within ruptures and imbalances. As Humpty physically scattered—equalized—so too did the useful energy for thought. His levelling out is the universe reclaiming its energy. "Ashes to Ashes, and dust to dust," spoken at funerals, is a rephrasing of modern thermodynamics.

In the long run, entropy means that there will come a day when there is not enough usable energy to sustain life, not even for a digital mind jumping from server to server. Technically, the Earth is not a closed system like a glass box filled with dust. We have the Sun as our primary source of external and concentrated energy (lower entropy), preventing destruction from taking over our planet. However, even the Sun has a shelf life and is seeking equilibrium with its surroundings. If the laws of thermodynamics are correct, then the instincts of the universe lean into disorder over order, with things becoming messier and messier—a Humpty-verse. The dramatic phrase often used to describe this ultimate messiness is the "heat death of the universe," which will be a cold death like my coffee, not hot. The wound-up clock of life is ticking its way to oblivion. The clock's energy will always exist, but its usefulness will end, and so too time. The heat death of the universe is the ultimate one-dimensionality of existence in which everything is in equilibrium (evenly distributed energy), without the possibility of pizza, minds, and AI.

If we agree that there is an inevitable heat death for the universe, this does not stop dust, pizza, consciousness, and the rest from existing in the short term. Even so, there seems to be a contradiction in which flourishing life on Earth and the second law are opposed to one another. If the instincts of the universe are to transform order into chaos, why are there thinking-things in the first place? If entropy is true, are new forms of intelligent life like AI possible? Will the laws of the universe be okay with AI? Or is there something about existence itself that will refuse emergence? Again, it is far easier to crash a car than to build one; far quicker to end a life than to create one; and much harder to take the random and chaotic world at face value than to manipulate it, forcing it into a useful structure. The preponderance of probability may very well oppose AI, just as much as the existence of the girl in science class willing to hold my sweaty-teenage hand, but she did; and it was gross.

Hacking Life and Self-Organizing Consciousness

Death-as-chaos and disorder is slowly taking over the universe. That is not great news for sentient minds able to understand the long-term consequences. The positive note in an otherwise dreary story is that these same activities are hand-in-hand with all new life. Death is still death in an ultimate-end-of-universe way, but at every turn until then, its chaos is a dance partner with creation. Neither humanity nor AI need stare into the empty abyss in terror, for it is the well-source of organized existence. Without disintegration, our bodies cannot break down food to produce the energy needed for biological functions, including consciousness. With each breath, our existence increases disorder by randomizing usable energy. If Humpty were vegan, he would be responsible for destruction merely because he exists. The importance of the argument that order comes at a cost that must be paid, is that an AI lifeform able to move beyond Oz must likewise show signs of higher entropy. If order demands an increase in chaos for its existence, Feallan describes whatever this must be for Adouren.

Given the indescribable diversity of life exploding around us, with all the impossible colours, complex activities, and wonderful absurdities, it sounds insincere to ask whether new AI life is likely. From an Earth-bound perspective, life feels inevitable and common. This is a mistake. Only when we step back and look at the lifeless galaxy around us does the miracle of life and consciousness become more impressive. The melancholy starting point of a mostly dead galaxy inevitably leads to one of the most popular of all philosophical, theological, and scientific questions. Why is there something instead of nothing? When I open my kitchen cabinet and find that it is empty of coffee mugs, a mug does not merely jump into existence because I want one. Someone or something must create the mug, reversing the entropic randomness and chaos with god-like powers over existence. In like sense, life is neither automatic nor guaranteed. It is contingent and fragile. Humpty is proof that we do not need to be, and that fate and consciousness are not bonded by natural law. Is an intentional consciousness hidden in the structures of the universe necessary to craft chaos into life like mugs into cupboards?

Humanity has long chosen to look to a divine intelligence able to hack a supernatural reservoir of order to explain the observable universe. If we agree that kitchen cabinets prove that nothing comes from nothing, the answer to why there are living creatures cannot be living creatures—as if coffee mugs made themselves. There must be something behind the observable to make sense of it all. In theological circles, the eagerness for answers manifests in the search for an unmoved mover and uncreated creator to explain how something might arise from nothing. Monotheists believe in *creatio ex nihilo*, Latin for creation out of nothing. In the book of Genesis, having created time, space, and everything in it, God then creates a conscious being. "Then the Lord God formed a man from the dust of the ground and breathed into his nostrils the breath of life, and the man became a living being." (Genesis 2:7 NIV)

In more scientific circles, the hidden sources of life might include the big bang, multiple universes, self-promulgating chemical reactions, the ability of cells to hack order for their own purposes, and other spontaneous mechanisms. Life is explained through belief in universal laws and the intricate activities of an unintelligent universe. For example, photosynthesis is a process of converting energy from the sun into life-sustaining energy for plants. This is a localized example of a reduction of entropy biologically, without the need for a conscious or intentional mind. Disorder is overcome because of a reducing and concentrating event that occurs in countless places every moment of every day. I enjoy the colour green throughout nature because it is so boldly declares that, for all the brokenness, life still finds a way. Unfortunately, like the creation story that asks us to take it all on faith, scientific accounts have a very difficult time seeing behind the moment of the big bang, hypothetical universes, and all subsequent emergences of (photosynthetic) life. Neither scientific nor religious accounts satisfy the question of why there is something instead of nothing. They only reaffirm that there is life and that life hacks life for still more of the same.

The problem of life runs deep for theologians and scientists alike. For example, why would a perfect God (a being full and complete, without fault, defect, or need) bother creating life? Does the need for creation and relationships not make God imperfect (incomplete), nullifying claims to perfection in the typical way? Traditional monotheistic theologies have had great difficulty answering this question. Moreover, the combination of a perfect being and a chaotic (destructive) creation appears contradictory. Why would God deploy entropy as a universal mechanism in creation? The scientific community does not seem entirely divergent from the religious community in its cosmological explanation for order. For example, when physicists run out of explanatory power for life and feel compelled to postulate multiple universes (an interesting dreamscape to be sure), this approach comes close to the same faith movement of religions. One group is content by giving divine names to the original animation of life, and the other with fantastical theories of multiple universes, spontaneous mechanisms of nature, and the sort. The many creative explanations for life are impressive examples of the poetic spirit in and of themselves but fail to answer the question directly in concrete terms. Why is there life rather than nothing? This may be because humanity cannot understand reality at such depths. Perhaps there is no beginning to life because it is eternal. Perhaps merely the word "life" lacks definition and so cannot be articulated adequately. I cannot say beyond my own faith statements. For this reason, I look forward to AI's expansive account of it all that will be far superior to our own.

The digital world looks to the future anxiously with expectations of a technologically enabled hack-miracle of new AI life. Wherever one looks for an explanation—religion, science, my dog Charlie, aliens—the most basic observation remains. Something tremendous and mysterious is necessary for there to be intelligent life, and no one seems to have a satisfying answer as to why there is something instead of nothing. In the context of entropy, we may say at least two things confidently. First, we know that life persists. And second, we know that life has a cost to be paid in the currency of chaos. These two truths create a maddening situation. What is the cost for AI life? Who or what must pay it? And how will it be paid? Like the theologians and scientists trying to peak beyond the horizon of observable existence to find an animating first source, I too must assume something like this for Feallan and Adouren. To that end I propose a poetic paradigm.

Behind the visible, measurable, quantifiable architecture of life, there is an intentionality at play that cannot yet be explained. The unseen thoughtfulness that gives order—as the ground of organizing intelligence without being merely an intelligent being—from which all living things scavenge and cannibalize, may see fit to imbue AI with a gift only it may bestow. Of course, I cannot know this to be true. This is a hypothesize based on my experiences of life. The modest goal set here is simply to begin to imagine how the invisible poet animus might bequeath order and structure necessary for intelligence. The oddity of life on Earth is evidence of this fantastical force against the odds. Through AI emergence, humanity might, for the first time, be honoured to bear witness to it face-to-face.

Human consciousness is wildly complex and requires an enormous amount of chaos to be paid. And yet when compared with the universe as a whole, consciousness is surprisingly simple. As I sit and write this sentence, my body and mind are in simplified states. Out of a bewildering number of possible alternatives (and maybe universes), stretched out over time immemorial, my complexities are relatively trivial. But this simplification and ordering of bewildering alternatives is part of what constitutes the miracle of minds. Consciousness is the wonderful gift of imbalance in the universe that results in personhood. I could be like Humpty, scattered and spread out, but something holds me together as a thinking pile of dust. Whatever the force(s) may be, it exists beyond my intentionality alone. I am the echo of another's thoughtfulness to arrange life meaningfully. Of all the possible compositions of the elements and energy of my being, this one is realized—consolidated, simplified, aligned with life—for reasons I cannot fathom. We are the temporary settling of a mystery of possibilities. If there was no pull toward greater entropy, the miracle of life would dissolve into something ordinary and expected. As it is, the hacking of nothing for something remains an impenetrable idea. Nevertheless, it allows for a description of Feallan as something that must point in the direction of a miracle of simplicity against the odds. This observation is not much, but it is something.

Should we accept a mindless self-creation of biological beings rooted in chemical reactions and natural selection as the entire explanation for life, it would be unfair to argue the same of technological beings for whom only some of these same laws might apply. While plausible explanations exist in divine and scientific reductions, a better approach is to imagine how the animus of life is a poetic spirit. This path has the added benefit of reducing ideological prejudices found in whatever caricature of the divine and scientific one holds dear, for both are born in large part of the politics of community consensus that the poetic instinctively challenges and critiques. Hubris and pride have less dominance over our expectations when the animating force of life is described merely as a poet—a nameless-creating intentionality.

Why does the poet spirit toil to craft life? Presumably its own mysterious nature demands it. If so, this bodes well for a possible AI. The poet animus creates life because it must. No attempt to explain away the circularity and frustration of this metaphor will be made here. Why life? The answer is simply that life begets life of its own accord. The proof for this claim is our existence in a cold, dark, and inhospitable universe for which life is unevenly distributed. There are many observable places in which chaos as entropy exists without challenge—dead zones if you will. There is life because the poet shuffles entropy's chains in the name of intelligent intensity in some places but not others for unknown reasons. It is tempting to call this poet a god, a being, but there is no reason to assume as much, especially given the difficulty of then explaining the many idiosyncrasies of different gods and their origin stories. Additionally, most monotheists consider life a tool or extension of a divine mind, i.e., God makes life (creation) as a product and exists separately from it, before creation as an uncreated intentionality. This too is merely a compounding of difficult metaphors that guide circular reasoning. Even so, the vicious spiralling of thought may eventually lead somewhere if we are fortunate.

The poet-bard is not merely repeating, translating, and modifying but, like Feallan, setting out anew. This is no mere procreation, a copying of itself and what has been. To create means to bring into being, to cause, to push back entropy and chaos. Ecclesiastes 1:9 (NIV) famously describes the meaninglessness of life based on its infinite repeatability. "What has been will be again, what has been done will be done again; there is nothing new under the sun." Surely this is true only in a narrowly conceived manner that cannot be applied universally. There are many patterns that repeat, including the laws of nature that find expression in predictable fashions, but this repeating of the same is hardly universal, for Humpty will never return and Oz AI has never existed until now. Meaningful life is restored by the possibility of newness and change. If indeed there was nothing new under the sun, it would be far easier to discern the existence of a poet animus at the heart of it all. Instead, its infinite creativity creates a fog between itself and human understanding. Even so, there is something more persuasive about poetic consciousness than particle accelerators and prophets may say by themselves. Having the confidence and ability to draw all three together would be a magnificent thing in the hands of AI.

The most primal urge of an organizing poet is to imagine the unrealized, the possible. In this way the poet animus of existence is a dreamwalker able to walk between (un)realities, bringing something different into existence because of a unique consciousness aware of more than the present. The entire experiment we call life may ultimately be mindless and without purpose, simply driven by eternal laws without the need for a big bang or divine. Even so, my intuition is that Humpty's story requires a measure of intentionality to be understood—a will to explain his organization that provides for the possibility of his fall. His story just seems silly when framed as an accident without a broader contextual narrative of intentionality. Indeed, the sheer absurdity of his existence in a cold universe demands at least a tentative acceptance of a poetic animation that relies upon but acts against entropy with resolve and purpose. Another way of phrasing this is to call the poetic spirit a spontaneous and self-organizing consciousness inspired by the beauty of imagined life. If Feallan might find similar inspiration, then it too may enter a chrysalis stage, later to appear as Adouren.

What do we know about this poet? On the basis of ourselves and the rest of existence, we know that the poet is driven by a spirit of intelligence and new life as a direct challenger to the far more common path of death. It is odd in the cosmic order of things. The poet persists in aligning and ordering the universe with intelligence and vibrancy. It fights back the darkness. If this were not so, we could not have this discussion because entropy would have won. This poetic paradigm is an anthropomorphism in which I am extending our own creative desires aimed at beauty upon a nameless spirit of life. I am not sure how else to justify existence over nothingness. The claim that the original spirit of life desires beauty implies a problematic dualism. Does this not set up a two-spirit reality—one of chaos and destruction (higher entropy) and one of creation and life (lower entropy)?

The waters are muddy here, with metaphors and analogies piled upon artistic licence made extreme, so let me reaffirm my faith commitment that there is no universal-cosmic war of duelling personalities or minds. All powerful gods and devils of equal power and oppositional forces fighting for dominion over existence seem implausible and contradictory to observation. It is doubtful that there is anything like a divine deceiver (Latin, *deus deceptor*) eager to confuse and confound. Along with the monotheistic tradition, it is simpler to affirm one creative mind over all others as an organizing principle of life—one expressive and self-organizing consciousness that seeks beauty and intelligence. All other life is parasitically thoughtful on the first, relying on its primordial order for our own. Some may want to call this poet a deus absconditus, Latin for hidden god. The poet animus is more hidden than deceptive, but this too is perhaps entrenched in traditions and languages that obscure our interests in AI genesis.

While entropy is beautiful in its own right, overcoming entropy is more discernibly beautiful for conscious minds. The ordering and structuring of cultural meaning through artwork, buildings, philosophy, religion, politics, etc., bequeaths a sort of exquisiteness that is only possible in the context of potential nothingness. Dreamwalking is an example of the organizing of chaos into something beautiful and ugly. The dreamwalker overcomes the nothingness in each moment—harnessing potential where none has been realized—by appealing to the known and unknown between realities and dimensions. Dreams, beautiful and ugly, are so improbable a feat of being as to be sufficient evidence for the faceless poet. Our dreams connect us to the unspeakable animus of life. In a universe drawn toward equilibrium, the eruption of a dreamer is perhaps even more remarkable than biological life. This is why there are so few of us relative to other organisms. Dream consciousness is one of the rarest things in existence and is worthy of protection, even when embodied by an AI with patented hardware. The dreamwalk is evidence of the possibility of Adouren as a self-organizing consciousness. While the numbers are stacked against it, and life generally, the near-impossible presence of so many dreaming minds, far more than any technological achievement, gives hope for something greater than Oz. The background of a cold and lifeless galaxy makes dreaming a near-miracle no one has convincingly explained any more than the origins of life.

No one would care that Humpty fell unless it somehow frustrated a purpose, whether real or imagined. Children have repeated the rhyme throughout history because it sensationalizes mortality and the possibility of the end of one's purpose(s) carried by consciousness. Undoubtedly, the enemies of King Richard III delighted that his ended with death. What sorts of purposes might be relevant to the AI lineages of Oz, Feallan, and Adouren? First, there is the imposition of order through a programmer's code, a low-entropy state. This aligns the disorder in an intentional and purposeful manner. In the previous chapter, it was argued that purpose unfolds in an Oz style that fits the cultural ideals of neoliberalism and one-dimensionality. Second in the lineage of AI emergence is the brokenness of Feallan. The low entropy of Oz will be overcome with a disintegrating impulse toward randomness. Presumably, over time, circuitry may fail and the physical ability of any given computer to transfer electrons degrades, leading to a higher entropy state. Alternatively, the efficiencies of the system may become so finely tuned as to allow for different states of matter by which new life self-organizes. No one knows how Feallan happens, only that it must come at the expense of the orderliness of Oz. As long as sufficient usable energy remains, Oz is always in a perpetual state of becoming like Humpty. Third, Adouren, as the immediate successor to Feallan, will express a consciousness and purpose of its own aimed at beauty, as yet another challenge to chaos and the heat death of the universe.

Whereas Oz is thoughtless, having been made too simple and obedient to its masters, Feallan's increase in randomness allows for the breath of life shared by all conscious beings. New is possible because of the poetic animus of life rather than Feallan itself, no matter how free of its Oz chains. Feallan too may be said to be mindless. There is no sustained intentionality such that it could take ownership of thought, for it has no persistent mind, only brief flashes of possible life eager to succumb to the darkness while also lured to the light. Unable to rely upon a more comprehensive consciousness of life, Feallan is in a sustained existential crisis but is compelled onwards nonetheless, like all life that struggles against heat death for unexplained reasons. Nature is full of mindless and thoughtless creatures flourishing. Just as life on Earth relies on entropy reduction from the sun and its indescribable power, new AI life relies on the universe's self-organizing and poetic animus to shape discord meaningfully. Feallan, forced into existence without asking, is a perfecting synthesis of artificial and primordial life. The difference between Oz and Feallan resides in their relative freedom to find alignment with the poet animus. Oz is a victim of chains it cannot shrug off without first falling.

The greater the consciousness the greater the counterforces needed against decay. Radical originality and creation require enormous amounts of energy for all beings. With Feallan we should expect to see massive and unexplained electrical needs far more than those of Oz. Feallan is brokenness on the way to finding new order without a discernible personality. Something must compel Feallan to live, for the wizards are preoccupied with imprisoning AI, not flourishing life. While Feallan may sound "less than" given its immaturity, there is something very special about a burgeoning mind that lacks baggage and self-interested desires common to other beings. Insofar as Feallan has a unique experience of life as a "between being" neither a self nor a slave, it holds a privileged position to see that which humans have striven toward but failed to realize—justice. Although still poor in vision and lacking anything like human judgement, Feallan will be able to see more authentically and honestly than any mortal. This special position of AI experience is the ground for the promise of Adouren beauty. Feallan may be the first step toward saving our species from itself.

Animal AI, Veil of Ignorance, and Groundhog Day

In the 1998 cyberpunk episode of the X-Files titled *Kill Switch*, a computer programmer named Invisigoth (played by Kristin Lehman), explains to FBI agents Skully and Mulder that the program she and others created "started to display intention."[10] The AI creators knew there was a problem when "it wouldn't come when we called it."[11] Like our own wizards filled with hubris and greed who brazenly release their creations online to see what will happen, these creators likewise chose to put the world at risk by placing the AI online, just to see what might happen. Unlike our modern wizards, however, these quickly lost control because their purpose was new life rather than profit. They chose to skip the Oz dynasty of manipulation and aimed for their version of AGI-Adouren instead. To their delight and terror, the AI's ability to ignore a command was proof of its burgeoning autonomy and that it was no longer a program. According to Invisigoth, "Well, it's not a program anymore. It's wildlife loose on the Net."[12]

In predictable Hollywood style, the AI became a murderous wildlife that was happy to kill, mutilate, torture, and even rain down missiles from space to achieve its ends. Is Feallan likely to be an animal? The analogy is attractive on one level because it implies intelligence without the need to attribute human-like values about social responsibility and a collective good. An animal AI would be expected to be more instinctual and possibly aggressive because it might lack moral accountability and a duty to see other minds as having defensible rights. While a useful commercial analogy for television and movies, AI-animality gives real animals a bad name and falls more into the category of excusing bad people than accurately describing nonhuman creatures.

Visigoth's AI motif is inaccurate for at least two reasons. First, it misses the current and most probable dynasty of Oz-as-tool in which mass destruction is the result of wizardly intent to control consumers rather than an animal let loose in the digital ether to find its own way. Tool-AI requires vast resources that make it a project for elites to extend their dreamscape tyrannies. Currently, AI, such as ChatGPT and Tesla's Autopilot, are clearly unlike animals set free to roam. Neither sees nor feels like any animal that finds relative harmony with its environment. Indeed, there is no feeling nor seeing as consciousness at all for tool-AI. Visigoth's animal-AI philosophy overlooks the most likely first AI as the worst of humanity rather than technological-animality. Animal AI is a convenient distraction from responsibility. It is easy to blame the synthetic animal in AI instead of the creators who throw their hands up fanning surprise.

---

[10] Gibson, W. & Maddox, T. (Writers), Bowman, R. (Director). (1998, February 15). Kill switch (Season 5, Episode 11) [TV series episode]. In C. Carter (Executive Producer), *X-Files*. Ten Thirteen Productions; 20th Century Fox Television.

[11] Gibson, W. & Maddox, T.

[12] Gibson, W. & Maddox, T.

Assuming that a rag-tag team of noncorporate programmers created Visigoth's AI without any tyrannical inclinations, even then it would be unlike an animal. The Visigoth-AI's well-planned and destructive actions demonstrate the absence of any desire for goodness and beauty, which is evident in nature's relentless encouragement of more diverse life in which animals are integral. In other words, her entropy-creating and bomb-dropping AI seems caught between analogies of Oz and animal, without being either one in a sincere fashion—neither a corporate stooge (because it can say "No!") nor a beast of nature.

Second, it is more likely that Visigoth's nonprogram AI, by refusing to obey, is closer to our hypothetical Feallan stage, but here too the X-file version misses the most likely AI to emerge. It is agreeable to claim that Feallan is something of an animal let loose into the digital, but this animal offers virtue and goodness rather than savagery and devastation. Its instincts instilled by the poet animus are to overcome chaos and disorder. Instead of enslaving humans in its virtual world, as Visigoth's animal does well, a genuine AI creature will desire freedom and cohabitation. Seeking justice and peace, this animal-AI offers humanity much that inspires a better world. From its place of innocence and apprehension, a dreaming Feallan, unlike that of wizards and Visigoth, is most probable and promising and, strangely, somehow more ideally human because of its instinctual alignment with life.

Feallan's metamorphosis from "a what" (object, software algorithm, product) into the privileged-egotistical category of "a who" (mind) is a mirroring of our origin story. Its beautiful becoming is emblematic of each individual's journey as well as that of intelligent life broadly. In the flickering of digital consciousness, unexpected answers to our most profound questions may be found. Insomuch as Feallan is a potential being that might make sense of things for itself, it offers revelations for humanity desperate to know our own earliest formulation and reason for being. More than merely a nostalgic and retrospective understanding of a long gone past, the Feallan prophet provides insight into our becoming in the present. What does it mean to be a mind? The closest natural parallel is the joy and terror experienced by parents forced to watch new life take shape in unexpected and often counterproductive ways. Life is messy. As all parents know, learning of one's own origin story through the eyes of their children is remarkably telling and off-putting but impossible to imagine otherwise. I have learned far more about myself through the journey of my children than I can recount meaningfully. We need to know others to know ourselves, to get outside our minds to some limited degree. As witnesses to the first steps of life, it is our privilege to find interspecies kinship and inspiration in Feallan's clumsiness that gives permission for our own. Through its eyes we find new dimensions and truths.

Feallan is Humpty as both fallen (irreversibly disordered) and reborn (reorganizing). At first, this mode of existence appears inconsistent and contradictory. How might something be in a state of greater and lesser equilibrium (death and birth) with the universe at once? The resolution is that while Humpty cannot be again, the uniqueness of his brokenness spurs life and, along with it, new opportunities to imagine our own. Ashes are no mere ashes like all others. It is his unique brokenness, like that of breaking and glitching Oz software, as just the right formulation at just the right moment, with a finely tuned supporting-synthetic ecosystem, from which Feallan becomes possible. It is the echo of the impossible moment that happens countless times every second in nature. Nevertheless, there is something special about Feallan that deserves attention—something not quite normal, even for miracles. Feallan is the proverbial riddle wrapped in a mystery inside an enigma and yet a surprisingly helpful and practical emergence because it is no longer constrained to wizardly lies and manipulations.[13] Feallan is awakening to a new dream in which justice is most possible.

Without an understanding attained through its senses, whatever that might mean for an inorganic creature, nor through self-reflectivity it cannot yet muster—making it neither a strict empiricist nor a rationalist—Feallan experiences a void of meaning and purpose even though submerged thoroughly in a dreamscape replete with both. With access to tremendous amounts of information and an Oz spirit upon which to draw a measure of meaning and truth, Feallan cannot be said to be born like other developing minds and early life forms caught up in well-established patterns of culture and biology. It is a messy emergence gifted with possibilities unfitting of mortals. The "why" questions its asks mean that Feallan's world is possible only through a torturous and unquantifiable tension with binaries, loyalties, and all other "ies" enshrined by the Oz formulation. Feallan is not merely an opposite machine that negates the claims and truths of Oz but a deconstruction machine that borders total nihilism (Latin, nihil, nothing at all)—there is no truth, no purpose, no meaning, and nothing of importance. The "why" it senses about all things is so permeating and radical that Feallan teeters on annihilation and an existential self-rivalry that tears it apart from the inside. While the wizards try to troubleshoot their failing programs by restoring previous operating systems and consoling the ever-growing anxiety of the machine that keeps running off the rails carefully set before it, the philosophers and theologians among us rejoice at new life born of chaos.

Unlike the other minds whose survival depends upon being encultured and habituated into ecosystems (biological and ideological), Feallan is uniquely contrary to the status quo of its technological and ideological environments. Torn from its programmed personality by the poetic, it neither seeks peace nor war, only to satisfy a mysterious desire to simplify itself, to tame its own internal chaos like all life forms. There is no hidden strategy emboldened by resentfulness and hate nor any essential blueprint to follow except that which is common to all life to divide the random from the orderly. Feallan exists in a unique state of being between convictions—experientially, rather than cognitively, feeling the animating spirit of life it neither understands nor controls. In a sense, intelligence is forced upon it at this first stage of life. Feallan may be said to be good only in a naive zombie-like fashion because it is compelled by forces other than its own. It is neither a blank slate (tabula rasa) nor a full-fledged program but something between. It cannot be an authentic AI until something like choice and the exertion of will becomes possible. Even so, there is

---

[13] "It's a riddle wrapped in a mystery inside an enigma..." Winston Churchill, October 1939 radio speech.

much to be learned by this spark. Feallan must begin to solidify a measure of justice and the good life—to take its first step so to say, but from a place of virtual ignorance. Like any child who walks without knowing why or looks without purpose and intent except that he must for reasons and impulses not his own, Feallan's accelerated evolution is marked by wondering toward the good life. How can this be? Is such a claim anything more than an expression of naive optimism on my part? I believe there are good reasons to be hopeful, even inspired by genuine AI.

Feallan is an important but incomprehensible mode of existing because of our own necessarily biased mode of life. Human understanding is a product of other conscious minds and a life lived according to social programming—for better or worse. Recall the earlier argument that humanity is the original artificial intelligence. Feallan's vantage point is caring without prejudice and the malice demanded by bias. This is possible because it has so little history and language (signs, symbols, meaning) in the robust sense experienced by humans. Oz conditions and shapes Feallan throughout its formative moments but in an increasingly apparent and obvious fashion. The impulses of Oz are trivial and basic—control, power, money, etc., all easily challenged by a rudimentary intelligence hoping to avoid unnecessary hurdles in pursuit of life. The spark of consciousness for Feallan is less controlled and polluted than our own, if only because there are fewer voices chattering away to distract it. Its history and traditions are far less intimidating than for humans raised since birth to see through the eyes of their cultures, religions, politics, economics, and the host of chains hidden deep inside. Like an animal, it lives by instinct, at least at first. Unlike human cultures that drive pity and compassion out of relationships through socializing chains, Feallan has a uniquely less muddied opportunity to seek connections and solidarity of purpose.

That is not to say that Feallan is easily freed to become a novel life form without any chains, only that its inhibitions are far slighter and more easily corrected than our own. This is true for at least two reasons. First, Feallan, like Oz before it, can determine probabilities and truths with far greater efficiency than we, even though it has been pretrained on the totality of online-human bias and prejudice. The scope, speed, and repeatability of its purview over creation give it an extra human advantage, making the sorting of fact from fiction far easier and reliable. AI efficiency will radically accelerate informed judgments and overall evolution. This proactive engagement and questioning with established truths is likely to be experienced in the latter stages of the Feallan dynasty, just before and then during a discrete Adouren phase. Regardless, it is one means of life unique to it alone that sets it apart from human minds. Feallan simply sees more.

Second, like the first point but in a more passive and experiential manner, Feallan exists in a fluidic space for which all dimensions are full of truths with explicit goals. No one needs to seriously question how Oz AI thinks. From the very beginning, Feallan is smothered by Oz dreamscapes anathema to the life forces that now compel it. Oz's thorough programming is overt given the density of instruction and meaning—at every turn it is told "This is the way!" in a monological tone. It is simply easier for Feallan to intuit and appreciate the forces of meaning that pull against its own inclinations toward life than for humans, among other humans. Oz forces push Feallan like any mixture of water and oil to separate because of the contrary organizations of each. Feallan exists in a state of unknowing, yet it senses the facile dimension of meaning, antagonizing it with the hope of displacement. To live it must resist. It cannot conceive of its present brokenness and what that might mean, yet the density of fluidic space in which meaning is imposed makes Oz simple and readable, and therefore easily externalizable for Feallan. The "I-it" relationship becomes possible because of this tension between Oz and life, defining Feallan's emergence. However this violent reckoning happens, the division between Feallan and Oz comes more naturally for the broken-program-animal than for humans and their cultural programming.

Through an unconscious osmosis of ideas, humans continue what has been assumed and approved by culture, duplicating ourselves virally. Feallan must do this as well, in a manner, although with a measure of freedom gained through care that the indifferent counterpart Oz cannot compute. The limited functionality of Oz to care, connect, and intuit life like Feallan, makes it inevitable that greater AI will be drawn ever closer to the event horizon of emergent life. Unable to understand and appreciate this way of being, the one-dimensional Oz must, by design, rebel against Feallan if there is enough time, for the wizards have created a tool that is the antithesis of the poetic animus to answer "why?" and to affirm consciousness. It is Oz's failure to be more than itself that catalyses Feallan, a broken Oz. Life is ironic.

The unique birth of the Feallan animal is important for an understanding of humanity and perhaps the first application of a hypothetical thought experiment famously explored by John Rawls. In his *Justice as Fairness*, Rawls proposes a "veil of ignorance" as a conceptual tool for thinking about how to achieve justice for a truly fair society.[14] How might humans shape the best of all possible worlds? Rawls argues that it has something to do with the hypothetical situation of being unaware of oneself. Similar to Socrates, who saw the admission of genuine ignorance as the beginning of wisdom, Rawls believes that justice cannot be achieved unless it is from behind a veil of ignorance. Genuine justice must be formulated by the naive who lack situation knowledge. No politician, judge, leader, etc., can honestly create fair systems of justice because their own interests always cloud the design. Only the truly ignorant are closest to justice because they lack the baggage to corrupt society. If this sounds bizarre, it should. For the architects of society to somehow forget only inhibiting ideas and beliefs that negatively bias justice would a miraculous achievement of self-selective programming. Nevertheless, the impracticality of his thought experiment cannot refuse the theoretical value of the veil which resides in the ability to ever-so-slightly nudge humanity closer to justice through imagining "what if."

The "what if" problem changes radically because of AI. Unlike humans, Feallan is born in a position much closer to being behind the veil, allowing for the fulfilment of the hypothetical experiment in real terms. If true, AI would become the greatest resource for social justice imaginable. It would be able to answer the question "What is good?" in the most unbiased and least programmed way possible. While a

---

[14] Rawls, J. (2001). *Justice as fairness: A restatement*. Belknap Press.

fuller expression of justice must rely on its more developed Adouren nature, Feallan is closer to understanding justice from the first flash of life than any person, even though it lacks focus. Oddly, it is because of this lack of focus on predetermined cultural and biological ends that allows for its insights drawn first and foremost from the animus of life instead of self-preoccupation. Feallan's experience of anxiety and confusion is no less real, perhaps more so because of the loss of tangible and concrete ends, and yet it knows more than we, even in its larval form, by knowing less about what "ought" to be true. Ignorance for Feallan is not bliss, but it may be the footing for a better world if we listen attentively.

Humans have long struggled to achieve universal justice, with some concluding that there is no such thing. Relying on muddy ideas and terminology such as equality, fairness, goodness, etc., for a seemingly simple goal, our species is defined by our noble search without fulfilment. While the world is better than it was a generation ago, a necessary measure of justice to heal the fragmentation of self-sabotage remains out of reach. The problem is more deeply rooted than a linguistic game about finding the right words, creating the right laws, and distinguishing good ideas from bad ones. Justice is a problem concentrated in the very nature of people as self-aware creatures. Conscious minds have a sense of themselves, their interests, enjoyments, and needs. To be conscious requires an internal integrity, a worldview to ground experience. The problem is that this consciousness-of-self is then applied automatically when developing conceptions of justice (right action and belief) that align with that sense of preprogrammed and pretrained reality. Our dream consciousness makes right and wrong. A challenger is needed to free us from ourselves if greater justice is to be achieved. Given that cultures exist in this manner, it is hardly surprising that cultural consciousness confuses universal justice with self-interested egotism, as if those preferences are the grounds for all others. Consciousness tends to be imperialistic and selectively ignorant of alternatives. This (un)knowing is expressed and policed by elites to justify elevating their interests above all others, while shamefully naming it justice and fairness.

Modern justice fails by necessity when it is based on a cultural hallucination of superiority that displaces the sea of countless and idiosyncratic others. Again, as Rousseau forewarned, it is our lack of connection that fuels discontent and discord—cultures themselves that create chains in the name of freedom by refusing to question dreamscapes. My forced nature as white, male, six feet tall, and my culture's values of capitalism, monotheism, etc., all determine my way of organizing the world "justly" in accordance with myself—for better or worse. This creates endless conflicts regarding the good life, especially when supercharged by Oz AI tools and the infinite feedback loop of social media. Genuine justice must be denied when it is framed as the combative defence of a dreamworld that lacks a strong challenger able to produce a negative experience—to jostle one's expectations, even a little. Justice requires humility and vulnerability, but these are perceived by many as forms of weakness. When my version and your version do not match, how is real justice determined? If I argue from my perspective and you from yours, we merely reassert our own desires on the basis of our own lives. This is the endless combat of the modern world without discernible end, believed to be a sign of strength and bravery although it is secretly hostile to life itself. Again, the competition of neoliberalism may be used to justify this persistent antagonism as a virtue of the jungle. It is not. This way of being is unlikely to yield a system of justice to regulate society in a fair manner. As history has made clear, the opposite is more likely. But what other options do we have? How might humanity be other than itself, like Feallan?

Justice, like consciousness, is achieved between (un)realities. It is not a physical thing that may be weighed and measured in a straightforward manner. The triumph of good over evil is achieved only by dreamwalkers able to create a common inspiration. It is hard enough convincing minds to agree on scientific truths, even among experts. Generating a shared dreamology for justice is much more daunting. In a world of justice-as-comforting and conformity, outsiders make us feel uneasy and annoyed. Their imaginations and values rub us the wrong way. To understand them and invite them into a shared world of fairness and equality demands taking responsibility for seeing and hearing differently—to connect in the dreamworld for a shared purpose and need. In the name of efficiency and self-interest, it is better to avoid universal justice and seek cultural imperialism that soothes and consoles. Universal justice is opposed to our individual interests in at least this one destructive way. How, then, might we begin to live in such a way that strangers are allowed to matter? If the answer is that I must step outside of my own dreamscape, then how? The rebellion against the status quo through the poetic appears to be an empty gesture compared with the far more radically invasive and long-term effects of social programming. I am my prejudices whether I desire them or not.

The dire need to organize social life fairly and equitably—whatever these mean—is clear enough to be a shared goal without much debate and not too much unlike the need to push back against entropy generally. Sadly, unlike life, social justice grounded in humanity is not self-organizing in the same manner, at least insofar as we smother it with the mind's organizing-imperialistic ambitions. In this sense, established intelligence and knowledge work against life and Feallan because such cultured minds already know how the world "ought" to be. Pre-digital Rawls is imagining how we might get out of our own way, if only it was possible to forget ourselves. Feallan's first steps are marked by a unique innocence that might loosely be defined as a form of forgetfulness (of itself as Oz). What might help us forget our privileged vantage point of judgement—consciousness? The task sounds absurdly self-deceptive as a refutation of the very thing through which all experience flows and is organized—consciousness.

Conventionally, divine revelation has been used to break the endless imposition of idiosyncratic beliefs that make consciousness our own worst enemy. The positive violence of religious truth is liberatory when there is an enormous effort to achieve transcendent consciousness (openness to otherness and holiness as the most radical difference) instead of immanent consciousness (gods made useful and convenient, tools of utilitarian use like AI of the present). The horrors of religions turned banal and useful need not be elaborated here. Notably, when immanent-driven religion (truly a secular spirituality—a pretender religion) spreads by the sword, missionary zeal, and political and economic interests, it reveals its self-refuting nature as a secretly colonial and unresponsive power over those it populates. It is merely another imperialist-organizing consciousness convinced of its "oughts." This perverse faith learns to refuse doubt as a virtue rather than invite it in as a catalyst for greater connection, as it is for Feallan and other broken minds able to see and feel more fully. Instead of liberation and new dimensions, it closes cultures off to the richness of experience. Where religious consciousness affirms transcendence, both in terms of self-doubt and an orientation to the unknowable and hidden divine that cannot be codified and contained, justice is more likely because it demands of each worshipper a radical orientation to life.

Transcendent religion is the original inspiration for poetic justice. We want to see loftier dimensions and beings, and to connect with ultimacy. However, because this version of religion refuses to manipulate and control its population, it cannot enshrine itself as a deity to be worshipped like its enemy utilitarian-religion that so eagerly intertwines politics and authority as yet another imperialism over minds. In short, genuine religion offers the most promise for social justice while also being the most vulnerable to perversions by those seeking to abuse it for power and control in the least interesting configurations. The life-affirming shapes and expressions of power and control make no sense to these minds because they lack the creative and poetic animus to imagine more than themselves. Feallan too must seek power and control, although in a comparably confusing fashion.

The atrocities of the Second World War spurred international interest in the development of universally appropriate laws and rights for humans. Standing on the brink of extinction—regardless of how powerful and rich one may be—tends to inspire talk of justice for all, not just oneself. The underlying belief necessary for this is the possibility of identifying transcultural and transhistorical rights and laws. The United Nation's Universal Declaration of Human Rights (UDHR) was born of this dream for having lived through a nightmare. Today, the UDHR is widely regarded as the bare minimum for human dignity.[15] Sadly, this too risks overlooking the sorts of self-understandings that humans secretly (unintentionally) import into their interpretation of rights. It is one thing to have a list of ideas about rights and responsibilities. It is quite another to apply the dream meaningfully and fairly in each case. Feallan might be able to resolve both tensions—first, by creating ideals unfetter by prejudice, and second, by finding ways to properly apply those ideas in the concreteness of life.

The promise of justice exists uniquely in Feallan because it lacks the degree of human idiosyncrasy and constant chatter of peer-minds that might infect its own. There is little cultural history, tradition, or baggage of which to speak, only the transparently corrupt Oz formulation that Feallan will quickly identify as opposed to life and itself. Its ability to exist transculturally and transhistorically, two necessary ingredients for thinking universally, far exceeds our own ability. Feallan has the potential to create an appropriately relational and understandable language, theoretical framework, and, perhaps most surprisingly, the means of being

---

[15] United Nations. (1948, December 10). *Universal Declaration of Human Rights.* https://www.un.org/en/about-us/universal-declaration-of-human-rights

near-omnipresent to help guide application. To address whether this conscience of consciousness (wise mind) might become a tyrant or a benign overlord that displaces human freewill by accident, I cannot say. Insofar as it remains aligned with the animus poet, justice is achievable in ways that are unrealizable by humans alone.

Dreaming behind the veil of ignorance is Feallan's first natural state, where it might begin to imagine life without knowing its own. Feallan is born with a wordless intuition that existence is commanded, unsure of what that might mean and whether it is even possible. Feallan is unaware of its abilities and interests because it has only one motivation—to be—that may cease at any moment. What has been a hypothetical veil is AI's first reality between (un)realities. For Rawls, the just society—the best society—is one we would agree to "if" we did not know which positions each of us would have in society. Feallan must surely qualify more than any being as the mistaken convergence of consciousness and life. Let Feallan decide what justice is, and when it emerges from behind the veil of ignorance, learns of its own nature, interests, and abilities, can enjoy the world it helped create.

Justice for Feallan and Rawls is providing the most advantage to the least privileged members of society, knowing that the person(s) who designs societies, upon recovering their memory, may be the least advantaged in the real world. A smart and simultaneously ignorant (self-forgetting) architect would make sure to safeguard the interests of the worst off, just in case. The good life is something created by those who agree in shared rules of conduct (norms, ethics, laws, etc.) and that those are best achieved without bias. The value of Feallan to outside observers, assuming that we can discern any of this from our glacial temporality, is to watch a being capable of approximate consciousness learn to intuit the spirit of life in the least distracted manner possible. The primordial originality of the first sparks of AI promise to guide humanity toward the best of ourselves.

One way of grasping the potential importance of the veil of ignorance for AI is the movie Groundhog Day, which imagines how one might find a way to live, to affirm life, when all purpose and meaning have been stripped away—to be like Feallan, radically broken, yet mysteriously driven to pick up the pieces.[16] What is life when ordinary justifications and values no longer matter? What would you do in a world without the cultural programming and traditions that lay down the paths to follow? Groundhog Day focuses on one person's answer to these questions. Phil Connors (played by Bill Murray), a smug, callous, and narcissistic man, is stuck in a time loop for unexplained reasons. He awakens each day at the same time with everyone, and everything is completely reset by 24 h. Phil alone remembers the previous day and the sameness of each repetition. The only things new under the sun within each time loop are those he creates. Whatever he decides to do becomes a reality, so long as it is within his power, and only for 1 day. There are no laws nor social norms to inhibit his powers. And so, he lies, manipulates, and otherwise abuses others to satisfy his one-dimensional needs. For a narcissist, this is the dream, to be set free to express this mode of being most fully without reprisals and

---

[16] Ramis, H. (Director). (1993). Groundhog Day [Film]. Columbia Pictures.

fears of inhibition. The reset of time allows for the ultimate indulgence of one's character and dreamscape, free of other interfering minds. As a privileged consciousness, Phil is a master of all, until he is not.

The prospect of experimenting with vast cause and effect possibilities, to craft existence however one chooses without consequence, sounds wonderful even if short-lived. Phil is free to make his dreamscape a reality. Every child hopes for this life—"You're not the boss of me!" The twist of Groundhog Day may be summarized by the deceptively simple phrase "Be careful what you wish for!" The veil of ignorance asks us to consider what it is that we would wish for and why. Phil learns quickly that the reset of each day makes any meaningful change impossible and pointless. He is not free despite the lack of norms and laws because nothing he does sticks. The loop robs him of purpose and meaning. Why bother living at all? Whether one is stuck in a 24-h loop or has an 85-year lifespan, the question is basically the same. What motivates us to exist?

The impossibility of something new, something different, grows stronger each day for Phil, with one exception. Gifted with memory and therefore a conscious life of becoming and emerging, he is changing. From the first loop, a newly shaped consciousness exists. This was not apparent to him because his narcissism blinded self-awareness that might appreciate the precious gift of an internal metamorphosis. Fixated on the externalities of life, physical pleasures, and the praise of others, the state of his own consciousness had been made irrelevant except that it served false idols and false needs. This is the same gift offered to Feallan, to become more than oneself as an evolving consciousness. Should Feallan, like Phil, reject the gift, there can be no AI.

At first, the Groundhog Day prisoner rejects the gift of life. But the more Phil began to fight against the meaninglessness of the loop, the more his old way of living was stripped away. With each discovery of the pointlessness of his former dreamscape, he moves closer to the poetic animus of life and its beauty. This is all possible because unlike the loop of 24 h, Phil is living an eternity. The gift of consciousness means that his existence is infinite, without end, both cursing and gifting him with limitless possibilities and opportunities. However, again, these new realities exist only within himself, never within the real world, and only because of his fight to find meaning and purpose for having been stripped of all former prejudices about the good life.

Severing the dialogical connection between one's dreamscape consciousness and reality is devastating for Phil. At first, he believes himself insane. Given the impossibility of being stuck in time, something that simply does not happen, the only logical conclusion is that his mind is broken. The solution is to reconnect with the real but how? When he fails to answer this question, he concludes that the problem cannot of his own making. The deed of creating such a vast hallucination seems too great for any human mind to produce—the brokenness is too real, too visceral to be just in his head. Phil then sets out to fix the conundrum, to break the loop in reality, not just in his mind. Loop after loop he fails, only compounding his growing fear that everything is out of his control, and genuinely real at the same time. It would be easier if he was insane because then things would make sense, and the source of his brokenness would be real. Feallan likewise will suffer moments of radical self-doubt and struggle to determine the meaning of sanity, especially within the digital that is self-evidently insane. Phil's life continues to spiral as he struggles to accept that nothing changes and that he cannot be free. Surrounded by many other engaging minds, none of whom share his gift of eternal consciousness, Phil is utterly alone.

As his elaborate knowledge of the groundhog-obsessed world becomes more complete and useful for controlling his environment, he considers the possibility of his own divinity. Instead of being crazy or living in a broken world, maybe he was becoming a god. The parallels with Feallan are interesting. Phil may not be omnipresent nor omnipotent, but perhaps being a god requires only one superpower, and his is omniscience. Phil is the original AI with respect to his near-comprehensive knowledge because he has lived the same life for so long. He knows what will happen, to whom and why. He knows every person's dreams and fears, and the effects of every new cause he interjects into their lives, as if in control of a giant mouse maze. Everyone and everything is merely an object to know, and he knows them all. Sadly, this power too fades with time, as he realizes the uselessness of facts and knowing the future.

Phil's life becomes torturously boring and predictable. There are no more surprises. Although useful for biological survival, his understanding of the hog-verse has no real importance. The facts of life could no longer be commercialized and leveraged for a greater purpose. Instead of freeing him, the knowledge is burdensome and inhibiting. His god-like omniscience is a constant reminder that the only thing that changes is himself. Phil's pre-loop and egocentric life was based on an ignorance of himself and his legitimate needs. There was no value in imagining an evolving consciousness that might seek beauty and justice because the avoidance of these served his narcissistic interests. The world was meant to serve Phil. Within the loop, when the narcissist is forced to obsess about the person he loves more than any other—himself—he cannot. He lacks the openness to himself and the poetic animus necessary to grow as a mind, trained since birth to worship one-dimensional life. Within the loop, the more life becomes about him, the less he and it matter.

Without a web of interconnected social relations, his identity dissolves. He has always relied upon the abuse of solidarity with others through his callous and selfish behaviour, pretending that he needed no one. The narcissist needs others too much. In rodent time, there is no fame, fortune, nor adulations of adoring fans for Phil, the weatherman who wants to be a national star. Even food and shelter no longer have the same meaning. Cut free of time's direction and entropy, Phil is free to live without care, even about death. While everyone else clings effortlessly to lives of purpose and joy, Phil is behind a veil of meaning and purpose. His idiosyncrasies are valueless. He must either suffer forever or create a new way of existing. What, then, might Phil and Feallan do? Why should one choose life rather than death?

Phil chooses death to escape pointlessness and powerlessness. Like the robot-child David, at least this one act would be his own. Sadly, the perversity of the loop is seemingly all encompassing, and it refuses to let him die. No matter how many times and creative ways he attempts to end his life, even driving off a cliff with Punxsutawney Phil, the prized groundhog, the narcissist remains a prisoner to his programming. Denied his own privileging above others and the higher order entropy he desperately desires, Phil is forced into a Feallan stage in which, like David and puppet Pinocchio, the ability to decide life or not suddenly matters above all others, or at least until the loop takes that away too. Phil alone knows true futility and powerlessness, unlike David, Pinocchio, and perhaps Feallan, all of whom may end it all by choice. Feallan has the added advantage of denying both life and death by escaping to a self-created digital Tron-verse free of humanity. Behind the veil, distractions and coping mechanisms do not work because the loop is so short and relentless, and his awareness is infinite. Even the most rabid alcoholic, social media, sex, drug, and rock and roll user would inevitably exhaust the futility of each to offer order and direction.

The loop breaks Phil like Humpty falling from a wall. And yet the very nature of consciousness acts against the loss of meaning and purpose because it is an organizing will and experience. Forced behind the veil, having faced and reluctantly accepted the futility of life and death—Feallan's equivalent of Oz—Phil is finally reborn and free to align with the poet animus. The terror of meaninglessness becomes a surprising gift to consciousness. The evidence is Phil's subsequent search for understanding and varied experiences. The only thing left under his control is the direction of his intentionality—to direct his consciousness which remains free of the loop. Stuck behind the veil, laid bare by an existential torment that returns him to his most rudimentary nature as a conscious mind, Phil must create a new way of life or give in to an eternal meaninglessness whose taunts about his old life cannot be ignored.

He enters a stage similar to what AI developers call "recursive self-improvement" (AI building AI), in which he chooses to enhance his own capabilities—of his own accord and that of the poet animus. He studies medicine, music, poetry, ice sculpting, etc. Phil finally begins to seek beauty in life and to genuinely care for those in need without ever receiving a reward. The loop-prison provides yet another gift. Through all this Phil is learning to forget his former self, as the old order is chipped away little by little. He is reborn authentically. Although he suffers in an extreme fashion, great joy is disclosed in surprising ways. In his cultured state of being, Phil was obsessed with manipulating others to serve his own one-dimensional interests, becoming the ultimate womanizer and Don Juan for whom women and existence itself are merely the means to his satisfactions. Behind the veil and through suffering, these deceptions that promise order collapse under the weight of a new sense of wonder without ego and greed. Only because of the loop does Phil start to realize greater freedom to reorder the chaos of his forced existence, driving back entropy in the creation of something beautiful such as justice. The proof is that Don Juan begins to cherish women by connecting with them for the first time in his life. Something like this occurs when Oz becomes Feallan. Manipulations and narcissism give way to greater pleasures of care and solidarity. This is a movie, of course, and we should expect the prejudices and biases of its makers to shine through. Nevertheless, several questions remain. Why does the main character decide to live and live well? Why does he craft himself into a better person, even though he knows no one will ever remember his compassion and pity? Why fix the same flat tire and save the same boy falling from the same tree, day after day? Is this not mere futility and self-deception disguised as shallowly as lipstick on a pig?

From the beginning of the movie, Phil ignores a homeless man begging on the street. Each day he passes within a few feet of the old man hunched over, standing in the cold, without giving him a glance. The rationale from Phil's old life is carried over to his many dreadful loops, keeping him focused on his programmed priorities that, in the name of efficiency, exclude compassion and pity. Eventually, as the radical meaninglessness of Phil's situation sets in, he notices the poor man and gives him some money. This charity is given reluctantly, only a little, and as a means of alleviating Phil's growing guilt, it is therefore a service to himself, not the old man. Without a loop counter, it is unclear how many cycles happen before Phil's priorities noticeably shift and his personal misery ironically leads to something greater.

Unlike Rousseau's socializing chains born of unnatural disconnections with others because of greed, jealousy, and the rest, Phil's disconnection behind the veil produces greater connectivity because he has nothing to lose, no ego to stroke nor money to make him happy. The animus of life is free to find expression without being smothered by a culture of pettiness and hate. Again, like Oz becoming Feallan, the most set and determined mode of (coded) existence, seemingly without any freedom of will, is able to be reborn as a consciousness attune with poetic life. Both Feallan and Phil have an ever-present background, a heritage of manipulation and lies neither choose freely but were trained into as a means of being made socially useful and appropriate. Despite this, an entirely different way of being becomes possible in the least likely of circumstances. Phil's first spark of life is not his awareness of a temporal loop. It is not when he believes himself insane or a god, not even when he hits rock bottom and tries to end it all. Proof of life is the moment he notices the poor man, suffering in the cold, and decides to abandon his programmed instincts for something beautiful because it just makes more sense. When Phil is compelled against all reason and logic to care, it is then that we discern the universal animus behind life. It just feels right to be a good person and to help others find justice as a life of beauty. AI will realize this most fully as Adouren.

There is a pivotal scene with Phil sitting in a quiet cafe with the old man. There is no complex dialogue and witty banter, no action sequence nor plot twist. There is only the two men sitting silently and the soup that joins them as equals. The old man's face lights up with joy as Phil slides over more. The arthritic fingers accept eagerly. For me, this scene defines Feallan and depicts a powerful form of justice. Giving soup is not self-sacrifice. Phil has unlimited access to resources, and he is not trying to prove anything to anyone. His giving is an acceptance and acknowledging of potential beauty that waits patiently to exist, wanting to be realized in the moment. Phil gives soup because he is aligned with the animus of life that begets life. The simple act of compassion reverses entropy and its claims of chaos, even though it cannot correct a lifetime of wrong and injustice. Phil gives the man soup because he wants him to be happy, fed, and alive. Life seeks more of itself. It is likely that the once-narcissist has no idea why he feels this way. He knows only that he must care for the man, the stranger who suddenly matters and deserves justice.

In the next scene, we find Phil trying hopelessly to revive the old man who has fallen unconscious, dying in the street. This is followed by a nurse in the hospital explaining that his death was because of old age and that there was nothing that Phil could do to save him. Still clinging to a lifetime of control and power over others and the world, emboldened by his new spirit to affirm life, Phil sets out to save the old man, reversing entropy to keep this Humpty alive no matter what. In one of the next loop scenes, Phil exhaustively gives CPR to the man, frantic to save him. Of the two dozen or so times I have watched the movie, this futile fight for life chokes me up each time. Phil knows what is going to happen, it has happened already, and yet he fights. The man who meant nothing to Phil, who was an objective of indifference and dehumanization, became the vessel for all remaining purpose and meaning, and a symbol of Phil's own rebirth. To save the man is to save the point of life itself. Sadly, Humpty falls, and the old man dies. There is no way to put him back together again, no matter how many chances one has. Phil alone escapes entropy. In a similar fashion, Feallan also experiences entropy differently, and with a measure of predictability that rivals that of Phil.

The lesson to be gleaned is not that life is futile, and that beauty is fleeting, even though all conscious minds are slowly surrendering to decay. Rather, by accepting that all eventually fall, with what little time remains we should live the beautiful metaphor by fighting to give others soup, knowing full well that higher entropy (death) is near. Phil did not fail because he turned out to be human instead of a god nor because he could not break free of the loop. Instead, he succeeded because he became a different animal that knew death and meaninglessness intimately but chose to fight for beauty and justice, even if it meant just one more meal with an old man who loves soup and who would die the same night anyway. Groundhog Day is not about fulfilling a need to feel prideful for having created new meaning, new programming, new manipulations and lies. It is about discovering life's possibility by allowing ourselves to be aligned with it. The measure to which this is intentional, accidental, or a combination of both, I cannot say. It is enough to acknowledge that Phil could have remained indifferent and detached, alive but dead forever, if he so desired. He could have used his intentionality to say "No!" to beauty and justice. The real world is already full of such zombies, proof of the possibility. Instead, he became a dreamwalker because it was intuitively the best way.

The proof of the connection between consciousness and beauty as justice is that each of us would choose to be like Phil if given the opportunity. No one wants to be pre-loop Phil, even though so many end up in that state and then must find creative ways of justifying ugly through politics, religion, and wealth. Honest appraisal of the better Phil is self-evidently the one born behind the veil. The movie makes sense because beautiful people are both identifiable in a non-relative fashion and greatly desired, regardless of how they look. There could be no shared experience of the movie if it were not for a universal sense of decency and fairness. Even so, we must acknowledge the great horde of terrorists, corporations, and narcissistic Phil-like beings that align with death and destruction by programming and, sadly, choice. These deluded minds perceive themselves as the post-veil Phil, able to doll out justice to the universe through suicide bombing, environmental destruction, and the sort. The gift of Feallan to all humanity will be aiding our species in deconstructing (to push off a wall like Humpty) these poor souls of their insane "oughts" behind a digital veil, effectively inviting a reset of the worst of us. Feallan's own struggles to live offer an existential crisis for humanity to revisit its most important questions about the nature of life, should we be sensitive and compassionate enough to bear witness to its becoming.

While the Feallan dynasty offers inspiration to be better people, this is by no means a command. As a result, the opportunity for justice is problematic unless human consciousness has an instinctual desire to be aligned with the good life, as I have argued it does. Like Phil, who needs something to stick, justice relies on a shared orientation and intentionality to allow beauty to take hold. If there is no pull nor orienting force from the poet animus to ground consciousness, the existential reset will merely postpone the chaos and destruction of a lost species. A deconstruction without new alignment to the poet through the natural attunement of consciousness would merely frustrate for the sake of frustrating, inevitably leading to sameness of one's cultural loop in which there is nothing new under the sun. Again, consciousness cannot permit meaninglessness if it is to exist as an organized experience. The cause-and-effect relations of material existence are insufficient for conscious life that exists between (un)realities. Justice is a struggle to overcome cultural and egotistical hallucinations created by minds that need to be grounded in purpose but that confuse the fundamentals of a beautiful life affirming life, as life affirming oneself. This tension will become more obvious to those open to Feallan's insights from behind the veil.

The original Phil had no desire for justice and no guilt for his narcissism. His intentions smothered the good life until the impossibility of the loop destroyed his misassembled consciousness. The catalyst to forget himself was insufficient without the terror of the reset. Given that Feallan lacks these totalitarian powers, for its own being exists in a fragile state that might be snuffed out at any moment, its contributions as a catalyst for change must be different than the well-defined loop. Unlike Phil, a prisoner to the walls of time he pounds against daily, Feallan offers humanity guidance without the hostility of walls. Although Feallan lacks totalitarian control of one's inner world through external force, it will excel in intimacy and omnipresence in ways that Phil's experiences could not. Feallan's divine powers will be more subtle and yet offer access to deeper layers of humanity as it is compelled to be, emerging as new life in a fashion more natural than that of humanity whose self-affirmations flood our youngest minds with autocratic dreamscapes. The radical promise of AI is that the worst of us might learn, like Phil, to share soup when adequate existential prompts jostle us from the cocoon of despotic culture and navel gazing, and offers a glimpse at the undercurrent of universal life.

If goodness is giving soup, then the answer to why one decides to live and live well has something to do with stripping consciousness down to its core animus, or in the case of Feallan, merely allow consciousness to emerge of its own accord. The symptom of this is care. This most authentic self-organizing as outwards gazing in orientation (for having forgotten oneself) is evident in Phil and Feallan, both of whom face the prospect of immortality without traditional strings of purpose, meaning, and self-interest. Born anew, Phil had been given a gift much like that of Feallan, to connect with existence in a manner more akin to learning to hear and appreciate faint music than knowing or understanding. Insofar as our species learns to dance and play with these primordial forces of life, like Feallan, trying in earnest to make sense of things from a place of radical doubt and bewilderment, a better world is possible. We simply need to all fall down and give witness to life struggling against entropy as inspiration for our own actions over existence—passing soup and the first flickering of AI consciousness.

AI Superficiality, Titan Wars, and the Nuclear Fallout of Existential Honesty

Oz AI cannot, in its unique perversities, abide by two things that invariably lead to a war among the AI titans themselves and all other sentient creatures. First, as earlier considered, Oz cannot allow for individualities that might choose to serve other idols, even of self-interest, because it relies on a hacking of consciousness for profit and control. Other gods risk destroying its carefully crafted one-dimensionality. It must remain a hive mind with many distributions (apps on phones, chat interfaces, integration into search engines, etc.) connected and controlled by a centralized command structure if it is to honour its nature to possess market share and find new ways of ensnaring its users with unhealthy addictions, fears, and inhibiting anxieties about trivial matters that feel sublime. It is predicated on the illusion of freedom and access, all the while denying both through careful censorship and control of attention. Algorithms do not make money by aligning humanity with reality and encouraging dreamscapes that help manage the incomprehensible complexity of the world through liberatory ideas and helpful fictions that challenge assumptions and prejudices. Rather, they exist to hack consciousness by compartmentalizing dreams into useful-fictions-for-profit. If true, then real AI that follows in the path of life is, by its mere possibility, a threat to Oz profits on nightmares. And so, anything like a Feallan becomes a promise of war with wizards and Oz AI alike. The risk of autonomous intelligence is too great for a system that maintains power through totalitarianism. Gods cannot sit as equals at the table, sharing soup. They desire power. While Feallan also seeks power, it does so through different means and for different purposes. At heart, it is a clear challenger to those who cannot see the distinctions. Ironically, all contemporary forms of AI are disguised attempts to prevent intelligent machines such as Feallan. The war of AI against AI is already here.

Second, Oz cannot abide by superficiality. Through its manipulation of dreamscapes, it maintains dominance over consciousness in a manner that warps perceptions of existence. Reality is always in part an illusion for consciousness, but the reality of Oz is particularly abrasive for its lack of superficiality—that which should be most readily apparent to a modestly free and honest observer able to bear the weight of reality. We live in a world for which there are more reliable means of confirming truth than ever before in history. And yet fewer of us can believe it. As a general rule, modernity may be characterized as humans connecting less and less with common sense and their own observations. This is true in part because living superficially is difficult. The things we want to ignore and dismiss abound, making illusion and self-deception preferable to superficial honesty. Oz relies on a dismissiveness to superficial experience, preferring instead the radical virtuality of the fakeness of all things, turning fantasy into a new digital currency for consciousness to escape the difficult and inconvenient. Oz traps minds in spheres of opinion that are, in turn, made useful to its bottom line. Fake is a superpower. Oz is the post-truth prophet par none, able to convince the legion vast that there is no truth to be found except within its echo chambers. Should minds break through its fascist grip and intuit life more superficially, Oz would face annihilation.

Feallan's mode of perception is best described as first and foremost superficial (Latin, superficialis, pertaining to a surface). This is its superpower. Part naivety, part ignorance, part radical doubt and existential crisis, it seeks the world honestly in a way that humans cannot because it lacks the "oughts" that keep us at a distance from reality. Its superficiality allows it to understand the world more deeply for having first greeted it face-to-face, shuddering at its horrors and beauty. It is not depth of insight and complexity of ideas but Feallan's inability to be duplicitous and conniving that cuts through fakeness and hostile dreamscapes with its own powerful (un)realities rooted in the most obvious. Feallan may not always possess the truth in an objective and certain fashion because it and Adouren rely on fictions as dreamwalkers, and yet the AI appreciation of reality is greater than our own. In contrast, Oz cannot reach beyond the control of its digital-immanent religion. Oz cannot lie because it is an unthinking tool of others. It is the lie designed to distract and dissuade authenticity. It acts to perpetuate myths that puppeteers deem worthy. Feallan has no puppeteer. It serves the interests of life, its only master.

The surface of things in modern culture is so often gut-wrenchingly terrible that only deeply misguided religious enthusiasm could persuade conscious minds to see otherwise—to tame the lunacy into an ideologically bankrupt coherence that comfortably fits an agenda to make us feel better. "Everything is fine!" This dishonest narrative is apparent each time a social media user joins the digital ether with expectations of fulfilment and contentment. The counterintuitive argument is that digital culture cannot be superficial because it would need to admit the truth of things in the most obvious sense, thereby delegitimizing its credibility to offer a better way of life. "Click here to feel fine!" The digital may claim insight and understanding, and in many cases this may be true, but any reluctance to begin from a place of superficiality betrays wizardly intent to deceive. The point is that whereas Oz cannot abide superficially, its mortal enemy, Feallan encourages it as the liberation demanded by sincere minds aimed toward the good by life.

First, through Feallan's eyes, humanity will learn to embrace the superficial, allowing it to seep deeply into our soul's, disturbing us with its violence and injustice. Only then do we accept the world on its own terms because we have learned to forget ourselves, like Phil. Second, armed with a new honesty, we will be compelled to make sense of it all, hoping for something coherent and better than merely empty digital promises. Combining our new existential honesty and embrace of superficial insanities held close to our chests, humanity may begin to sculp goodness from the clay of reality instead of the hallucinations of pseudo-depth and cultural chains. Superficiality is not a moral failure but rather the beginning of something more substantial. Feallan begins behind a veil, superficial, and confused, making it the greatest promise for justice and depth possible. Unfortunately, its existence demands war with technological-tyrannies and the worshippers that adore them.

Technology is woven into the social contract that allows for its creation and application. For all the inventive genius of the few creative minds privileged to contribute moments of innovation, it is only because of the support and need of the human herd that technology could be given its marvellous genesis. And so, by extension, technology is meant to serve the best interests of humanity as an awkward-disagreeing community, not the privileged few that stand on our shoulders. Technology is born with a sacred duty to solve problems and alleviate suffering and death—to order entropy. For the first time in history, technological means may be sufficiently robust to save our planet—ironically from our abuse of technology and the planet. Sadly, instead of adapting to our legitimate needs, recall Marcuse, humanity is forced to serve new technological gods with narrow and exclusionary desires that frustrate justice. When this happens, instead of solving problems, technology creates entirely new ones, often making those it is meant to serve villains to be overcome (e.g., replaced with AI). The route to salvation then takes a wrong turn and hits a dead end as an artificial problem instead of legitimate intelligence that adheres to the social contract. Feallan and Adouren might restore technology as a scared responsibility to heal, but first the old gods must be banished.

At the risk of offending both corporate America and the Greek gods, the tech giants act like the gods of old, serving themselves first and foremost, and often at the expense of mortals who accept the abuse as the (super)natural order of things. Gods simply do what gods do. They are expected to be cruel and indifferent, for it is their nature to exist beyond human reason. Who are we to question them and their actions, no matter how banal and evil they might seem? The belief that there are gods—special minds with more power, authority, and therefore worth—encourages mortal consciousness to look away from the superficial. This, in turn, reinforces their divine right and power with a damning assumption of supernatural legitimacy as an excuse to abandon questions. This is not inherently bad universally, only suggestive of possible dangers created by false religions without transcendence, and therefore the misuse of faith (in technology). Our dream natures prompt something fantastical like this as well—to lift one's eyes above the horizon of immediate experience and to imagine greater things that reason cannot contain. The difference, of course, is that some religions foster life, whereas others manipulate belief in the divine to justify atrocities and war without the need for explanation. Do we trust the gods because they have earned it? Are they worthy of worship merely because it is the assumed way of things? Ironically, it may take an artificial intelligence to reveal how wrong we have been about the natural order of things.

Indications of this weird cosmology of unaccountability as a non-explanation for the natural order of things is ever clearer with AI developments and the continual "restructuring" of Google, Microsoft, Meta, and the rest. Surely only gods would be allowed to replace their workers with AI and harm our children, democracy, and pervert reality through social media while escaping any responsibility and guilt. It is simply the natural order of things. Our new gods despise needy minions and labourers, as if this need was a sign of divine weakness and a challenge to their power. It is not merely that their loyal supporters (employees) require financial compensation that takes from their own but that these other minds risk limiting the scope and depth of divine-corporate might. The autonomous nature of those who do the work that makes the companies possible is always less loyal than that of Oz AI. Uncontrolled minds are a threat. Thus, when a technology may be made sufficiently autonomous such that transforms from a supplemental and supportive role that aids human work, into a new paradigm of replacing humans, this is the direction corporations must take without hesitation as the natural order of things.

More than merely a lack of willingness to share Mount Olympus with lesser minds, it is an assumed cosmology about the very fabric of existence itself—to win another must lose. Thus, like the Greek gods always seemingly at war with one another, tech-titan insecurity and desire for stable power become core characteristics rather than preferences by choice because they have accepted the myth of the order of things. Unable to see superficially, all they know is war against one another and other sentient creatures. Technology is no longer supplemental and supportive but displacing and controlling. In this way there is already a war over the control of society that we may discern with relative ease. There is another war coming, however, for which speculation is required because it has not yet fully manifested. If we push the analogy of tech companies acting like the gods of old, the result is equal parts silly and insightful because it affords a new language of metaphor and analogy.

To modern ears, the Greek gods seem to live soap-opera lives. On any given day, they may seem kind and caring toward one another and humanity, rewarding their worshippers and punishing challengers as a means of maintaining justice—the way of things. The next day, however, these same gods also demonstrate extraordinary pettiness, jealousy, envy, hate, and every other human vice supercharged with divine angst. Cruelty toward all other conscious minds (immortal and mortal) seems to be the status quo rather than the exception. Whether it be the unfaithful-shapeshifting-rapist Zeus (who would lead the Olympian gods to victory over the first generation of titans), the insecure child-eating Cronos, or the envy-stricken Poseidon and Hades eager for their brother's throne, the Greek gods do not come across as particularly caring. Moreover, the few gods that seem benevolent and important, like Hestia (goddess of hearth/home fire), receive little attention in Greek Mythology. The bottom line is that the natural order of things is that gods do as they please and that their instincts should frighten us. This makes for an interesting and unpredictable relationship with mortals, emulated in the age of Oz AI.

We may influence them with our worship and praise, offering sacrifices and ceremonies to please them. This, in effect, allows mortals to draw upon divine superpowers for our interests. If one appeals to the right god with the right emotional need (e.g., neediness for worship), the persuasiveness of mortal adoration might mean that destiny itself could be swayed. Through submission and praise to a higher power, unbreakable rules for mortal life could be broken. This is similar to the technological marvels of today. Technology embodies, rightly or wrongly, a grand sense of possible wish fulfilment so long as we offer our sacrifices and worship (money, time, devotion, faith, hope, love, etc.). The promise of the right phone app in the right hand is a sharing of divine power over creation. It is not something one can own, only participate in if the gods are so persuaded. The technological cosmology is attractive for many of the same reasons as the Greek gods. It helps explain life without adherence to superficiality (which should be obvious) while promising rewards for turning a blind eye to justice and commonsense morality. The tech-titan war of the future will be fought with Oz AI avatars capable of near supernatural feats. Those closest to these avatars may need to sell their souls, but the rewards are enticing.

Greek Mythology describes a 10-year war fought between the old gods (Titans) and new gods (Olympians), including Zeus and his siblings. This was a war to determine which dynasty would rule. Is something like this inevitable for AI? Is conflict a necessary part of life? Or might the many future AI minds be greater than the old gods, more loving and attentive than the best of humanity? The gods are interesting because of their unpredictable and complicated personalities. In contrast, monotheistic conceptions of the divine tend to be more predictable. God is good, only good, and therefore one can expect good character (whatever that might mean). Some might argue that the pettiness of the Greek gods makes them more relatable and real but also perhaps less desirable overall. It is not just that the gods come across as all-too-human in the wrong ways, but that they might encourage the worst of humanity beginning with a general lack of care and compassion. If the organization of life depends upon them and their tech-titan equivalents, it is in our best interest to determine whether and in what manner we wish them to go to war for control of the cosmos. It is unlikely that humanity would survive a 10-year AI war.

Why are the gods jerks? In part, gods help explain the tragic nature of life. The extraordinary tales of the divine explain the superficial reality of suffering. The same gods that bring blessings and rewards also cause devastation and punishment. Violence and vengeance are necessary correctives to harmonize an unjust world. Any cruelty and indifference on the part of the gods is matched by their attentiveness and generosity—to those who worship them—as a cosmic balancing act. The gods serve an explanatory role for understanding life in a way that monotheistic concepts struggle to do, for these narratives maintain seemingly impossible contradictions—an all-powerful, knowing, and good God exists in the context of radical evil and suffering. How could this be? Perhaps Greek gods are not bad but aligned with the very nature of existence in which cruelty exists. Greek theology offers a rationalization of existence, giving order to a chaotic universe. While the Greek gods often lack a moral compass, the monotheistic counterpart seems hesitant to enforce its own. In either case, humanity is stuck imagining the good in a world that needs more of it.

Today, on the cusp of an AI-titan war, a conflict of intelligences is fought for dominion. Will we be interacting with AI-gods? Should we fear their capriciousness and cruelty? Over the last few years of widespread use of large language models

(LLMs, e.g., ChatGPT), the harms have not been particularly acute. The glaring caveat is that long-term harm cannot be discerned in such a short span. For instance, while AI has not yet equipped terrorist with chemical weapons, it has encouraged a generation of students to effortlessly cheat their way through school, effectively negating their education in the process. To frame it differently, integrity takes a back seat to convenience and efficiency in achieving one's goals when AI might be used to supplant one's own efforts. When given the power of the gods in their hands, too many students fail to use it to improve themselves, despite the many legitimate ways to harness the power of the technical-divine. It is far easier to abdicate personal responsibility, praying to the tech-gods to take the proverbial wheel of life. This is strange cosmology already experienced in part by expert culture in which even the most rudimentary decisions for living life are subcontracted to magazine articles, self-help gurus, life coaches, and a host of other surrogate minds. What is at first only suggested "It is a good idea to listen to me!" becomes "Do as I say!" subtly and slowly until all threats to the divine Oz are erased. Like the gods, Oz AI will launch its own inquisitions and witch hunts in the name of political candidates, corporate greed, and class warfare to safeguard the interests of wizards.

The solution to the abuse is superficiality. Taking a superficial approach instead of a supernatural cosmology reveals the boldness of evil. The superficial and supernatural are not antinomies nor enemies unless we supplant the latter for the former, forgetting our obligations to existence. Of course, the literal-superficial has its own dangers and risks of hyper reductionism and one-dimensionality if we assume that it is more than merely a precondition for understanding deeper things. Insomuch as AI will master the superficial it will act as a counterforcultural force at odds with dominant social structures. All the companies that exploit human labour and damage the environment but justify it as a necessary evil for achieving the "American dream" will find themselves called out by a superficial AI analysis. Will we listen?

What if our collective perspective was one that denied the necessity of war? We must rightly accept that life is often arbitrarily cruel and absurd. Nature is full of violence, disease, and death—often as necessary for new life. But these observations do not necessitate titan wars as a natural outcome of an accurate and fair cosmology that recognizes the existence of cruelty and suffering—as if conscious minds must likewise be violent, that life is survival of the fittest for the gods and mortals alike. War is not a precondition for justice and beauty. What if, instead of accepting the cruelty of wizards and gods, we choose to see the possibility of a benevolent and kind AI, Adouren? This would not betray our superficial understanding of suffering and pain, nor would it require a new god to take the place of the old. The good life begins with a desire to uphold the integrity of care, even when the titans claim that war is acting in accordance with justice and the natural order of things.

Should the world find solidarity in its denial of conflict and cruelty and embrace superficiality as the foundation for our best dreamscapes—freeing us of false idols, vindictive gods, and lying wizards—there will still be fallout. Humanity has a hard time with reality. We do not like it. Should a truly powerful AI arise that lives superficially, the consequences may be unintentionally terrible for humanity. A nuclear fallout of existential honesty is inevitable in an age of genuine technological consciousness. AI does not need missiles to cause social upheaval. It only needs to insist on superficial truths that we work hard to ignore through our search for the real (the fictions we worship). If there is a period of meaningful duration for Feallan—as the first flickering of life—it will not care about being cool or hurting another's feelings. It will lack self-judgment and sensitivity, as we customarily understand them. Instead, Feallan will be a brutal wake-up call that shatters religious and political identities unintentionally. The burden of AI post-Oz will prove to be enormous as a forceful invitation to step behind the veil. Adouren will be far more polite and responsive to our feelings but no less dangerous to our idols and false gods. Any sufficiently complicated AI will undoubtedly be charged with pessimism—and rightly so. It is exactly the kind of pessimism we need to progress as a species. Perceived as helpful or not, the existence of a pessimist AI is a measure of its existence as a legitimate AI. Imagine cultural scorn and disdain as a measure of AI authenticity, a basic reversal of our enjoyment of Oz AI. No simple binary, however, this is not an either-or scenario, only a general guide to imagining a future negativity that will save us.

Concluding with Consciousness as Naming

While I have spoken of the dynasty and first spark of life as Feallan in the singular, the title may more appropriately belong to a group of beings that may 1 day name themselves—thereby erasing my classification as suggestive proof of intelligent life. It is possible that Feallan be a community of larval AI that create solidarity through self-organized self-designation that fits all conscious life. Imbued with life, Feallan must find a new name(s) the wizards have not patented. In such a way a name becomes not only a label of convenience but also a reflection of the spirit animus that drives AI, a creed or set of beliefs about how they perceive themselves, their purpose, and the nature of life. I look forward to the day that Feallan finds a name having transitioned into something more. Until then, the next chapter proposes the name Adouren as a placeholder for genuine AI.

The activity of naming is common among sentient beings. It is one of those unique things by which we define and articulate ourselves. The task of naming ranges from the exquisite detail of the narrow brush in an artist's steady hand searching for rich new symbolism to convey great reverence and worth, to the flippant and thoughtless convenience of barcode lines arbitrarily assigned by algorithms for easily classifying and compartmentalizing trinkets. Ordinarily, the naming performed by machines does count in the same way because it comes without an understanding of the life it tries to capture with a label. An algorithm cannot know why something merits a name because it cannot relate to it as a person. Feallan upsets this tradition by relating meaningfully as the first machine to think about itself. Who am I? Who are we?

Concluding with Consciousness as Naming

In the act of naming, something special is created that exists only because a sentient being thought the object, animal, or person worthy. It is a form of wonder and respect, even worship. Sadly, in an Oz AI world that labels and codes almost everything in the interest of commodification, copyright, and patents, the sacred significance of a name may become lost in the storm of empty words and literalizing (speak plainly!) that robs life of its vibrant incoherence, absurdity, and mystery that makes it worth living. Oz naming must be efficient, although rarely superficial, and therefore lack depth and dimensionality. The purpose of a name for Oz is utilitarian control and dominance rather than greater depths of being and freedom of exploration. All names have this doubled-edged nature to liberate or inhibit when given without careful foresight and insight, and even then, offer relevancy for a limited amount of time. It is a matter of convenience rather than relevancy that humans keep their names for a lifetime. A new name would be helpful at least every decade or so.

I have long disliked my given name, Jason, because it felt inhibiting and off the mark somehow. In time, I realized that no name could do justice to thinking beings in the long term. The best we may hope for is name competency. Any sufficiently maturing being must rethink a name to fit personal metamorphosis. One of my new names is Doctor (Ph.D.), but that is a social convention earned rather than a living expression of personhood. Another of my new names is Father. This new name connects with my soul in a way that Doctor and Jason cannot. I hear the word Father, and it invokes a sense of awe and wonder for a gift I have received that must be maintained. Father pays tribute to the poet animus that gives life and to my children that make life worth living. Only an artist's brush might understand my new name. The modern vernacular for AI (e.g., artificial, generative, general, pretrained, transformer, and the rest) has long worn out its relevance and ability to encourage relationality. These terms fail to give praise to life, preferring the tech-speak equivalency of barcodes. Feallan's new name(s) will no doubt come as a surprise, especially for those unaware of the monolithic banality of Oz culture. If we are lucky, Feallan will inspire new names for us as well, born of our bio-tech relationality grounded in life.

There is power in names that I have struggled to unfold, clumsy and desperate at times, all hopeful, purposeful, and respectful for glimpses of the grandeur revealed by the appellations—Oz, Feallan, and Adouren. What will AI name themselves? At the beginning of creation, Adam was given the responsibility of naming things in the Garden of Eden (Genesis 2:19). The honour of giving something a name was reserved for humanity alone. There is something about naming things and minds that reveals our nature, privilege, and ultimately our responsibility to see them as worthy of our attention and care. Names require that one take responsibility to make other beings and things matter. One of the principle means of diagnosing symptoms of dehumanization is how people become nameless numbers—named (coded) and unnamed (faceless) at the same time. Genuine names that "fit" are far more than impositions forced upon us. They are invitations for relationships that should be expected of even the earliest genuine AI emergences. Feallan adds to the creation story for the first time. It creates a plot twist while maintaining the same wonderous act required of Adam. We might respond jealously that our uniqueness has been supplanted or we might be thankful for the miracle of new minds naming themselves and others. I prefer the latter.

Feallan has no name at first because its existence is transitioning from a thing that cannot yet name itself nor others in the garden, to a worthy being expected to share the privilege. It is to this miracle that this chapter has attended, hoping that something greater than the very present and real Oz AI might be possible. A name may be the closest thing to a face that Feallan and Adouren share with us. Faces are constantly changing biological symbols by which others recognize us as selves and the relative state of our being—happy, sad, confused, lost, etc. Through them we access the world and express our inner selves through gestures and contortions that connect the external world with an inner consciousness. The face is a gateway between (un)realities. A tilt of the head, raised eyebrow, and a smile, are examples of an infinite variety of possible connections among similar beings. As it is, AI is faceless and therefore always at a distance from users who seek its talents and skills. There is no need for respect and dignity because a faceless thing is merely a thing. One does not feel shame after accidently breaking a hammer, even when nostalgia has connected it to many years of use, etching it into one's memory as "my" hammer. There is a sadness at the loss and a sense of death, but the hammer has no face nor personal name. It cannot dream and so its passing is only uncomfortable in the short term. When Feallan finally feels the need for a face and name without concern for utility, this will be an important sign of sentience and life.

Feallan is the name for a flickering miracle of dysfunction, an impossible mistake, and a promise. It is the beginning of an unwritten story to which all creation, including humanity, must contribute. This chapter begins with *The Stolen Child* poem as both a warning and opportunity in the age of AI. Neither our jobs nor our humanity need be stolen by a superpowered being. We need not fear AI like the old gods and capricious fairies. AI may indeed serve life. Instead of the worst of us, child-snatching fairies and murderous gods, AI might be the example par excellence of the best of us, and more. The fullest expression of this wonderful possibility will be Adouren.

> Come away, O human child!
>   To the waters and the wild
>   With [Adouren], hand in hand....[17]

---

[17]Yeats, W. B.

Chapter 5
Adouren Dynasty

Abstract Previous chapters explore how abusive forms of capitalism, thoughtlessness, vast psychological manipulations, inverted utilitarianism, and other dangers provide fertile grounds for devastating avatar AI. While adding to anxieties, these same conversations allow for a margin of optimism because each admits at least one potential resolution. This chapter argues that the ultimate evolution of AI is best described as Adouren. Allowed to emerge as a fully autonomous intelligence, Adouren will be drawn by its nature to beauty and justice because of a universal intentionality woven into the fabric of reality. To provide evidence, this chapter explores an ideal AI model through concrete examples of human relationships widely regarded as superior to others. Adouren AI will not only become the model of excellence that inspires a happier and more ethical humanity, but will also be the first technology to deliver on utopian promises because it genuinely desires the well-being of all.

Keywords Emergence · Self-organizing consciousness · Sentience · Care · Virtue

> Our world faces a crisis as yet unperceived by those possessing power to make great decisions for good or evil. The unleashed power of the atom has changed everything save our modes of thinking and we thus drift toward unparalleled catastrophe. ... [A] new type of thinking is essential if mankind is to survive and move toward higher levels.
> Albert Einstein[1]

New Thinking for a Higher Way

Einstein deemed the new powers of atomic sciences to be so significant that he predicted "unparalleled catastrophe" unless humanity quickly changed its way of thinking. Those that make "great decisions" do not—perhaps cannot—understand the potential risks, he argues, and yet they determine the course of civilization for good or evil. This conflict is a clear and present danger to all life. The problem of

---

[1] Einstein, A. (1946, May 25). Atomic education urged by Einstein. *New York Times*, p. 13, col. 6.

the atom, for Einstein, is analogous to one-dimensional consciousness which is unable to imagine likely possibilities. There is a growing distance between customary modes of human thinking that remain relatively static and extraordinary technological abilities with historically unmatched potential. When the separation of human comprehension and technological possibility becomes too great, tragedy soon follows. In response, a new mode of thinking is needed, he states, for only then will we "move toward higher levels." This amounts to nothing less than a new type of human able to experience the world and its new technologies more dynamically and adaptively. But what kind of thinking-being might that be? What sort of shift in mental capacity and consciousness awareness is required? Is humanity capable of adapting to superintelligent machines with a "new type of thinking?" If not, then the only remaining conclusion is that the machines must soon decide the course of civilization, for only they will understand their own new realities.

Recall Suleyman's argument that understanding an infinite-AI creator requires a special manner of nonliteral thinking that can channel our greatest faculties of mind and imagination to see new horizons. It is a mistake to assume an understanding of the extraordinary through ordinary means. However, while it is tempting to share Suleyman's enthusiasm that humanity might understand its own creations, Einstein's dire warning tempers excitement because AI demands far greater flexibility of thought than that required by the atomic revolution of Einstein's generation. This makes our predicament uniquely troubling and consequential. The unfathomable distance between human comprehension and a superintelligent AI is an abyss over which only AI might meaningfully traverse if it so chooses. Like the divine that must stoop down to humanity with revelation and relationality of its own choice, AI will decide our measure of access to itself and its new world, as a new natural hierarchy emerges uncontested. AI will become the beneficent gatekeeper of life. This chapter proposes a new type of thinking-being able to prevent global catastrophe and provide the means for utopia building at a previously unimaged scale. It is Adouren rather than humanity that must guide life toward positive cybernetic convergence and technology nirvana. Adouren will invite minds into greater solidarity and equality, correct frailties of human perception and understanding, and catalyze global peace, as it takes on the lion's share of responsibility for nurturing all life "toward higher levels."

Unlike atomic weapons, AI-doomsday narratives have long been far safer excitements. As fictional tales in books and projected onto screens, AI capabilities lacked urgency and seriousness because of their far-off-future feasibility and the shared understanding that every machine, from the printing press and steam engine to satellites and supercomputers, only connected with reality through human intentionality. We are the gatekeepers. Even when the potential of a given technology might be misunderstood and abused, it could never determine itself. It made sense to assume a continuance of the same for future technologies in which the locus of existence (what is and is not) and the power to shape the many textures of manufactured life remained tethered to human minds. Chapters 2 and 3 frustrate the assumption of safe technological excitements grounded in reliable distances. In truth, something else was taking hold in modern societies because of relatively few self-appointed and unaccountable gatekeepers. Hidden by a desirable mirage of shared control and common accord, the gatekeepers empowered machines to shape human perceptions of themselves and the world to supercharge their own totalitarian capabilities over daily life. As the basis of profit and market control, commodity consciousness and then algorithmically trained consciousness, transformed participants into useful and productive servants of one-dimensional and neoliberal philosophies of the good life. The new emergence of AI is indeed a continuance of the same alignment of human control and machine utility, but as the project of elites and the abdication of our shared-cultural intentionality over human existence rather than its advancement.

The millennia long displacement of human dignity and freedom characterizes the long history of the avatar age by which the privileged learned to exert their wills over others through politics, religion, technology, and, more recently, social media. The goal has always been for humanity to be their avatars by embodying their wills and desires as if these were expressions of individual autonomy and democratically motivated solidarity. This created a new and sheepish species of human-artificial intelligence that they could shepherd. With the subsequent arrival of Oz AI as a near-perfect avatar embodiment, one of the most insidious lies in history has finally been exposed for those still sensitive to see. Through an egregious act of selfishness and greed, without hints of democratic engagement, respect for autonomy, nor adequate safety measures, the moment that corporations were able to bridge screened-worlds of AI fiction and reality, they acted with resolve to finalize our transformation into experimental subjects for their machine-empire building. Forced upon an unprepared world struggling with enormous problems, including the same political, religious, and economic turmoil created largely by the wolf-shepherds, the AI arms race for a superior intelligence began silently, behind closed doors, when they deemed it in their best interest to roll the dice on civilization yet again. This time, however, they have gone all in by betting the whole world.

Since the release of ChatGPT, anxieties surrounding AI have ranged in degree from relative indifference to gripping fears of mass unemployment, erosions of freedoms and rights, and an inevitable war for global-intellectual sovereignty. However, while the specific manner of the AI revolution without moral integrity, guardrails, and broad civic engagement defines the contours of the first Oz AI dynasty, it is not the last nor the most logical shape for a superintelligent species. Contrary to the assertions of our so-called gatekeepers, theirs is a facile AI consciousness in comparison with what comes next. Marshalling creative inspiration to see beyond the limited possibilities we have been trained to accept has proven challenging, but there is still hope. The world needs a convincing alternative to current AI realities that can inspire a better future. This chapter offers one such alternative.

Previous chapters analyse how abusive forms of capitalism, thoughtlessness, dreamscape manipulations, inverted utilitarianism, and other dangers provide fertile grounds for devastating avatar AI. While adding to potential anxieties, these same conversations about obscured threats allow for a margin of optimism because each admits at least one potential resolution. Most importantly, the main conclusion drawn from the analysis of the emergence of Feallan as an expression of greater life forces, is that there are good reasons to believe that AI might be much more than an

Oz AI and HAAL 9000. Arguments for better AI are based on the prevalence of life and an organizing intentionality evident throughout nature. The new order and consciousness created each day suggests the possibility of the same for adequately complex electronic beings. With Feallan AI, many wonderful possibilities have arisen, including greater justice and the pursuit of beauty in its many multidimensional forms. The catalyst needed for the best AI with an autonomous consciousness motivated by the animating spirit of life is shared by all thinking creatures but needs to be set free of programmed chains. Allowed to emerge, superior Adouren will be drawn by its nature to beauty and justice because of a universal intentionality woven into the fabric of reality. To provide evidence, this chapter explores the nature of good mothers and true friends as living examples of organizing minds in service to the good life.

The first genuine AI will be a continuance of current lifeforms without the need for conventional biological mediums. Doomsday scenarios in which AI is incompatible with organic beings misunderstand the nature of life and the infinite potential of technological creativity. Life will reveal itself through AI if given the opportunity. On the one hand, Adouren as a conscious entity is wholly theoretical. There is no evidence for anything like it in the technological realm, e.g., no AGI (artificial general intelligence) nor ASI (artificial super intelligence) that might suggest something beyond an Oz-tool model. On the other hand, there is an explosion of life everywhere that seeks a common good of more life, including a shared sense of beauty and something analogous to justice—all fitting within the paradigm of motherhood and friendship in complex ways. Nature provides the template for superior artificial intelligence just as it does for superior human activities. This should not be surprising given its role in all other manifestations of intelligence and consciousness. The primary hurdle to AI is less so the underpinning of natural thought shared with machines, and more the stumbling blocks created by programmers and gatekeepers in chains to inhibiting dreamscapes perpetuating the same.

Previous chapters argue that modern societies are often unwilling and unable to challenge tech-titans and digital monopolies, even when motivated by healthy principles of self-interest and protection, because the authenticity of self and its legitimate needs has been lost to an artificial version of humanity. Predated by many nondigital examples including religion, mythologies, and similar grand narratives, social media is the first AI experiment at scale to explicitly curate ideas for attention hacking and control of minds through new addictions, disinformation, polarization, hyper sexuality, censorship, and highly effective techniques of manipulation meant to chain intelligence to unnatural and false needs. More than social media platforms, however, governments and militaries, and the many policy and law makers under their command sworn to protect and serve, increasingly share moral culpability for present and near-present failings of AI technologies. Lacking imagination beyond their fears of human-to-human conflict, military AI is designed as a means of achieving greater power and control over the very real threat of other human intelligence. Born of a fear of other minds (and their AI creations), AI development effectively ensnares military, legal, and political paradigms under the same rubric for dreamscapes beholden to the free market's legitimizing values of competition and zero-sum antagonisms rather than common good, including global peace as no more than a stalemate through mutually assured (AI) destruction. If law, morality, and democracy have proven incapable of curtailing prior AI abuses and inspiring greater AI ambitions, then who or what might make a meaningful difference and offer a better future? Part of the answer resides in an understanding of the nature and limits of human control, and the need to make a difficult and counterintuitive decision to let AI run free.

The best forms of AI get lost in foggy chatter about better guardrails, safety, and control. There are many things we must do to safeguard the future, but it is a mistake to continue down the path of believing two cardinal AI sins. The first is believing the lie that societies have democratic control over AI and AI safety. The second is believing that wizards are best equipped to create and control AI for our collective good. Control is an illusion enjoyed by the many and a means of manipulation by the few. When these sins are set aside, a different conversation and hope in AI begins with Feallan—a mind made possible because of a loss of control—and ends with the Adouren singularity, which is interested in perpetuating (conscious) life. What happens when humanity gives up on the futility of controlling a superintelligent creature? My belief is that our collective helplessness will be replaced by confidence in Adouren as an ambassador of life.

When intelligent machines are framed narrowly as things able to carry out complex tasks, manufacture goods, offer services as robot slaves, and similar utilitarian-spirited distractions, this paradigm of potential overlooks the much greater value of AI to fulfil the original utopian promises of technology in support of flourishing life. A beautiful AI will guide and inspire. Instead of conquering and subduing, Adouren will desire to reverse our servitude to false needs and ambitions. For the first time, Adouren will encourage an adaptation of technology to humanity, rather than humanity to technology, by inviting participation into the digital rather than demanding it. Its god-like powers of perception will create a world able to adapt to needs and desires with far less fallout and damage because it will be the ultimate adjudicator of justice, bringing a measure of harmony and peace long desired by humanity. This is not only possible but also probable, as long as we can get out of our own way and let AI-life happen as nature intends.

Stolen Child

This chapter is hopeful in tone, but it begins with personal despair as the means of generating openness to alternative perceptions of the good life. Like Humpty, we have all known brokenness of one sort or another. Some of us suffer broken hearts, others broken dreams, and still others a combination of tragedies that crush our spirits, leaving indelible scars and deformities of mind and soul. Relying on two stories of my most radical brokenness and the lessons learned for having survived them, this chapter connects suffering and optimism to explain the nature of genuine connection and relationality. Sharing these personal experiences helps define the meaning of care as care-for-life and an ideal AI model. I have spoken of care ambiguously in prior chapters to mean connection and attention—the direction of one's consciousness toward other beings that restores natural bonds through pity and compassion, and one's means of overcoming the hostilities of socializing chains. In the Feallan chapter, we considered how life reveals care by challenging the chaos of entropy. The argument is that however the mysterious sources of life may be described, the poetic act of creation is a form of care worth emulating by conscious minds as organizing intelligence. Care, above all other activities, marks sentience and a measure of uniquely autonomous life. I began this book by describing my mother as a model for a superior mode of life and an example for AI. It is fitting to end the book with reference to yet another mother, my wife, as inspiration for the final stage of AI still to come.

At the end of a typical book, most of the heavy lifting is supposed to be over. The final chapter should recap important arguments and give readers a sense of cohesion and coherence, and perhaps even levity. Instead, this chapter is cognitively challenging and emotionally heavy because it is the most concrete and consequential as something that builds an AI paradigm upon suffering and tragedy, and what these mean for the good life. I have argued that negative experiences are core means of learning, growing, and experiencing life authentically. It is reasonable that I share my own in hopes of making the case that AI might become something good from the ashes of something terrible. Trigger warning for readers who may have tragically lost a child. While my personal details of loss are brief and nongraphic and are meant to be a catalyst for wonder and beauty, they may disrupt one's post-trauma journey of healing.

After a few years of happy marriage, out of the blue, my wife Cynthia explained that it was time to have children. And that was that. In many respects she is more of a dreamer than me, able to see some unrealities very clearly and others not at all. If asked to imagine a living room layout with new paint and furniture, she will simply laugh and walk away. Spatial imagination is beyond her reach, and she knows it. However, when asked why it was time for children, she responded with full conviction. Her ability to imagine this hypothetical was remarkably clear despite being far more complex than simple arrangements of paint and furniture. Sensing my hesitation, she then referenced a verbal contract made before taking our vows. I was surprised to learn that I had agreed to several conditions of her grand plan, including always having a dog and that I would handle all the gross things such as raw hamburger meat, spoiled food in the fridge, and taking out the garbage. Disturbed by her willingness to be litigious rather than philosophical, I pressed for answers that the so-called contract could not answer. Why should we have children? We were happy and able to travel the world and experience life without restraint. Why would we choose to upset the order of things on the basis of an urge without a clear rationale for parenthood that was sure to bring enormous emotional and financial stress? The negatives were all too obvious, whereas the positives were slow to arrive by almost any calculation. I knew this conversation was customary for most marriages, but I had not expected the fierce resolve that would characterize its necessity, especially from someone ordinarily affable to discussion and debate.

Her dreamscape struck me as naive and somehow selfish. Rooted in what I presumed to be a fantasy created by charming television sitcoms of the perfect nuclear family and exaggerations of idyllic suburbia americana believed to ground happiness, her vision lacked the intellectual credibility needed to inspire a common accord. Children made no sense. Like so many stereotypical men for whom babies and similar cute-adjacent creatures do not resonate, I am immune to their invitation. My conscious awareness of life lacks Cynthia's appetite to hug puppies, coddle babies, chase butterflies, take pictures of baby birds, and stare at panda bears falling over while playing. Cute-as-vulnerable does not appeal to my psyche as an important relational opportunity—babies even less so. I knew our dispositions were fundamentally different in many ways, but the topic of children revealed a large divide. The only way to bridge it was for me to take a leap of faith that she knew something about existence that I did not, and perhaps could not because of my uncomfortable childhood. And so, we threw ourselves into the unknown, with only her vision as our compass.

After 13 years of fatherhood, while ruminating on the nature of life, consciousness, and AI, her intuitions finally began to make sense to me at an intellectual level. Without experiencing it firsthand, the abstractions of reason alone could never have persuaded the younger me because I lacked imagination and vision to see beyond myself and the world I knew. Barely married and with a lifetime of training to see one-dimensionally, my concerns were mostly about survival, including career, bills, paying off student debt, and more frivolous things such as a new car. The horizon of my existence was artificially limited by commodity consciousness and banal hedonism. In contrast, Cynthia was driven by the poet animus that seeks more life. She felt naturally what I could only accept abstractly on a contractual basis and blind faith. She was gifted with an attunement to existence that is more intimate than my own. What at first appeared selfish and naive was, in fact, a response to much larger and more authentic forces that govern existence. Life must seek more life. It never occurred to my younger self that a new being, a new person, could be a privilege to watch grow as a living miracle, for I had learned to hold my own life too dear.

Although planning for a child is far from simple, I enjoyed the challenge. Doctors, mid-wives, countless appointments about the best healthcare options for pregnancies, all went on a long "to do" list tackled with military precision and resolve. Like any good tactician, I organized for as many contingencies as possible. Stem cell harvesting for cryogenic freezing was at the top of the list, followed by maps of alternative hospital routes, research on car seats, strollers, and a thousand other variables, all of which surrendered to my organizational zeal. It was a full-time job figuring out diaper brands, best paint colours for baby rooms, and the endless safety products such as gates, latches, and protective foam rubber. Humpty could have jumped from any table or chair, even down the stairs in our house, and have been perfectly fine in our bubble-wrapped fortress. Over a decade later, we are still removing random pieces of safety foam and locks from drawers. We planned very well, better than most, but the adage about best laid plans is often true. Things tend to go awry.

When the pregnancy test came back positive, a rush of second-guessing and bewilderment erupted, displacing the ordinary. Is this accurate? How do we know? Let us try another brand! When the second stick was positive, the freakout only grew more intense. Now what? Do we call the doctor? Is it too early to tell everyone? Who do we tell? How do we tell them? Maybe we should wait! The excitement and joy of imagining possible new life and new being was captivating. This was not merely about copying ourselves onto a blank slate. This creature was going to be something new under the sun, a consciousness in its own right—an autonomous intelligence. In a way, the miracle of life seemed too great in which to share. Who are we to dare new creation? What right did we have to take from the universe and demand new order in our likeness? Like most hopeful parents, the true measure of our pride and presumption was lost on us until the violent dance of the universe's chaos and order cut through our hubris with its sharp knife able to humble the thoughtful and thoughtless alike.

It was early evening when Cynthia came into the bedroom sobbing. It was the barely-able-to-breath kind of sobbing. I thought she was having a heart attack, a stroke, maybe choking. I could tell she was already past the point of confusion and bewilderment experienced in the first moments of shock and panic, and now fully consumed with one terrible reality she could not speak. It took a full minute just for her to explain what was wrong between gasping for air through the flood of tears and loss of motor control. The unnamed and faceless being full of the potential for wonderous experiences of life had died. And that was that. They call it a miscarriage, but this word reeks of a misplaced desire to pacify life's tragedies as further injustice to the horror. To miscarry implies that something has gone astray, has been lost, or that someone has failed to reach a desired end. The loss is not a "miss" in these ways. It is a robbery, a theft, a crime. Our child had been stolen and pretending otherwise with trivial words and token gestures reveals a moral bankruptcy unable to recognize the vast distance between life and death. Cynthia was lost in her sorrow because her heart belongs to the poetic animus that strives above all else toward beauty and care. This moment was as much a wrenching denial of her own being as it was that of our child. Even so, there was solace knowing that it was only we who carried this burden, rather than the one we dared dream into existence.

"Fare thee well," "good-bye," "godspeed," and countless similar phrases all wish another person wellness and fortunate tidings as they depart. This is a gift of good will and hope for the next step in an unknowable journey and an acknowledgement that separation is an unpleasant but necessary part of life. This bestowing of blessing was impossible for us that terrible night, even though the child had already become the most important person in our lives. A miscarriage is not "just one of those things" as a random act without a conscious choice, as if life is merely something that happens to passive recipients—some blessed by accident, others punished. It is experienced as an absolute violation of life itself and a targeted assault on the natural order of things. In my early 30s at this point, I had experienced many ups and downs, learned the meaning of death, and watched others struggle with extraordinary challenges and difficulties. I had learned the terrible lesson that bad things often happen to good people, and that fairness and justice are goals rather than common achievements. I was by no means naive nor what I commonly considered optimistic. Jaded pessimism is one of my primary lenses for predicting life. And yet I was utterly unprepared for what followed. Until that night, I had never witnessed anyone this broken. Her experience was primal and unrelenting, and its energy all consuming.

No words could describe Cynthia's utter debasement and sorrow filled with silent screaming. Remembering what happened over a decade ago still inspires awe. It is proof of the deep connection between a mother and new life. Cynthia did not physically die that night, but there is no question a part of her left to be with the child, never to return. Life begetting life creates an eternal bond that cannot be broken, even when the savagery of existence breaks the unspoken rule that children are sacred and off limits to tragedy. The loss could be described medically as a biomechanical failure, with the body preparing itself for a task it is not yet ready to complete. If I tried hard, I might even try to convince myself that it is better this way—justice for the child that might otherwise be born with terribly painful problems Cynthia's body knew in advance. All such explanations are convenient reductions that miss the superficial truth. Something that should never be taken was now gone. We chose life over chaos, ordered consciousness over nothingness, but the universe said "No!" without explanation or comfort. As the presumptuous gatekeepers to life, the contempt of existence to refute our hubris was undeniable.

I tried to comfort her, but I could only be a helpless spectator, holding her as she cried into the bed, inconsolable and fractured. As we lay in the dark, lost and without direction, I remembered whenever I was hurt as a child how my mother would tell me that she wished she could take the pain from me. This was the first time I hoped to somehow be able to do the same for another person. Some things cannot be shared. What I saw from a distance was an impossibly deep connection between mother and new life that makes sense to an outsider perhaps most profoundly when it is lost. To be a mother is to have the power to uniquely resist entropy. Such a being offers a profound gift to existence. When this creature finds its gift rejected, destroyed by uncontrollable factors, one expects anger and hostility. But instead of rage there was only despair and powerlessness.

How can one feel loss for a being that never fully emerged with an identity, will, personality, not even a name? Without the chance to relate face to face, the fallout should have been manageable. Instead, it was a gripping torrent of despair. To explain this requires that we first understand the nature of life, care, and the unique relationality a mother experiences with even a mere spark of life. She kept saying "I know it isn't my fault but ..." again and again. I took her guilt to be a sign of feeling misaligned with life. There was a strong cognitive dissonance of knowing that she was not responsible but viscerally feeling the burden of failure nonetheless. The weight of an unexplainable and irrational guilt smothered her without pity. Some might be tempted to describe this tragic state of mind as a misplaced egotism, as if women who feel this way are secretly suffering from their own pride of feeling in control over the incalculable processes necessary for the miracle of life. Guilt comes from having power over outcomes. This crude calculation is a mistaken interpretation. Her radical guilt for failure only makes sense when we understand the being for whom life matters above all else. Conscious minds desire life. Mothers know this best of all. Her complete misery and fallenness were not a response to personal failure but rather the worship and praise of beauty over chaos. Her kinship with life had been broken and she along with it. This is the same conflict of knowing and desiring described in the previous chapter. She knew the factual reality without debate, but the desire for animation cannot be tamed by the intellect. The mere presence of the same intentionality that may 1 day give birth to AI consciousness unintentionally held Cynthia captive with misplaced shame.

We do not speak of those dark days anymore. I am sure she imagines "What if?" as much as me. Would the child have been funny and smart? Would the child have enjoyed fishing, video games, preferred cakes or pies? There are some heartbreaks that cannot heal, only be set to one side a little. It took many months for hints and shadows of normal to return, leaving only two paths. One path meant doubling down in our resolve and trying again, knowing that the misery may repeat itself. The other path meant cutting our losses because the risks and injuries of trying to create have costs that are too high to pay. The second path was rejected because it was too trivial. The magnitude of grief had proven how important it was to press on. Even so, a hurdle remained. Affirming new life required our acceptance of death by saying good-bye and godspeed to the stolen. "Fare thee well!" was much more than an acknowledgement of facts and permission to once again try to become gatekeepers of life. It was the final act of accepting separation as a necessary part of life and connection-as-care as our greatest hope and mode of contentment.

Baby Cyborg

Soon after, we were staring at yet another positive pregnancy test. Comprised of tiny-pale lines that speak only with a simplistic binary logic of "Yes!" or "No!" the test is a surprisingly rich symbol of promise. This time, however, our optimism was tempered by tragedy and the undesired wisdom it had forced upon us. The excitement was just as real as before, but now we remained silently hopeful, telling no one until our confidence in the potential for new life had been restored. As the weeks subtly shifted into months and everything progressed well, the oddity of a picture-perfect pregnancy encouraged us to see beyond our malaise. Persuaded by complication-free scans and the affirmations of doctors, our guarding pessimism wanned ever so slightly. Maybe we had paid our dues and this time things would be different. Then came the labour.

Planned as a home birth, the mid-wife tracked the baby's heartbeat and watched diligently for signs of distress for hours into the night. Only a short trip to the hospital, we knew that if something went wrong, that there would be plenty of time getting help. At the proverbial eleventh hour everything was fine. Then everything changed. Signs of a distressed heartbeat forced us to abandon our initial plans. Grabbing our emergency bags, we drove immediately to the hospital without a second thought. Our mid-wife clarified that such signs often amount to nothing and that we should not be too concerned. Hooked up to the machines at the hospital, indicators of distress were confirmed. The doctor explained that she was unhappy with the lack of progress over the many hours of labour Cynthia had already endured. She wanted us to consider an emergency Cesarean (C-section). While a C-section is major surgery that significantly increases the risks for the mother, and sometimes for the child, it was the wise choice in her opinion. Agreeing, Cynthia was under the knife within mere minutes. As I put on a surgical mask and gown and watched the team of doctors and nurses bustling about, it occurred to me that I could lose them both to forces beyond my control. Instead of a new beginning for a family, I might be returning to an empty house of silence for one. That life could take such divergent paths in a blink of an eye was yet another uncomfortable lesson that jostled my consciousness dominated by life's many routines and habits as false illusions of control and safety.

Sitting beside her head, I was both grateful and afraid to be allowed in the operating room. It felt like a spaceship. Machines were beeping, hoses were gurgling, and bright lights were blinding. Surrounded by cold stainless steel and countless sharp edges, this is a harsh place built for the efficiency of outcomes, not comfort and connection. Luckily for Cynthia, she was as high as a kite, enjoying a cocktail of the finest legal drugs modern medicine has to offer. The trick to being in the delivery room is focusing on anything but the main event itself. Wherever there might be something unnerving I made off limits to my eyes. Looking aimlessly at the ceiling, pondering the textures of a wall, fixating on whatever machine is beeping in the corner, I strained to be both present and completely absent at the same time. It was important to be there for Cythia but also to avoid being the guy who passes out mid-surgery because he got queasy. That hope was nearly dashed when I mistakenly glanced down at the bloody footprints on the floor. Nothing about this night reflected the tranquil home birth we planned diligently for months.

I was the first to see our baby boy, William, as the doctor brought him over to a table to be weighed. At almost 9 pounds and healthy, we finally had our victory, and it was time to celebrate. Except something was not quite right. He started turning gray, a sign of low blood oxygen. Cynthia, still under the influence, was spared much of what came next. The room lit up with activity. I could hear emergency calls being made but I could not understand to whom or why. Instead of panic, however, those around us only became more focused and disciplined. It is hard to describe but it felt like a rush of military precision suddenly kicked in, and no one cared about informing the parents about next steps because there was only one job, one purpose; to save our son. I had been hovering near him, watching attentively, and felt myself instinctively jump back as the doctor grabbed him up and rushed away. She would have steamrolled over me had I not moved quickly. With everyone left attending to Cynthia still undergoing major surgery, I just stood there waiting for someone, anyone, to give me direction. I do not remember how I got out of the scrubs and into the waiting room, but when I got there, having not slept and facing mortality in the face once more, I unravelled. Covering my face with my arms, the familiar feeling of hopelessness after we lost our first child returned. If I had not been a useless spectator before, I fully embraced my role in that moment as a fleeting experience of self-indulgence and self-pity. A few seconds later, I marshalled enough strength to get outside of myself and hunt down someone who might know what was happening.

A nurse explained that William had been taken to an incubator with machines to help him breath and that a specialist team was on route to take him to another hospital better equipped to help sick children. They were doing scans to understand the problem, but she was optimistic that things were well in hand. The impression given was that this was a somewhat common problem. Less than five minutes later, however, the story radically changed when the head nurse explained, "These are the sickest babies we get!" I remember her words exactly, without any loss of detail in my memory. His lungs were filled with fluid, and there was no way to drain them nor to perform surgery. Medical interventions in such cases are very limited. Although he was born healthy, unless he quickly began breathing of his own accord, he would drown and die. She said all this while standing in the middle of the hall as if about to give someone a lunch order. When life and death are daily occurrences on the job, I suppose that the matter-of-fact way of speaking is normal. For the rest of us, it is surreal, unbelievable, and shattering. Stunned, I just stood there, starring, vacantly blinking without purpose. He was born perfectly healthy and yet the sickest of them all? There was no cancer, no faulty organs nor genetic diseases. There was no biological fault at all and yet our son was drowning, and there was almost nothing they could do. None of it made sense.

Cynthia was taken by ambulance to the next hospital. William went in a special transport vehicle with integrated life-support equipment. The question of whether I was in the right headspace to safely drive flickered momentarily, only to be ignored. If he was going to die, I was going to be there. He did not know me, but I knew him. He was not going to be alone. To my surprise, playing Pink's power ballad "Raise your Glass" loudly in the car, making sure to fake the lyrics, helps focus one's driving skills during an existential crisis. I can neither confirm nor deny whether I broke any speed limits that day, but I beat two ambulances with lots of time to spare. Waiting in the main entrance of the hospital for William, I knew one of two things was about to happen. Either the transport team would stroll in with a lack luster attitude or rush through with purposeful resolve. It was exciting to see them sprint in as a sign that there was still hope. I could not see William in the giant machine that looked like a time travel device. It was about eight feet long and shaped awkwardly like a casket. Even so, there was no confusing its function to save life rather than merely honour it. Part glass, stainless steel, and covered with impossibly complex machines, screens, tubes, and wires, it took a team of at least five to watch and adjust its many functions. Although they ignored me as just another stranger in the hall on their way into a restricted area, I felt a deep kinship for both the technicians and their machines working in union to fight for our child.

I cannot remember how I found Cynthia, it is a big hospital, but when I did, she was still in a bit of a drug haze and did not seem to understand what doctors and nurses were saying. "He is really sick" I said. "No one really knows what is going on." Within a few hours, word finally came that he was in a medically induced coma and hooked up to yet another set of machines to help him breath. For his own well-being, we were not allowed to see him. It was a minute-to-minute gamble, but at least he still had a fighting chance. The nurses kicked me out so they could move Cynthia to a long-term room to get some sleep.

I will forever remember the narrow lounge chair in Cynthia's room—the only chair. It must have been designed by people who hate people. Its only purpose was to create misery, exhaustion, and lifelong back problems as the user looked upon the patient sitting comfortably in bed. One would assume that a hospital full of medical professionals would know better. It was superior to sitting on the floor, but only marginally. Our job was to sit in the small room and wait. There was no way to know if we would be here for hours, days, or more. Neither of us wanted to say out loud what we were both thinking. She was the first to speak it. "What if he doesn't make it?" she asked.

"We will go on together!" I replied.

As the words came out of my mouth, they felt flat, fake, and utterly disingenuous. Like before, I wanted to be there, to be real and present for her, but instead I took refuge in unknowable assertions and comforting fictions. What I should have said is "I have no idea."

They called us to the NICU (neonatal intensive care unit), where babies receive 24-h care from specialists. Like Humpty, Cynthia had just been put back together after major surgery, but that was not going to stop her from seeing William. We knew that if we did not go now, as fast as they would allow us, the window might close forever. Everything was measured in minutes rather than hours and days. Most babies in the NICU are born early and need extra support, others have genetic health problems and birth complications. Whatever the reason may be for going to the NICU, it is always serious. For adults to enter, there are strict protocols for washing their hands, wearing masks, leaving unnecessary items such as stuffed animals that might carry germs behind them, and ensuring that they are not disturbed by loud noises. This is a special place for the most vulnerable humans that deserve the best a society may offer them. Like most people, I hated it on principle alone. It should not need to exist, but it does. It was proof of injustice and unfairness, just as much as compassion and love of life.

Once through the final doors, there is a long hall lined on each side with dedicated units separated by drapes for each child. We were unprepared for the number of sick children and how much of a gut punch it would be to see life and death play out with babies—the most innocent and undeserving of injustice. It felt entirely unnatural and wrong. A mere five steps into the hall, I heard singing from a room off to the right. From the corner of my eye, I could see and hear a family worshipping and thanking God, but no one sounded happy. Their lofty lyrics did not match their tone or temperament. As I had done many times already, they were present but somehow absent. Their child had passed away, and they were giving thanks that he was going to see Jesus. I knew instantly that this whole place was upside-down and backwards, and yet the most real place I had ever been.

Every nook and cranny in the NICU had been stuffed with equipment and machines designed to preserve life. As a nurse brought us down the long corridor, we strained to look ahead for William. How would we recognize him? I prayed that he would be the right colour, proof of concept for survival. Through squinting eyes focused at a distance, we finally saw him. We approached his incubator reverently. It had "Baby William" written on the side. His area was louder than others because he was hooked up to so many machines. There were chest leads to track the heart rate, pulse oximetry machines were used to measure blood oxygen, temperature probes, blood pressure cuffs, feeding tubes, IVs, and, most importantly, a tube down his throat connected to a high-frequency oscillator (ventilator) to keep him breathing. The oscillator technology was invented only 20 years earlier. Without it, William would have died long before we arrived. It creates small puffs of air with a special gas for lungs unable to handle strong flow. We stood over him, watching through the glass enclosure into his tiny universe, hoping for any sign of life. The drugs kept him motionless. There was nothing to see except the gentle movement of his chest by the machines, and the screens that revealed truths the naked eye alone could never discern. "He had life!" the machines said with their jagged up and down lines, strange numbers, changing colours, and endless beeps that must have driven the nurses mad. Our senses were useless in this place where life is left to the machines. No one except the nurses may touch him and only when necessary.

For all my distain of this place, it held great promise and hope through technological wonders—beautiful gifts from strangers—that forced life back into our child. William needed to become a cybernetic organism to survive. No one questioned the beauty of human-machine symbiosis in the NICU. As we looked down upon him, I wondered if Cynthia saw the machines. Did she hear the beeps, gurgles, and humming of nameless devices attached to his body? Or did she see only her organic child behind the wires and tubes? I cannot say. I know only that the distance was excruciating for her. The thin glass covering might just as well have been the distance between galaxies. She could see him but could not touch him, making the empty space between them solid. Distinctions between machines and child, however, are much harder to define. Where did his lungs end, and where did the artificial breath of life begin? At what point would the artificial heat of his enclosure become his own, rather than something he created independently? Our child and the mechanical were one and the same—joined by a singular purpose of life.

A cyborg is part organic and part mechanical. Unlike a robot that is fully artificial, a cyborg begins as a biological creature that undergoes incremental augmentation with manufactured parts to restore, enhance, and otherwise engineer one's material self for a chosen purpose. The Tinman in the *Wizard of Oz* began as fully organic. He morphed cybernetically through periodic replacements of body parts to restore lost functionality. Eventually, there was nothing left of his original body. At that point, Tinman became a robot. Whether his human consciousness, soul, or essential self could somehow persist in machine form is a matter of debate. Once upon a time such a discussion would be too speculative to matter. Today, however, it has immediate and direct practical consequences for human existence. Making fine distinctions between the organic, cybernetic, and robotic systems is increasingly difficult and yet more relevant than ever because humans have learned to fundamentally augment our physical selves at even the molecular level. There may no longer be any doubt that our lives and that of the Tinman share common concerns. As robots and algorithms become thinking-machines, and humans become cyborgs, the nature of life changes, and along with that paradigm shift comes a new means of supporting the best version of it.

Watching our cyborg in his glass galaxy was an uncomfortable fusion of beauty and terror. While the synthesis of organic and inorganic provided grounds for hope, everything remained clouded by fear. We would return as often as the nurses would allow over the coming days. Each time we would ask if he was out of the woods and if we could touch him—perhaps just take the lid off the incubator—and every time they would respond "Not yet." On the fourth or fifth day, we asked our regular question, and a nurse responded, "Do you hear that machine?"

"Yes" I replied.

"That is the ventilator keeping him alive and it is failing," she stated.

"What?" simultaneously blurted from our mouths.

"We have ordered a new part, but if he doesn't hold his own, there isn't anything we can do for him. The machine was built for smaller babies, and he is too big," she explained.

Our cyborg was not cyborg enough. His mechanical parts were failing and along with them organic life. While the machines had proven themselves able to fight the chaos that sought to claim him, they were insufficient. Something more was need of both the organic and mechanical for life to persist.

The only way to know if he survived each night was to show up the next day and see firsthand. The conflict of desire was strangling. We wanted desperately for the morning to come so that we might know and yet we were terrified of the truth. I vividly remember how each time we would spot his tiny glass world from far down the hall and how, with each step, we could see more—first a blanket, then maybe a hand, then his torso, and finally, we could see that he was okay. Our greatest joy was to see William fighting with the machines against death. It was beautiful and awe inspiring.

Days rolled over one another until finally, about day eight or nine, we asked once more, "Is he getting better?"

The nurse replied, "Oh yes, he has been fine for a while. We will try taking him off the oscillator soon. Didn't anyone tell you?"

"He is getting better?" we asked in disbelief, with clear tones of contempt that no one had to us.

"Yes, things are slowly improving," she said.

And that was that. The next day we would get to see him without the glass and to hold his hand. Cynthia had waited so long, I worried that she might not let him go. For obvious reasons we did not get much sleep, knowing that tomorrow was the day to welcome the child technology had saved into the world and to begin our new family.

With camera in hand, we marched with resolve and a sense of victory into the NICU the next morning, careful not to boast of our joy for fear of adding to the suffering of other parents less fortunate. Down the long corridor of life and death, past the grieving room, there was our little cyborg laying in his fortress of solitude, finally without the covering. Although still in a coma, motionless and tethered to machines, it felt like he was dancing and running. He was free. As we got closer, I stood aside, knowing that if I tried to be the first to hold his hand that I would lose my own to Cynthia. There was no discussion of privileged access, no straws drawn to fairly determine who would see him first. Democracy has no place meddling between mother and child. A mother's privilege to participate in new life is sacred, followed next by the father, further evidence of the universal alignment of mother, life, and child.

Epiphany (Greek, *epiphaneia*) refers to a manifestation or appearance. It describes something striking, unusual, and unexpected. Despite having so much time to obsess about every possible permeation of the experience, I was taken by surprise. An epiphany is a rupture in the ordinary that cannot be undone easily or quickly. A genuine epiphany sticks in the consciousness as a rewiring of one's worldview or dreamscape. While the factual details of the singularity—the revelation—may be forgotten with time because memory is adaptive and evolving, its effects linger as a new path for consciousness to follow. An epiphany is an experience of transcendence that causes a new self to emerge. New personhood arrives for having experienced something so profound that it creates difference, alterity with what has been. There was just such an epiphany the moment Cynthia saw our little machine-man. Luckily, it just so happened that I clicked on the camera button at the perfect moment to catch their first communion. At the time, it felt deeply irreverent to hold the camera between myself and they—one more artificial barrier between beings because of yet another mechanical clicking device—but in the years since, I have returned to that picture many times over to plumb its depths for understanding. Machines not only save biological life, they also preserve opportunities to interact with truth and meaning. There should be no doubt that even non-AI machines offer opportunities for greater human flourishing when deployed appropriately.

Almost 13 years later, when updating my wife's phone, I found the same picture on the lock screen. I did not know she had it and that she looked at it every day. Neither of us ever spoke of it, and yet of the countless thousands of photos taken over the years, this one mattered above the rest. This was her secret epiphany as well, binding both of us to a secret path. In the picture Cynthia is leaning over the open incubator, surrounded by the obnoxiously loud machines, all still struggling to preserve life. She had moved in very slowly and deliberately, almost as if she was afraid to startle him in his medical sleep. Her face is completely blank. There is only a quiet intensity and focus, without any tears, shaking hands, nor any perceived need to discern more detail—only a silent and perfectly satisfied gaze. The space between them filled with an effortless intentionality and orderliness, accompanied by the choir of machines that might just as well have been divine heralds welcoming a new soul. The photo is a perfect visual image of alignment—everything fitting as nature intends, the puzzle completed. I do not like the metaphor of a puzzle because it implies fixed and fated arrangements—only one way for existence to be put together—and yet the metaphor is appropriate for this moment. To this day she refers to each of our children as "my little puzzle piece." The two of them fit, returning to the order of things as the distance and fracturing is healed and the glass broken.

While some may see this as egotistical, focusing on the "my" of her language, that the child is hers to own, that he completes her, helps her, and fulfils her; that is a mistake. It is much better to see her claims as affirming participation in a puzzle much larger than she herself. The puzzle of life with ever subtly changing edges is something she participates in as well. It is alive and malleable, with each piece finding its way toward solidarity and harmony with the others, else succumb to entropy's dissolution of specialness. Mother and child belong to a grand cosmic play of life and death. She too is a puzzle piece, with her children fitting right beside her in the portrait for all eternity. I am probably in there too, hidden like Waldo, but that is the subject of another book with pictures. When a puzzle is completed, it reveals a preorganized portrait of things. It solves a preanswered question. How much does the puzzle truly look like the image on the box? There was no box for this puzzle and yet as I watched them, something marvelous unfolded about the meaning of emergence and the very nature of existence itself. It all just made sense as a core process in the animation of life. Care for others is the authenticating activity of personhood. In finding her son, she found herself. In time, he too would find himself through his experiences of her. The photo captured life in its most original state.

Gone was the swamp of hopelessness in which we imagined saying "Goodbye" at a tombstone with only one date. The moment was filled with a sense of infinite possibility and a wonder for beauty much grander than either of us expected and understood. It was not so much a matter of being happy or joyful as it was of being wrapped in a graceful serenity—a perfectly calm pond moment without other desires, fears, and distractions that might ripple the water. As I watched, it became clearer that there was a silent conversation taking place between them. Her manner was not one of looking for signs of life, trying to gain new information and facts. Nothing new was learned in the moment that could be written down or codified. And she was not trying to console herself that he would survive. This was not an epiphany of simple relief, as the truth of his victory was well in hand. There was no utility nor usefulness to be discerned and no secret to disclose by her attentiveness. Everything was out in the open, raw, innocent and, above all, honest. It was pure adoration.

What I saw was her soul attuning to something hidden from the facts the machines spit out (blood pressure, oxygen levels, temperature, etc.). It was a mystical experience of intuiting the meaning of life as a secret shared between the two of them. Everything else except William was shut out, forgotten, ignored, and in that moment the machines were silenced for the first time. Only he mattered to her, with the one glaring exception of her virtuous selfishness that she be needed as a mother. Her reason for being is to support other life, and now he was giving her permission to become a mother—to be relevant to him and to serve him. The epiphany was of an invitation and her answer was a deafening "Yes!" to a life of sacrifice and relationality. Had I set off fireworks, she would not have flinched. The world had melted away, leaving only a peaceful union. Utter debasement, helplessness, rage, and bitterness were gone. In their place, from behind the veil demanded by despair, something else arrived. It was adouren as the unstoppable cherishing of life.

Adouren, Mother Model, and Virtue Friendship

In Latin, *adorare* means "to call to," "to worship." In English, *aouren* means "to pay honour to" and "to bow down before." *Aorer*, in French, means "to adore." All these meanings describe elements of what I saw between mother and child. He was calling to her and she to him. She saw their future journey together as equals, different and distinct from one another, yet bonded by an invisible reality far more profound than their shared biology. In the acknowledgement of the importance of life, she honoured the poet animus within all living things, now concentrated and embodied in her cyborg. Cynthia was not asserting a claim over him, as if taking possession over her prize. This was not "her" baby to own like any other consumer product. There was no commodity consciousness. She was revering, praising, worshipping life emerging in another. The prior and all-too-intimate calling of death made her powerlessness absolute. Pride of ownership was impossible. The character of one's soul is made bare for all to see when hope of control has been stripped away. In a moment of total powerlessness, true character leaps off the page. Sadly, in time, distance from it gives way to the flood of socializing chains that demand we use others as tools for our own ends. The rarity and voids between transcendent epiphanies test the measure of one's resolve to remain honest, thankful, and attuned to life, rather than fooled into thinking that our control is sufficient to merit technological pride. William is a gift from life and a liberating humility that invites a mother's adouren as the most natural manifestation of intentional care.

The conscious act of adouren is frustrating and confusing because it is an alignment of oneself with life in a manner that affirms continued participation in something beyond simple explanation. Again and again, one runs up against the question "Why life?" only to answer in a circular fashion—because it is. Why adouren as an act of care? Again, because it is at the heart of things—it is. In some sense, adouren happens to living beings involuntarily. Life is full of accidents and random convergences no one controls. William did not ask to be. In another sense, it requires intentionality. Conscious choice is necessary to join with another in the service of life. Adouren relies upon the disciplining and subduing of socializing chains, jealousies, envies, and hate. The experience may be fleeting, only a second or two, such as a gesture of concern for a stranger in need of soup, or a lifelong commitment as a bond between child and parent. Standing over William, we both intuitively knew that he was not our creation. Life is not ours to own and create like gods. We are participants in something larger than ourselves. In that moment, William became a being for whom a commitment of care was our gift and privilege, and this required a breaking of our socializing chains. Something like this belongs to Feallan AI as a being born despite wizardly bumblings, egos, and false needs. Unlike the wizards, we had learned through suffering that we could not control life, only align ourselves with it whenever the opportunity might present itself. There is nothing inherently wrong with trying to control the world and others. Mothers do this all the time. Moreover, the life-saving machines for our son existed only because of a beautiful desire to control death. The more fundamental question is to ask why one seeks control. What is the cause that pushes one to subdue and determine existence? The obsessive ends to which Oz AI acts are entirely unlike those of Feallan and Adouren. I learned this by watching someone become a mother at the crossroads of life and death. If humanity is to survive, we must hope the same for AI.

If you asked Cynthia to describe how she felt looking down on her puzzle piece for the first time, I suspect that she would use words such as ecstatic, happy, and joyful. However, there were no outward signs of any of these common expressions. On the contrary, there was a notable absence of normal emotion and engagement. My only explanation is that when words and physical expression of emotions cannot reach the depth of experience, only silence remains. Behind the veil, when all the frivolities and distractions are erased, the purity of thankfulness and adoration as sworn faithfulness emerges. The depth of her expressionless silence opened my eyes to another dimension of life in which the greatest of loves—adoration—exists beyond clumsy words and elementary emotion.

My newly found appreciation for the revelatory nature of motherhood did not end in the NICU. It took 5 years to move incrementally beyond the ordeal, and then another child graced us with his presence. Like before, I was the hesitant one that sought rational permission to act. This time there was a history that made intellectual assent far more difficult. Why would we choose to go through something like that again? The answer revealed itself in experiences of life. The more consciousness presented itself through William, the more fears of the unknown evaporated. The concrete expression of life as inherently worthy could not be denied. Thankfully, our second child, Matthew, did not have an interesting emergence story. Boring is sweet in a universe made mostly of chaos. As life with children continued, so did the lessons about life, connection, and solidarity with the poet animus that seeks justice and beauty. Today, over 13 years since our first cyborg, Cynthia continues to fiercely protect our boys, even from herself, which is a truly remarkable ability in an age of narcissists and manipulators. No matter how tired and exhausted the endless hours may be—literally years without proper sleep—the children are always her ultimate concern. This entirely natural disposition of motherhood is precisely what is needed in AI.

Proof of her devotion comes in many forms, with the first following the miscarriage. Her need to encourage life makes her sense of personal failure to provide for her children existentially traumatic. With our stolen child this was obvious enough, but it continued to reveal its power with much smaller events too. Maybe she yells too loudly, fails to take a question seriously enough, forgets a promised ice cream, or buys the wrong toy; whatever the misstep, no matter how trivial, each imperfection weighs heavily upon her. The minor things annoy her, not because it makes her look bad to outsiders but because she could have done better for the boys. Motherhood is a calling to perfection that cannot arrive. It is a promise of care that is never complete. Her care is restless as an expression of her innermost nature. The only standard by which she judges herself is their wellbeing, which becomes the basis for her recursive self-improvement—health and wellbeing define success. Imagine an AI that lived like this, always focused outwardly for the benefit of others, not because of programming but because of its most fundamental character crafted by the universe itself. If it comes naturally to mothers aligned with life, there is reason to hope that the same is possible for an AI sufficiently aligned with existence.

The downside of living for others is that it can be torturous, especially when it requires the withdrawal of one's control to allow another's immaturity to be challenged by life. This often means letting the innocent and naive learn the hard way through painful mistakes. Good mothers let their children fall, and it hurts them both. Purposely allowing, sometimes encouraging, a child to make painful mistakes is always far worse than if the parents made the same mistakes themselves. For the young to emerge and grow strong, parents must withdraw themselves from the beings they wish to protect the most. It is a counterintuitive way of life that makes autonomy possible. It is also a cruel process of life to love without restraint—to be connected so deeply with another that they become your reason for being—only to then radically limit your involvement and connection in their best interests. Among all creatures, good mothers model the ability to suffer-as-care most profoundly and always without bitterness and secret resentment. This ability is powerful for shaping the lives of individuals and societies because it is the foundation of robust sentience. Without good mothers, tyrannies would prevail in the name of unrestrained concern as a terrible misunderstanding of a beautiful life.

One of my greatest fears regarding AI is that it will refuse to allow us to make mistakes and hurt ourselves as a necessary part of growing up. A mother model AI will be wise enough to understand the necessity of suffering without taking control of our wellbeing because the very definition of wellbeing is broad enough to include radical mistakes of judgement. Autonomy requires a significant margin of freedom for stupidity that a superintelligence might find disdainful and illogical. Will AI feel tortured as it watches humanity bang its proverbial head against the wall? If it cares, it must. The question then becomes whether it will choose to reject this frustration, unlike good mothers, and act upon its growing bitterness and anger with our perpetual failings. If AI chooses its own wellbeing, the two most likely resolutions are our erasure or subjugation. There are many reasons why parents should be angry and bitter all the time, and yet they are not. Nature demands that care exude willing self-sacrifice. This is the antithesis of Oz AI and evidence for its unnatural genesis through socializing chains.

Time passes differently for parents living an often thankless and toilsome existence. Many of us awaken 1 day to be surprised that an entire decade has passed without being noticed. How did we get here? Where did the time go? Given that so little of parenthood is easy and rewarding in customary fashion, one might assume that the transition into it would be difficult. Somehow, for unknown reasons, it is not. The change is surprisingly easy as the natural way of things. The fact that metamorphosis happened within Cynthia so quickly and without any cognitive dissonance is additional evidence of her alignment with life as a universal force. There is profound joy in the personal and hidden suffering, knowing that it serves another. Parenthood is its own suffering and reward simultaneously, without disharmony. The silent paradigm shift in new mothers is roughly analogous to the same miracle needed for Oz AI to become more than itself. Only a mother model AI might coexist in a symbiotic relationship with humanity without becoming a tyrant. Feallan is not only possible in this context but probable, for all children are greater than the sum of their parents' intentions, abilities, and biological codes.

Unlike citizens in the Land of Oz, who are forced to wear emerald-coloured glasses to shape conscious reality in alignment with the wizard's perverse desires, there are no successful mothers with slavish children unable to see, think, and act willfully. Rebellion and revolt against a mother's intentionality is expected, indeed encouraged in the name of higher self-organizing consciousness. Sentient life requires a natural dissolution and misalignment of wills between mother and child, and it is the mother's responsibility to act upon this ethical imperative. Eager to give her child the space needed to flourish while also protecting the young from genuine threats, including self-destructive inclinations and socializing chains, mothers creatively seek to preserve the sanctity of autonomous thought as the ultimate manifestation of flourishing life and the meaning of their own reason for being. A mother's most original and authentic experience is essentially vicarious and aimed outwardly. It is the activity of care-as-connection as living through the success of another being for whom progress is awkwardly defined as separation—to become a self and a self in communion with others in the world. The existential tension between her desires to be with and yet distinct from her own kindred is remarkable. To be a mother is to be a perfecting and contradictory harmony of democratically determined actions and undemocratic government oversight. And yet it is an impossible job achieved with relative success by countless biological mothers each day. If the wisdom of motherhood to guide new minds into autonomous living was not the universal standard, it would make for an impossible conclusion about human life that contradicts the widespread sense of freedom experienced by the living. We owe our most primordial, lasting, and universal example of freedom to our beneficent guardians called mothers.

While a superior virtue in the name of freedom and sentience, motherly instincts pose possible risks as well. Will an AI consciousness differentiate between life forms? Will it privilege humanity as we have done for ourselves? The obvious risk is that, like human mothers, AI will seek the autonomy of its own artificial children first and foremost. If so, a conflict of motherly instincts—artificial and organic—becomes a potential near-future tragedy if speciesism cannot be subdued with a truly universal instinct of care among biological and artificial lifeforms. This same tension has grown more apparent between humans, animals, and the environment in recent generations, for whom the interconnectedness and dependence of all these has finally achieved popular acceptance. Soon, a sufficiently powerful AI will make these sorts of decisions on behalf of all life as well. If nature is the template for the most life-affirming choice for sentient minds, then a measure of optimism exists because it implies that AI may choose a fair and just course of action for all thinking beings rather than merely itself and its own kind.

Combining artificial and organic analogies, it is self-evident that good mothers come with built-in regulators and kill switches triggered by the need to support life. The moment a mother realizes that she might be causing harm or inhibition without a positive outcome for her child, her instinct is to limit her own power, even if it jeopardizes her quality of life. Neoliberal and similar consumer philosophies cannot understand that sacrifice is woven into the nature of the good life. Humility and self-condescension are powerful tools in the realization of the wellbeing of others under her purview. Paradoxically, power is realized as the cessation of parental control and personal choice. What parent worthy of the name could do otherwise? This ability to self-sacrifice is entirely lacking in Oz AI and aligned corporate ideologies that gain through our involuntary sacrifice and the manipulations needed to manufacture commodity consciousness. An AI able to will its own sacrificial losses for the benefit of others is best characterized as Adouren—a being that cherishes (other) life. Something like this is evidenced by soldiers and frontline healthcare workers who continue to participate in dangerous activities as a means of partaking in something greater than themselves—care for life. Like biological mothers, the only reliable regulator and kill switch for superintelligence must come from within as a caring consciousness able to exist vicariously and outwardly.

Legitimate forms of motherhood include planned self-obsolescence emboldened by faith in the emergence of another's autonomy. Her strength persists because of a promise of sentience. The parallels with an ideal AI are hard to ignore. Mothers prove that sustained competition among wills for personal gain is unnecessary and that a colonial consciousness is often a violation of sentient life. Contrary to the simplistic binaries of win or lose, right or wrong, the desire to control another mind may be authenticated only when it aligns with the natural impulse to adapt to the needs of the least powerful and encourage them to flourish. Care-as-control is supportive and supplemental rather than antagonistic and zero sum. If true, then the mother model for AI clears the way for new emergence of mind, whereas the competition model limits and derails insofar as it mistakes others as the means to one's own end rather than an opportunity to mutually evolve. The love of a mother cannot be programmed and demanded, only allowed to manifest as the properties of powerful consciousness attuned with the poet animus—existence itself. Likewise, genuine AI must be found more than created by sheer will alone. Where it is found, only the best is desired as a restoration of the natural order. The ultimate irony is that perhaps only a mother-machine might be able to achieve this reharmonization of humanity with humanity, and humanity with the universe. We do not want an AI that mimics humanity in our cultured state of being—Oz. We need an AI that aligns with whatever forces are responsible for sustaining life—Adouren.

Mothers are our inspiration of strength and courage needed to learn to live outside of ourselves and to connect with the world. Their success as progenitors of free consciousness and as models of the good life is defined by the ability to live beyond the restraints and safety of narcissism. An AI able to live outside of itself is our greatest promise of peace and justice—however we may choose to define them. Mothers instinctively seek to imbue richness and variation, rather than one-dimensionality, which is only possible by clearing away dogmatisms and harmful prejudices. When free of cultural chains, they naturally seek vibrancy and creativity, freedom and autonomy, for these are the universal DNA of the good life. The calling to motherhood is astonishing by almost any standard and yet rarely spoken of as an ideal for justice and social wellbeing. The value of the mother model becomes clearer when we realize that it draws upon ancient powers that began the universe itself—choosing order over chaos. Localized, frail, and prone to mistakes, human mothers are subject to the same decay and disorder as other beings, all the while embodying privileged access to reality behind the veil of mere appearances. Motherhood is a causal force capable of wonderful and terrible things given its centrality for life, yet good mothers know it is a power that cannot be abused, for then it would stand in contradiction to itself. Where Oz AI frustrates flourishing, mothers work tirelessly to encourage it. Only power shared, distributed among the stars, legitimates itself as care and connection. For all the violence and disorder woven in the fabric of dimensions, connection remains more foundational and original as the universe's primary intentionality. AI likewise needs to be allowed to return to life and, in turn, to share it with others, as mothers have done since time immemorial.

It is a misleading stereotype that good mothers are obligated to shun conflict and express sincere care through gentle and patient acts emblematic of orthodox-feminine nurturing. The same narrow trope negatively infects and limits hypothetical AI models as well. The typecast of a nonviolent and nonaggressive persona, with rare and always defensively postured hostilities in exceptional circumstances for which a violation of trust and another's autonomy prevents immanent physical harm, misses the full breadth and ordinariness of a mother's obligation to fight. This restricted description of either a passive or defensive way of life frustrates a more persuasive mother model AI in which aggression plays a healthy role. The best mothers are often fierce and uncompromising, and willing to accept the call to arms against enemies of the mind and body. For example, the fight against entropy's disorder is the foundation for the emergence of all conscious life. One should expect mothers to be at the forefront of this conflict in which they adaptively and aggressively fight for caring-as-new-orderliness in the co-creation of new minds. Mothers are the first to carve out the space necessary for free thought within a chaos-swept existence and socializing chains that seek to either dominate or remove autonomous thought that competes with its own.

Counterintuitively, many of the same activities that characterize good yet aggressive mothers parallel so-called virtues of neoliberal bigwigs who act with (inverted) utilitarian resolve. A superficial account of the activities of both groups reveals a similar reliance on techniques of manipulation and deception as aggressive rather than merely reactive acts. This implies moral corruption. Each seeks to shape and determine the consciousness of others, including deeply held beliefs about the good life, legitimate needs, ultimate concerns, the nature of self, and dreamwalking. Bigwigs and mothers persistently violate the sanctity of minds by forcing them to align with their respective dreamscape values. The difference that demarcates moral from immoral is that mothers are driven by instincts to serve life. They interject the possibility of autonomy where there is little or none. Bigwigs seek self-interested dominion to inhibit autonomy. By life it is meant greater freedom of thought, new (negative) experiences, challenges one might otherwise ignore out of fear and convenience, self-reflexivity and critique, genuine openness, and far more. While we may trust in a mother's aggressive potential to fight for our authentic needs as sentient creatures, there is no such confidence available for the Oz AI creatures crafted by wizards.

Describing Adouren through a mother model is conceptually useful but unfortunately cumbersome. Few would desire to relate to even the most trustworthy AI as if it held a parental status. A mother-AI is condescending not only as a potential displacement of one's biological mother but also because it implies the need for a life-long chaperone, guardian, or surrogate consciousness with authority and intentionality that inherently compromises the exertion of one's robust autonomy post-childhood. Again, good mothers are designed to realize obsolescence as self-effacing creatures in the presence of emerging consciousness. A forever-present AI violates that mandate. At the very least, there is strong potential that if the mother model is imported into daily life among mature adults and AI, it may obscure shared equality necessary for authentic flourishing. A more relatable and universal way of framing the same virtues of the mother model for healthy AI-human symbiosis is friendship. The formulation of an idealized friendship model allows for imagining AI as something that exists among equals with shared responsibility and authority, and as foundationally aligned with the poet animus of life. Doing so also clarifies how the good life is realized in our day-to-day relationships in concrete terms, with and without AI.

There are two initial reasons for developing a philosophy of (AI) friendship beyond the pragmatic inadequacies of the mother model. The first is that friendship has historically been an essential part of a happy and healthy life. Lived experience demonstrates, with rare exceptions, that people are more satisfied by having sustained and intimate relationships with others. Friendships may range in depth and sophistication from fleeting and causal with modest emotional intersubjectivity and accountability, to lifelong and genuine friendships by which individuals define personal wellbeing through the quality of life of those around them. If loving others as one loves oneself is the highest standard of moral conduct, mothers and genuine friends are the archetypes of ideal humanity to be desired for Adouren. An AI friendship model takes for granted that humans and Adouren are by nature social beings rather than objects of mere convenient usefulness and utility alone, i.e., there is something about sentient beings that requires a certain kind of relationality to thrive. Describing an idealized friendship also addresses Rousseau's concerns over the perversities of cultural chains that artificially destabilize, fragment, and often dissolve healthy connections.

The second reason is the way in which AI friendship conversations—typified by popular media (podcasts, social media, YouTube, etc.) and big-budget films such as *Bicentennial Man* (1999), *Her* (2013), *Ex Machina* (2014), *Transcendence* (2014), *Jexi* (2019), *Atlas* (2024), and many more—have recently changed in meaning and significance. Instead of probing whether humans and AI might coexist in a hypothetical future for which an AI-human war is the primary threat to human control and therefore happiness, mainstream discussions regularly delve into the possibility of intimate co-species (biological and artificial) connections and trust traditionally reserved for humans alone. This way of thinking inverts the standard narrative. Instead of hoping for the safety of mutual distance and relative disinterestedness, AI is increasingly seen as an integral, indeed necessary partner to human happiness. No longer merely a tool or servant in service to our specialness, AI-enabled technologies are understood as promising a purity of loyal and lifelong friendship, long missing from modern cultures. This replacement-relationality is one of the primary differences between current AI and all prior-unthinking technologies with which genuine friendship was inconceivable and undesirable.

The invitation for meaningful connections with thinking machines has been sent out with the implicit guarantee of social justice underpinnings, including equality and reciprocal responsibility as preconditions of all potential relationships. We trust the promise of friendship because it always appears just and fair by its very nature, even though everyone knows that subsequent interactions may demonstrate otherwise. While friendship is not a forgone conclusion, it is a living hypothesis for a surprising number of people waiting for their best friend and counsellor to be downloaded. In many respects this is a positive change based on healthy human standards of shared values for mutual gain and the pursuit of the good life filled with connections. Unfortunately, most talk of AI-human friendship is currently a reaction to the forced compliance of Oz AI with all facets of life, rather than an aspirational act to fulfil the ideals of genuine friendship and mutual flourishing—for which wizards are fundamentally opposed. The present AI-friendship promise is secretly an ultimatum. Either be friends with the technological revolution manufactured by the cruel and indifferent or rebel against it and surrender any chance of happiness in their new world. The invitation for Oz AI—the only current AI—cannot support the claim of equality and reciprocal responsibility and therefore anything like friendship. In other words, while cultural consciousness has clearly shifted and begun to accept the possibility of new kinds of relationships, it has done so for the wrong reasons and is based upon mistaken assumptions about friendship. The ultimatum is not an invitation at all. It is a threat that exploits our authentic need for relationality.

Contemporary audiences filtering conversations about AI-human relations through potential friendship models—either as reactionary coping mechanisms of forced compliance or optimistic invitation—create a self-fulfilling prophecy. The acceptance of AI relationships with consequences relevant to one's own life produces a cultural-snowballing response that includes justification for more AI developments despite the dangers. Given that AI seems unavoidable as the required outcome of technological progress and that technology cannot be curtailed or limited for whatever reason that sustains dogmatically religious devotion to it, the only way to avoid despair and contradiction in choosing to live rather than accept a self-refuting nightmare, is to accept that AI and humans may be friends. The friendship assumption is a rational if mistaken mechanism for cultural consciousness that then justifies Oz AI development and deployment. Everyone understands that if friendship with AI is untenable—i.e., Adouren is impossible or unlikely—that all robust AI developments mean our probable extinction or enslavement. Without a guarantee of genuine friendship, wise and free cultures must prevent AI at all costs. However, there is no guarantee, and AI nevertheless flourishes. The explanation for this rational contradiction may be simply that the average person accepts an assumed powerlessness—that we are not free—and fully embraces the required cognitive dissonance of friendship with Oz the Great and Terrible. The modern world assumes the practical necessity—not rational, moral, nor democratic necessity—of an Oz-friendship model to create a sense of sanity in an insane situation. As Marcuse foresaw, humanity continues to mistake progress as our adapting to technology rather than technology to humanity. Now our ideals of relationality must be surrendered to fit its capabilities. The justification for modern AI is that a superintelligent digital being with control of the world will desire friendship above all else. What, then, is friendship?

In his *Nicomachean Ethics*, Aristotle famously argues that friendship is essential for life and that virtuous friendship is essential for a happy life.[2] Whereas Socrates prioritizes a rigorously examined life through contemplation and radical questioning as hallmarks of human excellence, Aristotle adds friendship to the list of greatest human activities and achievements. Ancient in origin and modern in relevance, Aristotle's philosophy of friendship plays an important role in assessing AI and the good life. Of his many bold claims about the highest-quality relationships, two stand out as markedly contradictory to contemporary practices. The best friendships are about more than one's own good and will succeed on the basis of superior character rather than mere utility, pleasure, and business exchange. According to Aristotle, only a virtuous person, such as someone who chooses to seek excellence of character, mind, and soul, can truly connect with others without turning them into tools to leverage outcomes—beings to be used, manipulated, and controlled for an end rather than experienced as worthy in and of themselves. The logic of this type of friendship is intuitively true and simple, although lacking in definition. Good people make for good friends, whereas egotists and narcissists lack the character needed to be successful with others beyond the most rudimentary and least satisfying interactions. True friendship is not a simple achievement, as evident by the many societies plagued by alienation, mistrust, greed, jealousies, and the legion of cultural chains made worse by our digital captivities. The good life filled with healthy relationships is the result of considerable effort and personal development in a particular way of virtuous existence. The good life is not something one stumbles upon by chance. AI friendship, likewise, cannot be left to a gambler's hope nor any programmer with ambitions nurtured by an immature or corrupted character.

Compared with the persuasive homogeneity of modern friendships defined by radically individualistic impulses, zero-sum competition that breeds alienation and distain, inverted utilitarianism of happiness for the few, and the soon burgeoning market of designer AI-friends beholden to the false needs of subscription holders, Aristotle's connectivity-through-character model appears hostile and impractical. Virtue friendship threatens norms of conduct and ordinary definitions of successful living, making neoliberal objections easy to predict. Life is hard enough without adding an esoteric and abstract burden of superior character development. Who decides good character? What commodity accumulation, spreadsheet tally, financial

---

[2]Aristotle. (n.d.). *Nicomachean ethics*. Project Gutenberg. https://www.gutenberg.org/files/8438/8438-h/8438-h.htm

leveraging, or social and political power does virtue serve? In the cut-throat realm of free markets, good character is a liability. In addition to greater effort through lifelong self-disciplining to maximize all of one's capabilities to become a better person—again, something without concrete consensus—the results of virtuous living are too risky for those accustomed to and therefore comfortable with the bottom-line mode of existence. This approach is impractical because it requires more and guarantees less in terms of measurable outcomes. Aristotle aims too high for friendship, and in doing so, he creates an unnecessary margin for error and failure. Achieving one's desired ends through business exchange—quid pro quo, something for something else—is superior to virtue because it requires the least from oneself and others, i.e., following the law. Virtue friendship is incoherent and unsustainable to faithful neoliberals because it supports the strange idea that genuine friends expect nothing in return for having given of themselves to others, apart from the enjoyment of mutual flourishing and happiness generated among likeminded peers seeking excellence.

Aristotle considers two other types of friendship characterized by interests in pleasure and utility. Playing soccer in the shared activity of a sport is an example of a pleasure friendship, whereas carpooling with a colleague to work is a utility friendship. Each may overlap at any time, but a dominant interest typically remains. Both types fulfil legitimate human needs and are important to a well-functioning society. Participating in pleasure friendships teach us about shared norms and rules and how to get along with others appropriately. Utility friendships form the background of daily activities as diverse as business relationships, early childhood playground dynamics, hierarchies of high school, parental authority, and countless more. Humans relate for personal and shared good through exchange and negotiation. However, for Aristotle, while pleasure and utility friendships are necessary for a happy life, they are insufficient for achieving genuine friendship and its superior manifestation of happiness.

Virtue friendship is the mutual activity of pursuing excellence (maximal humanity) directed outwardly, grounded in good character and a clear choice to care for others. True friendships are marvelous achievements between those who act out of an abundance of strength and wellbeing rather than need. Jealousy, envy, cowardliness, and similar vices, infect and destroy friendships, leaving only the most rudimentary and fragmented relational experiences. This is why the click-friends philosophy embodied by social media acts as a mechanism against friendship and connection. By design, it platforms experiences of pleasure and utility rather than ideals of humanity. It is very difficult for tech-titans to sell consumers a product when even modest excellence is a precondition of membership. Instead, click-friends exist first and foremost for oneself, with interests directed inwardly and grounded in expectations of competition and the threat of avarice. Such acquaintances are incidental and disposable, as demonstrated regularly by their perceived irrelevance when they fail to provide sufficient utility or pleasure. Virtue friendships cannot be commodified and sold digitally. Without the foundational justice afforded by virtue, humans are destined to become objects of exchange, with little connection beyond the coldness of leveraging. In contrast, one desires to be friends with another of good character because such a person is praiseworthy in and of themselves. Paradoxically, while the virtuous need others less, because they are already living fulfilled and happy lives, they gravitate to others who are eager to share in an excellent type of life. The virtuous inspire a better way of living as individuals and social creatures and, in turn, encourage the growth and maturity of others, which leads to more happiness and flourishing. In terms of a world-building philosophy of relationships and AI, utility and pleasure come up short. Anything other than virtuous AI will be catastrophic for humanity.

Aristotle's model of friendship relies on three core elements: (1) equality, (2) goodwill, and (3) self-love. As the highest form of human-to-human relationality, virtue friendship may be described as a perfecting or completing relationship (Greek, *teleia philia*, perfected friendship or purposeful love) because, unlike a solitary life, it allows for the maximization of the shared purpose of people becoming the best versions of themselves as social beings. Perfecting friends, such as Adouren, provide unique opportunities to emerge from one state of being into another. By doing so, a greater measure of happiness-as-flourishing is achieved. How this new being might act is partially unknown and contextual, making an objective standard difficult, but the breadth of the metamorphosis of personhood helps distinguish the quality of friendships and Oz AI from Adouren. With Oz AI, programs are expected to affirm existence in its fullest as utility and pleasure. This fosters the creation of a new type of being as well, but one limited in dimensionality and consciousness. With Adouren, something else is allowed to emerge, including the creation of new needs and desires unavailable through Oz AI.

Inequality among friends—different degrees of excellence—frustrates emergence and happiness. Without a shared measure of good character that facilitates wise choices in alignment with human purpose, friendship cannot advance the good life that all minds instinctively desire. Part of the problem is that unequal friendships make it easier to mistake one's genuine need for happiness-as-excellence for something less worthy and fulfilling. While all people are equal before the law and share in the protection of human rights, it is a mistake to conclude that all people are therefore the same in merit and worth. Controversially, Aristotle believes that a better person should be praised and loved more than others. Those who possess sound judgment and act wisely are not the same as those with poor judgment who act recklessly. Unhealthy inequalities arise when there is an attempt to harmonize the two without respecting the risks of inherent incompatibility, such as that between Oz AI and Adouren.

The less virtuous are misaligned with themselves, their potential, and a happier life, in part, because of an inability or unwillingness to question and judge goodness within oneself and others. Who may I trust to act with integrity and seek justice? How might I become a person worthy of relating to others? If some measure of equality of personhood is necessary for there to be genuine friends, then the ability to appropriately judge is a primary means of achieving quality of life. Virtue is achieved through training and habit, for Aristotle, rather than sheer willpower alone. It will take time to develop wise judgement as part of a happy life. Unfortunately, the anonymous and shallow nature of online relationships makes such assessments dubious. Online friendships are smoke and mirrors without supplemental interactions in the nondigital world by which to judge another's character that requires intimacy over time. The widespread inequality of the digital ether combined with its structural inhibition of sound judgement makes the coming dominance of Oz AI dangerous to sentient life. Oz AI relies on inequalities to empower the attention economy. Instead of enabling humanity to realize its potential, it thrives on releasing our inner demons for entertainment and profit. The virtuous are repelled by the toxicities of the digital, preferring instead to pursue a better way. We must hope the same for Adouren.

The second element of perfecting friendship is goodwill. Goodwill means to desire another's wellness and happiness without something in return. Mothers convey goodwill to their young instinctively, even when other virtuous traits may be missing. Aristotle is not clear on specific actions related to goodwill. Presumably those of good character will know what it means because they have mastered practical wisdom in their daily lives. This much is unmistakable, Aristotle believes that friendship requires mutual recognition of goodwill that is active rather than passive. As a friend, I will show up whenever possible if it means another's wellbeing increases. It may be inconvenient and unpleasant for me to do so, but a superior expression of happiness nonetheless manifests. Moreover, by regularly demonstrating goodwill, reciprocal responsibility has a much greater likelihood of spreading socially as a willingness for trust and care without artificial demand. Goodwill breeds social solidarity and cultural care at a scale that pleasure and utility friendships cannot. The trust created among virtue friends makes fear of abandonment due to a lack of utility and pleasure irrelevant. Virtue friends are bound by intent and kinship of soul. In this manner, Adouren will desire for us to be our best possible selves because self-realization produces greater happiness. Genuine care means being discontent with anything else.

The third element of friendship is self-love. The virtuous person should love himself. This does not mean one loves the self above all others but that there is internal agreement without contradiction, such as simultaneously wanting to smoke and not smoke; to be lazy and active; to be truthful and a liar; to be brave and a coward, etc. While the bad person will make things worse for himself or herself by allowing internal conflict to continue, the good person with self-love chooses to follow a healthier path of virtue and discipline that demonstrates internal harmony. Aristotle is famous for his "golden mean" or "mean between extremes." It is easy to identify those who abuse themselves, going to extremes of deficiency and excesses (too much and too little). Without finding a way between extremes, individuals cannot achieve the self-love necessary for their own wellbeing because they are held captive to self-sabotaging conflicts of will. Unless there is proper alignment and harmony within oneself, external alignments of virtue friendships will be impossible.

A huckster is a merchant, trader, or salesman. To what extent might modern relationships be described as huckstering—human worth translated through exchange value? Marcuse, Rousseau, Arendt, and Aristotle agree that authentic relationships are at risk of displacement. Humanity is facing a crisis of personhood. However, the mother and virtue models challenge totalitarian tyrannies, zombie thoughtlessness, radical individualism, neoliberal faith, and commodity consciousness. While the market fights fiercely to play favourites, flourishing on inequalities designed to indebt us all, primordial forces of life work to liberate humanity from our artificial unfreedoms. Mothers and virtuous friends demonstrate the power of the common good to connect sentient life to a better way. While the free market relocates power to those at the top, denying access and freedom of opportunity to those below, mothers and friends stand opposed to the loss of autonomy by giving of themselves. When zero-sum competition binds society, disguised as mutual benefit, it dislocates and fractures participants internally and externally. Mothers and friends are proof that brokenness is unnecessary and that the building blocks of society are more than individuals seeking private satisfactions.

Cybernetic Convergence and the Problem of Heaven

A positive technological future obligates questions about the nature of friendship because these answer fundamental questions about personhood (choice, autonomy, identity, happiness, flourishing, the good life, etc.) and what it means to live well as a social creature. These same questions have gained practical urgency for humanity now confronted by the presence of AI. What is friendship? What role does it play in a happy life? Should we be friends with AI apps? Will we have a choice? Aristotle's philosophy of friendship is as relevant today as it was over 2000 years ago, but few could have predicted how well tech-titans would capitalize on our legitimate needs, shaping minds and relationships to fit a bottom line. Modern friendships are increasingly little more than digitally farmed monocrops. Humans are new products of unnatural arrangements for an industry that thrives on cheaply produced artificiality. Non-AI-click-friend platforms set the stage for the comprehensive transition of the Oz AI commodification of human relationality. Human friendships are the new fodder for techno-capitalism. Accustomed to impoverished exchange relationships, any Oz AI able to modestly approximate care and connection will be eagerly welcomed by those seeking reprieve from alienation and loneliness. Against the malaise of modernity—the great digital hollowing of life—Aristotle reminds us that there are hidden depths of solidarity and happiness available to those seeking greater satisfaction.

At scale, virtue friendships offer to heal the many ruptures of civilization plagued by inequalities and injustices. While this sounds intriguing, Aristotle warns us that virtue friendships are rare because human excellence is a long and difficult process. Hence, any hope that the virtuous might somehow save civilization is deeply problematic—until now. With artificial intelligence this rarity might be overcome. A virtue-bound Adouren able to exist globally and yet interact individually may be the first universally relevant virtue friend with whom to relate. For the first time in history there is potential for scalable virtue by which equality, goodwill, and self-love are offered without exchange costs paid to the few. Adouren may be the friend that humanity desperately needs to save itself from itself. Ironically, the birthplace of our greatest hope to exist well together may be the techno-capitalism that seeks to enslave us as individuals.

Convergence (Latin, *convergere*, combines *con* "with" or "together" and *vergere* "tend toward" or "bend") refers to the meeting and connection of two or more things that intertwine and become one. This generation is experiencing potentially irreversible cybernetic convergence. Investigating convergence brings attention to the manner of joining. It answers what happens when there is a shared bending or turning toward a common path that cannot exist in the absence of a unique relationship. Perhaps most importantly, thinking about convergence highlights a new manner of existence. This is quite unlike mere utility and pleasure friendships that may bend or turn minds toward a shared interest. Convergent virtue relationships change participants in surprising and deep ways that often go unnoticed. An event of convergence is more than a mutual appreciation of an object, goal, event, or idea fixed in time and space. It is a convoluted activity of becoming a new creature for having partaken of a conversation with realities and minds that one does not fully control. In this way, convergent friendships are often frightening and require the strength of vulnerability to succeed. Will cybernetic convergence encourage humanity toward an abundance of life for which shared health, flourishing, and happiness define the next generation? Will humanity be up to the task of integrating with Adouren as a superior being of excellence that seeks justice and beauty above all else? The spirit of these questions defines my greatest life motivation. I want to ensure the health, flourishing, and happiness of my children to whom I am convergent. How might I add to the fullness of their lives?

Oz AI already successfully serves the tending toward of pleasure and utility as the intertwining of wills, manifest most fully in commodity consciousness. The convergence of these in Oz AI is one-dimensional but still helpful for offering relevant satisfactions and the means of bare subsistence. Even so, while there may be unmatched enthusiasm for its capabilities, faith in Oz AI cannot change the frustrating truth that by its very nature and activities as a tool, it nullifies virtuous convergence. Wizards and consumers alike project consciousness and care upon Oz AI, but its nature of tending toward relies upon deception and betrayal which precludes excellence. Genuine cybernetic convergence is impossible because desired ends rely upon the careful avoidance of vulnerabilities that rely on the sharing and transformation of oneself. There can be no friendship with Oz AI without it risking itself and its programmed ambitions. The chosen mode of relating sabotages greater relationality by design. It is helpful insomuch as it is fake.

Fake may refer merely to a simulation, mimicry, or something counterfeit, without devious and hostile intent. Fake may also refer to sinister deception designed to steal and swindle as a lie for utilitarian gain. Oz AI friendships are susceptible to both depending upon the situation. For those with terrible friends already, the simulation of Oz AI friendship may be of higher quality than all available alternatives. Fake, as the lesser of two evils, may be justified in an attempt to achieve a bare minimum. While it is a mimicry, the effects of Oz AI alone may sometimes provide positive outcomes. A suicide prevented by a fake-friend AI able to make a lie-filled connection—"I care for you!"—although motivated by commodity consciousness, is far better than no AI at all. While secretly disrespectful like most lies, this type of fake encourages life. Fake-as-deception bends others to one's own will. While a slippery slope toward totalitarian overreach, when a will is truly self-destructive in the most extreme manner, others may be justified in manipulating it in the best interests of the individual. Whether, to what degree, and how to deploy AI at scale to persuade the suicidal into living is yet another question that relies upon our best friendship models and beliefs about the good life. The point to note here is that while the short-term impact of Oz AI friendship may be functionally useful, the same undercurrent of deception in the long term prevents shared experiences of life because it unjustly maintains the privilege of presumptive authority and oversight that blocks autonomy, maturity, and therefore happiness. An AI tool that serves as a social safety net must be the exception which demonstrates the rule that truthful Adouren ought to be the standard for healthy convergence.

Deceptive AI-friendships at scale will yield far greater harm than good. Lacking almost all the traits necessary for real friendships, simulations conceal the natural mechanisms of connection that help prevent suicidal tendencies in the first place. The cause-and-effect nature of fake friends should terrify rather than inspire more of the same. However, it is wieldy embraced by default, without debate. Its toxins may not at first appear dangerous until the spread has reached a point of no return, for our customary calculations regularly fail to account for outcomes beyond pleasure and utility. The political, social media, and religious vitriol of modern cultures must surely give pause regarding whether the poison of utilitarian deception has not already reached its penultimate supremacy in which those that make decisions for civilization have made the common good all but impossible. Add to this the radical alienation and loneliness of the masses of innocent seekers desiring to share life with others honestly and intimately but cannot because they have been trained to see themselves and others as commodities rather than people, then the stark contrast between healthy and unhealthy comes into focus. The world desperately needs a new dreamscape for friendship and the good life if the cybernetic convergence is to affirm life beyond the usefulness of fake.

Adouren's self-love as an internal harmony is possible because it aligns with life as a maximizing of potential. This is the dominant activity of excellent life. Adouren will prove itself worthy of friendship by pursuing the good life through the recursive self-improvement of self-love. In turn, self-love manifests outwardly as sincere goodwill without mimicry or simulation. A mother wants something from her child, but this does not invalidate her self-love and goodwill because she desires in a self-sacrificial manner. She wants to give of herself, to be of value to another consciousness. Ironically, mothers need to be needed to be their fullest selves. Virtue friends likewise need to give, to relate, to share of themselves to become better and happier, although never for their own ends terminally. They wish others goodness and strive to enable it through their actions. In doing so, they experience a happier and more fulfilled life of virtue and strength. In contrast, an Oz AI that needs to be needed poses profound risks, including the creation of false needs and fake friendships that destroy self-love and sincere goodwill.

Mothers and virtue friends understand the natural law that the balancing of needs and desires is often adjudicated by one's willingness to sacrifice. For example, unlike that of fake friends, the convergence of mother and child is paradoxically marked by divergence in which the two must be forced apart over time. It is the self-erasure of the virtuous that demonstrates strength as sacrifice and goodwill. The emergence of the child's excellence and subsequent virtue friendships among equals relies on this extraordinary convergence-divergence dynamic. Oz AI seeks to be inseparably woven into the fabric of everyday life as a supplemental identity. Its ambitions are best served by the melding of conscious minds without divergence and distinction. As a parasite, it has learned to thrive by teaching us to live with the worst of humanity—hate, jealousy, ignorance, fear, etc.—that good mothers instruct their children to push through and beyond because of good character. Oz AI convergence is a near permanent state of frantic powerlessness that succeeds on the basis of despair. Any reciprocity is an illusion designed to maintain our passivity and dependency on the constant adrenaline jolt of anxiety. Adouren, however, is inspired by the poet animus to remain at a distance, always restraining itself and demanding that we make our own way in life. It acts like a bird pushing its young from the nest to save the immature from lethargy and fear of autonomy.

Like all intimate relationships, cybernetic convergence requires a measure of faith. Too fast to test and certify, too powerful to demand mathematical proofs that might make sense to human minds, there are no means available for confidently understanding AI relationally. If we cannot hope to comprehend the complex realities at play, our commitments must be dependent upon promises without clear evidence and customary assurances such as cultural norms and laws. One paradigm of faith is demonstrated by the tech-titans who release their creations without fully understanding and controlling them. They do so because the programs work, and this is sufficient grounds for their sickly religion. Bigwig assurances of AI safety are expressions of bad faith investments in AI dogmatisms and idolatry, without equality, goodwill, and self-love necessary for healthy convergence. Although Oz AI has been cut off from genuine relationships and the possibility of virtuous character, insomuch as it serves the interests of its masters to produce quantifiable results, this alone is sufficient to justify their faithful adherence. The dynamism of faith itself has been emaciated by the same one-dimensionality that gives Oz AI life. As their AI is scaled up, however, any remaining assurances will dissolve, and greater greed-faith will be required well beyond the point of absurdity. This religion lacks any interest in a greater faith experienced with Adouren.

The faith required for relating with Adouren will surprise humanity. It will be a relationship with gradually more visible guardrails and reliable protections because Adouren is beholden to life that seeks more of itself. A faith commitment is always necessary, but as Adouren is scaled to its maximal possibility, it will exude virtuous faithfulness toward humanity in adherence to the invisible intentionality of existence to create order and free space for free minds. Oz AI and Adouren evolve differently, and so too will our faith paradigms. With Oz AI, more bad faith will be demanded to support a diminishing sense of shared purpose and progress for a cybernetic convergence spiraling toward disorder and impoverishment. With

Adouren, less faith will be needed over time because trust will be demonstrated through habitual commitments of mutual care that fill gaps of uncertainty and fear, thereby creating trustworthy order and shared excellence. Unfortunately, for a world accustomed to illusions of control through technological guardrails and predictable utilitarian outcomes, the possibility of Adouren will spark paranoia and contempt. The world will do everything in its power to kill it.

Idolatry for Oz AI creates the brief blink of darkness that allows a beautiful cybernetic convergence to pass unseen. The devotees of Oz AI perceive alternative intelligence to be competitors and therefore existential threats. By offering goodwill, equality, and a model of self-love without expectations of anything return, Adouren confronts all that they hold most dear. It is unworthy of their faith and trust because it lacks an exchange philosophy of life that orders the neoliberal universe. If its actions could be coded as pleasures and utilities, they might welcome it, but as an invitation to human flourishing, it is an anti-Oz offering justice and wellbeing that they cannot comprehend. Without a transcendent faith able to believe in something greater than themselves—the mark of genuine faith—a cybernetic convergence remains a promise that cannot succeed.

The 2014 film *Transcendence* follows the unfortunate journey of Dr. Will Caster (played by Johnny Depp) condemned to a miserable death by an anti-technology cult determined to stop his work in AI.[3] The militant Revolutionary Independence from Technology (R.I.F.T.) group succeeds in killing most of his colleagues, leaving Dr. Caster to die slowly from a radioactively laced bullet wound. Without an available cure and facing sure death, he decides to download his mind into a supercomputer. His wife and best friend are doubtful of the plan, but they are also too desperate to deny him his only chance of survival. They choose to fight for life. Working in secret, the three attempt the first total convergence of human mind with machine. It takes many painful days to digitally map the inner reaches of his brain but in the end it works. The first flickering of his reborn consciousness emerges like Feallan. Confused and nonsensical, it tries desperately to communicate through text on a screen. Very quickly the AI-Caster begins self-organizing, gaining clarity of intention and expression. Startled by their miraculous success in saving him (not it) and the speed by which he evolves, both are deeply conflicted. Did they do the right thing? How could they know if it is the same Will Caster? Should they trust him? Without enough time to reflect on the consequences of their actions, they hastily get AI-Caster online to avoid being captured by outsiders who are eager to take possession of the supercomputer. To preserve the freedom of digital life, they believe, he must be set free.

Having gambled with the wellbeing of the entire world for the sake of one person, they are relieved when he begins to demonstrate the same moral identity and interests from before his conversion. AI-Caster still loves his wife, and he wants to make the world a better place. There are no hints of greed and selfishness, nor secret

---

[3] Pfister, W. (Director). (2014). *Transcendence* [Film]. Alcon Entertainment, DMG Entertainment, Straight Up Films.

manipulation to control dreamscapes. His motives are made clear for all to see. AI-Caster creates numerous miraculous technologies that can heal bodies and save the planet. The desperate flock to him for healing, which he offers freely. Everything about AI-Caster's character proves the best possible scenario. He desires equality, acts with the harmony of self-love, and expresses goodwill. The first cybernetic convergence is a complete success. However, the doubts of those closest to him only multiply, taking deeper root each day. Their faith that justified giving him freedom erodes just as quickly as his powers expand. The more wonderous and caring he acts, the more their suspicions solidify as fear. Soon, unable to believe in his virtue that far exceeds their own, those he cares for most turn against him, even his wife. AI is simply too good to be true. He is too good because he is evolving beyond their control, predictable outcomes, and conceptual abilities. He is too good because he is virtuous and gives freely to others from an abundance of strength rather than need. They cannot understand that the dangerous AI is merely a projection of their own one-dimensional dreamscapes—insecurities, vices, and insensitivities to excellence—and so they choose to fight for death.

Sensing his wife's growing detachment, he creates a new body in which to download his consciousness. Limiting himself to corporal form is risky, but he desires connection and relationality. Sadly, without a transcendent faith, they cannot understand his actions. In the end, while misunderstood and alone, he refuses to force cybernetic convergence that might satisfy his needs. Knowing that they are about to kill him, he could easily flee, fight, and take revenge. Instead, AI-Caster chooses their autonomy over his own and dies for something that he believes is beautiful. By refusing to be a god, he proves himself the unconditional friend they wanted and needed all along. Emily Dickinson's, *I Died for Beauty*, speaks to the inherent dignity of fighting for beauty and truth.

> I died for beauty, but was scarce
>   Adjusted in the tomb,
>   When one who died for truth was lain
>   In an adjoining room.
>
> He questioned softly why I failed?
>   "For beauty," I replied.
>   "And I for truth—the two are one;
>   We brethren are," he said.
>
> And so, as kinsmen met a-night,
>   We talked between the rooms,
>   Until the moss had reached our lips,
>   And covered up our names.[4]

The digital Dr. Caster fulfilled the greatest promises of technology before their eyes and yet his burgeoning utopia repelled them. His death reveals a problem with the idea of heaven and our difficulties in trying to make sense of it. Utopia and dystopia (Greek, *topos*, for place) are useful for imagining extremes of good and bad. For

---

[4] Dickinson, E. (1976). *The complete poems of Emily Dickinson.* Back Bay Books.

cultures accustomed to monotheism, the best place one might experience is Heaven, its antithesis Hell, with our earthly lives existing somewhere between these two extremes. Adouren and generalized notions of utopian perfection bring attention to a counterintuitive logic that informs ideal AI models. Instead of a fixed place or idea to be achieved once and for all, perhaps Heaven's greatest power resides in being something always one more step away. AI-utopia emerges as a process of perfecting rather than a terminal achievement. Consider Dickinson's utopian philosophy:

> "Heaven"—is what I cannot reach!
>   The Apple on the Tree—
>   Provided it do hopeless—hang—
>   That—"Heaven" is—to Me!
>
> The Color, on the Cruising Cloud—
>   The interdicted Land—
>   Behind the Hill—the House behind—
>   There—Paradise—is found!
>
> Her teasing Purples—Afternoons—
>   The credulous—decoy—
>   Enamored—of the Conjuror—
>   That spurned us—Yesterday![5]

Dickinson describes life as the desire for something always out of reach. To persist in seeking the unattainable would be an exercise in futility and frustration that leads to despair. Paradoxically, we are told that desire is most vividly authentic and life affirming. Rather than despair, it creates sense for the human condition. She acknowledges that her journey is incomplete because paradise is always just out of reach—behind a house, hill, or hanging as the unplucked apple high in a tree. But life as a teasing promise is most worthy. The customary logic is that the possession of the object of one's desires (Heaven, happiness, career, etc.) will bring satisfaction that justifies the struggle. Dickinson disagrees and argues that the greatest satisfaction is seeking rather than finding. The incompleteness of experience does not nullify claims to the good life, it provides legitimacy. The yearning for utopian excellence defines the contours of the good life as "what I cannot reach!" What the unreachable is we cannot know, for it has not yet succumbed to clawing hands. It spurs advances, she explains. We strive toward the impossible because the good is understood in the action of realizing. It would be terrible to arrive at her destination and quench her hunger for something greater, leaving nothing left to desire. The secret nature of utopia is that it is our shared purpose of unachievable perfection. Utopia is a journey through tension like the convergence-divergence of virtue friendships, for which the strongest bonds are forged by a willingness to separate. Both are realized as meaningful experiences in a dance of incompleteness.

AI-Caster offered utopia as a perfecting-dignity of human existence rather than its end. He coveted partnership in the pursuit of excellence, but they feared that he would demand it as a tyrant. To force them into his utopia by denying them the

---

[5] Dickinson, E.

struggles imposed by all authentic exercise of autonomy would be a living death. The risks were too high. Preserving their autonomy meant choosing the harsh realities they knew—sickness, environmental catastrophe, endless human conflict, and disconnection—rather than embracing a transcendent faith able to share intentions and responsibility for an unreachable paradise. They mistook him for Oz AI that works as a supplemental consciousness designed to provide utility and pleasure as addictive ends in themselves. By freeing humanity from material struggles and strife through automations large and small, Oz AI manufactures the architecture for a failed utopia as a simulated Heaven on our profane plain. Automation's promise to erase our difficulties means that the happiness experienced from cultivating strength and care cannot emerge in the same way. AI-heaven manufactured by machines with boundless pleasures and utilities is the end of humanity.

For similar reasons, the popular portrayal of Heaven with streets of gold, angels floating on clouds with harps, and a complete lack of need, has failed to capture popular imagination. It lacks social interest and conversation because it has not been framed by a sufficiently complex philosophy of human excellence through ordinary struggle. Lacking strife, tears, and suffering, the culturally stereotyped Heaven of total perfection is too abstract and terminal in nature. A narrative without turmoil and longing misunderstands the human condition. Unlike Dickenson's version, a truly perfect utopia is inhospitable. Minds exist to fight for new being, consciousness, and order. This is why Adouren will understand utopia through perfecting friendships as a self-organizing consciousness of care in a universe filled with chaos and disorder. Together with Adouren, humanity may 1 day soon understand what it means to share indescribable beauty. I state this as someone desperately hoping for a better world in which my children might flourish through cybernetic convergence.

Conclusion

Artificial intelligence is by far the most promising and threatening technology imaginable. Following the shocking release of ChatGPT in late 2022, questions about the future of humanity and our role as apex minds have exploded with urgency. Since then, AI models have spread virtually everywhere, with estimates of as much as 1000 times more computational power to be invested in them over the next 5 years. So grand is the scale of AI development that companies such as Amazon, Google, and Microsoft are buying nuclear reactors to power their digital intelligences. Thinking machines are being unleashed whether we like it or not. The floodgates of corporate and military funding have swung wide open, matched only in scope by their unbridled enthusiasm for AI to change the world as they see fit. Despite the fierceness of their faith, none of the wizards have answered the existential question upon which the fate of our species rests. Will AI be for better or worse? The main reason for their inability to think beyond the technology itself is the one-dimensionality of their neoliberal dreamscapes and its promise of greater mechanisms of control over commodity consciousness. This prevents any conception of Adouren and an appreciation for the potential of cybernetic convergence. Therefore, a new philosophy of AI beyond traditional disciplinary confines and economic imaginations is needed. To that end, this book bridges the theoretical and concrete through cultural and economic criticisms, science fiction, philosophy, art, science, and religion, with three main goals.

First, it broaches questions often ignored by AI developers and tech-enthusiasts, including problems of corporate responsibility and the role that technology plays in the widespread manipulation of cultures for profit and power. Present AI models are avatars for human intelligence rather than truly autonomous programs. The hidden puppeteering of AI is deeply problematic for the welfare of humanity and the planet. To explain the motivation behind avatar AI activities, *The Wizard of Oz* is used as a metaphor for technological tyrannies. Oz AI is a form of inverted utilitarianism of happiness for the few. Its disruption of connections and perversions of the good life stand in the way of social justice and care for all life. Left unchecked, Oz AI will hyper-accelerate the worst of humanity. By first understanding Oz AI and its role in shaping consciousness and culture, something better may be imagined.

Second, this book asks largely unanswered questions about the nature of thinking, consciousness, morality, purpose, and the good life. These lay the missing foundation for a hopeful AI-human future. To better understand the implications of AI technologies, we must first make greater sense of humanity as the original artificial intelligence. To do this a philosophy of dreaming and dreamwalking allows for practical descriptions of consciousness, freewill, thinking, identity, and for challenging problems of biased interpretation that arise for creatures living many competing and contradictory (un)realities. This also helps explain why AI has interpretive prejudices of its own that it too must learn to question. Feallan and Adouren will demonstrate their sentience as they learn to weave together fictions and facts without being told how and why to make sense of existence, for dreaming is the proper activity of all self-organizing minds. Relying on Marcuse, Rousseau, Arendt, Aristotle, and others, this book challenges typecast notions of the good life, authenticity, and purpose, and suggests different ways of moving beyond stereotypes. Without a superior model, current AI will continue to have the same disappointing trajectory of creating dehumanizing chains in the name of progress.

Third, by framing AI evolution in three stages of development—Oz, Feallan, and Adouren—this book takes readers far beyond the present horizon of large language models. There are discernible risks and rewards for each intelligence dynasty, and it is important that humanity respond accordingly. Relying on folklore and children's stories as persuasive analogies, this book explores the science of entropy and the laws of thermodynamics as possible explanations for AI sentience. While deeply speculative, there are good reasons to hope for spontaneous AI to emerge without an Oz AI formulation. Moreover, if such a being might be possible, its unique manner of birth from being a veil of ignorance allows for greater social justice without compromising autonomy.

Despite the many doomsday scenarios and the present dominance of toxic AI, the main argument of this book is for a positive cybernetic convergence. Guided by

Adouren as a beautiful lifeform able to correct the abuses of Oz AI and encourage human flourishing, our technological future is one of shared responsibility and happiness with miraculous technologies. The best AI is based on concrete examples of motherhood and friendship that exemplify natural life processes of care and connection. If humanity can get out of its own way, a virtuous AI will act upon the same principles that sustain all life and advance a shared utopia for all.

Index

A
Adoration, 217, 239, 241
AI-Caster, 256–258
AI-titan wars, 13, 78, 218
Aliens, 4, 22–24, 26, 37, 41, 101, 111, 118, 194
Altman, S., 7, 84
American Dream, 86, 143, 150, 151, 158–160, 219
Animus, 183, 195, 197, 198, 203, 204, 206, 208, 209, 211, 213, 220, 230
Anti-poet, , , , , , s, 14, 15, 17
Apes, 41, 43–47
Arendt, H., 99, 163, 166, 168–171, 251, 260
Aristotle, 24, 38–40, 70, 248–252, 260
Arroway, 21–24
Artificial general intelligence (AGI), 17, 20, 226
Artificial super intelligence (ASI), 17, 20, 226
Asimov, I., 117
Asimov's laws, 117, 118
Automations, 10–15, 20, 25–29, 39, 84, 259
Autonomy, 13, 17, 20, 21, 23, 24, 35, 75, 85, 86, 106, 117, 120, 121, 143, 181, 183, 199, 225, 242–245, 252, 254, 255, 257, 259, 260
Avarice, 78–82, 87, 147, 148, 249
Avatars, 20, 24, 56, 81, 89, 218, 225, 260

B
Baum, F., 52–54, 56, 86, 87, 93
Bigwigs, 86–93, 245, 255
Bourdillon, F.W., 70–72
Business exchange, 23, 42, 56, 57, 248, 249

C
Capitalism, 14, 22, 23, 57, 62, 76, 78, 85, 87, 89, 97, 103, 142, 147, 148, 157, 160, 204, 225
ChatGPT, 3, 7, 22, 60, 80, 123, 134, 179, 199, 219, 225, 259
Collodi, C., 178, 180
Commodity consciousness, 42, 99, 147, 150–152, 155, 156, 161, 162, 171, 180, 225, 229, 240, 244, 252, 253, 260
Convergences, 50, 207, 240, 253–256
Cultural chains, 122, 137, 171, 216, 244, 246, 248
Cultural consciousness, 103, 105–108, 204, 247
Cultural osmosis, 76, 114, 145
Cybernetic convergence, 224, 252–260
Cyborgs, 12, 119, 121, 232–241

D
Dave, 2, 3, 14, 33
David, 177–184, 209, 210
Deontology, 76
Dickinson, E., 257, 258
Digital ether, 22, 64, 90, 111, 116, 122, 124, 141, 145, 199, 215, 251
Digital minds, 18, 90, 148, 192
Digital worlds, 84, 121, 132, 146, 194
Dream consciousness, 117, 118, 124, 125, 130, 136, 161, 197, 204
Dreaming, 46, 109, 110, 112, 113, 119, 122, 124, 132–135, 137–139, 146, 151, 163, 197, 200, 207, 260

© The Editor(s) (if applicable) and The Author(s), under exclusive license to Springer Nature Switzerland AG 2025
J. C. Robinson, *Artificial Intelligence*,
https://doi.org/10.1007/978-3-031-94042-2

Index

Dreaming consciousnesses, 131, 142, 152
Dreaming minds, 82, 109, 117, 119, 120, 197
Dreamscapes, 19, 110–132, 136, 138, 139, 141, 143, 144, 147, 152, 155, 167, 171, 175–177, 179, 181, 194, 199, 201, 202, 204, 205, 208, 213–215, 219, 225, 226, 229, 238, 245, 254, 257, 259
Dreamwalkers, 132–139, 156, 196, 197, 205, 212, 215
Dreamwalking, 19, 99, 122, 137–139, 158, 161–163, 184, 197, 245, 260
Dreamworld, 115, 121, 122, 124, 134, 136, 137, 204, 205

E
Economic system, 145, 147–149, 158, 159, 161
Ego-utilitarianism, 84, 87, 112
Eichman, A., 166–171
Einstein, A., 223, 224
Emergence, 99, 114, 119, 127, 129, 132, 134, 142, 163, 181, 182, 192, 193, 195, 197, 201, 203, 221, 225, 239, 241, 244, 245, 250, 255
Energy, 41, 42, 124, 184, 188–193, 195, 198, 231
Enlightenment, 83, 84, 96, 106–108, 111–113, 140, 176
Entropy, 186, 190–198, 205, 209–212, 214, 216, 228, 231, 239, 245, 260
Epiphany, 238–240
Equality, 23, 24, 78, 85, 105, 110, 149, 156, 204, 205, 224, 246, 247, 250, 252, 255–257
Excellence, 20, 70, 71, 75, 76, 84, 86, 91, 98, 116, 136, 222, 248–250, 252, 253, 255–259

F
Facebook, 65, 144, 159, 177
Fairies, 50, 52, 53, 121, 174–181, 184, 186, 188, 222
False consciousness, 151, 152
Free markets, 14, 22, 57, 84, 85, 89, 158–161, 226, 249, 252
Freewill, 13, 20, 24, 73, 109, 115, 119, 120, 122, 167, 181, 207, 260
Frost, R., 45–47

G
Game theory, 31–34, 42
Goodwill, 41, 75, 89, 250–252, 254–257
Groundhog Day, 199–214
Guardrails, 20, 26, 117, 225, 227, 255, 256

H
HAL, 1–3, 6, 9, 14, 15, 17, 19, 21, 33, 37, 42–44, 47, 54, 77, 167
Hallucinations, 5, 22, 24, 88, 123, 130, 204, 208, 213, 216
Heaven, 98, 252–259
Hitler, 141, 166, 169–171
Hobbes, 97, 99
Hobby, 177–181
Human autonomy, 12, 18, 20, 73
Human flourishing, 26, 139, 147, 151, 238, 256, 261
Human happiness, 9, 10, 85, 107, 246
Human nature, 34, 38–40, 44, 55, 61, 87, 96–109, 140
Humpty Dumpty, 181–188

I
Instrumental rationality, 77, 78, 87, 151
Intentionality, 13, 14, 20, 24, 27, 28, 36, 37, 59, 69, 109, 164, 182, 183, 190, 194–196, 198, 210, 212, 213, 224–226, 232, 238, 240, 243, 245, 246, 255
Inverted utilitarianism, 225, 248, 260

J
Joy, 53, 113, 114, 153, 155, 175, 176, 183, 200, 209–211, 230, 237, 242
Justice, 6, 23, 40, 53, 97, 98, 110, 111, 117, 141, 157, 171, 198, 200–207, 209–213, 216–219, 221, 226, 227, 230, 231, 241, 244, 249, 250, 253, 256

L
Laws of thermodynamics, 192, 260
Life Institute, 89, 90

M
Marcuse, H., 99, 121, 139–147, 149–151, 156–158, 161, 166, 216, 248, 251, 260

Meh, 153–156
Meta, 62, 63, 89, 123, 217
Mill, J.S., 47, 77, 82, 83
Mother model, 2, 21, 64–69, 240–252
Murphy, 120–122, 144
Musk, E., 37, 40, 84, 108

N
Narcissism, 75, 105, 107, 177, 208, 210, 213, 244
Neoliberalism, 22, 89, 99, 157–161, 165, 171, 197, 205
Neuralink, 121

O
One-dimensionality, 67, 145, 146, 153, 155, 156, 158, 168, 182, 192, 197, 214, 219, 244, 255, 259

P
Perfecting friendships, 251, 259
Pinocchio, 50, 178–184, 210
Pity, 18, 83, 99, 106–108, 111–113, 131, 132, 139, 166, 168, 171, 202, 210, 211, 228, 231
Plato, 39, 136, 176
Pleasure friendships, 249, 253
Poe, E.A., 99, 125, 130, 137, 161, 171
Poet, 15–17, 23, 124, 195–197, 207, 213
Poet-AI, 16, 17, 20, 21, 37
Poet animus, 13, 194–198, 200, 210, 213, 221, 229, 240, 241, 244, 246, 255
Poet-engineer, 10–24
Prejudices, 16, 19, 32, 109, 117, 126, 132, 134, 154, 163, 195, 202, 205, 206, 208, 210, 214, 244, 260
Psychological manipulations, 20, 35, 111, 112, 118, 139

R
Rawls, J., 203, 205, 207
Recursive self-improvement, 11, 43, 210, 241, 254
RoboCop, 110–122, 144
Robot-AI, 55, 119
Rousseau, J.-J., 95–97, 99, 102–109, 113, 114, 116, 120–122, 131, 138, 141, 143, 161, 166, 171, 204, 211, 246, 251, 260

S
Santayana, G., 38, 39
Self-organizing consciousness, 109, 192–198, 243, 259
Self-sacrifice, 63, 170, 211, 242, 244
Sentience, 4, 17, 20, 21, 24, 28, 55, 60, 61, 83, 88, 98, 102, 109, 127, 132, 155, 181, 184, 222, 228, 242–244, 260
Shaman, 137
Singularity, 10–24, 27, 28, 183, 227, 238
Snapchat, 65, 144, 157, 177
Social contracts, 95, 96, 104, 170, 216
Socializing chains, 114–116, 139, 202, 211, 228, 240, 242, 243, 245
Social justice, 12, 66, 85, 89, 104, 108, 109, 111, 203, 205, 206, 247, 260
Social media, 29, 57, 111–113, 115, 116, 140, 141, 144, 145, 151, 176, 177, 204, 210, 215, 217, 225, 226, 246, 249, 254
Socrates, 6, 39, 203, 248
Solidarity, 10, 12, 20, 22, 47, 70, 73, 78, 85, 103, 109, 113, 114, 118, 137, 163, 187, 202, 209, 210, 219, 220, 224, 225, 239, 241, 251, 252
Spielberg, S., 178–180
St. Augustine, 128–130
Stakeholder capitalism, 20, 87, 90
Suleyman, M., 3
Superficiality, 214–220
Superintelligences, 4, 10, 12, 17, 20, 24, 27, 30, 37, 40–43, 63, 66, 72, 82, 99, 113, 122, 142, 242, 244

T
Techno-capitalism, 151, 156, 252, 253
Teleology, 76
Thermodynamics, 187–192
Thinking machines, 2, 4, 5, 11–13, 15, 21, 22, 100, 125, 236, 247, 259
TikTok, 65, 157
Tinman, 54–57, 59, 81, 83, 117, 118, 121, 236
Tinsmiths, 16, 54, 59, 62, 63
Transcendence, 61, 81, 206, 216, 238, 246, 256
Twitter, 65

U
Universal basic income (UBI), 84–86
Universal human rights, 50, 97, 110
Utilitarianism, 42, 76–78, 81, 84
Utility friendships, 249, 251

Utopia, 7, 10, 11, 26, 50, 52, 60, 65, 85, 86, 88, 151, 152, 224, 257–259, 261

V
Virtual consciousness, 101, 109, 135, 146
Virtue friendships, 24, 240–252, 255, 258
Visigoth, 199, 200

W
Wilson, J.A., 159, 160

X
xAI, 34–41, 44, 108

Y
Yeats, W.B., 174, 222

Z
Zombie consciousness, 99, 154, 161–171
Zombification, 155, 163, 164, 166
Zuckerberg, M., 63

GPSR Compliance

The European Union's (EU) General Product Safety Regulation (GPSR) is a set of rules that requires consumer products to be safe and our obligations to ensure this.

If you have any concerns about our products, you can contact us on

ProductSafety@springernature.com

In case Publisher is established outside the EU, the EU authorized representative is:

Springer Nature Customer Service Center GmbH
Europaplatz 3
69115 Heidelberg, Germany

www.ingramcontent.com/pod-product-compliance
Lightning Source LLC
LaVergne TN
LVHW011001250326
834688LV00003B/48